Lecture Notes in Computer Science 12928

More information about this subseries at http://www.springer.com/series/7412

Islem Rekik · Ehsan Adeli ·
Sang Hyun Park · Julia Schnabel (Eds.)

Predictive Intelligence
in Medicine

4th International Workshop, PRIME 2021
Held in Conjunction with MICCAI 2021
Strasbourg, France, October 1, 2021
Proceedings

 Springer

Editors
Islem Rekik ⓘD
Istanbul Technical University
Istanbul, Turkey

Ehsan Adeli ⓘD
Stanford University
Stanford, CA, USA

Sang Hyun Park ⓘD
Daegu Gyeongbuk Institute of Science
and Technology
Daegu, Korea (Republic of)

Julia Schnabel ⓘD
Helmholtz Center Munich
Neuherberg, Germany

ISSN 0302-9743   ISSN 1611-3349 (electronic)
Lecture Notes in Computer Science
ISBN 978-3-030-87601-2   ISBN 978-3-030-87602-9   (eBook)
https://doi.org/10.1007/978-3-030-87602-9

LNCS Sublibrary: SL6 – Image Processing, Computer Vision, Pattern Recognition, and Graphics

This Springer imprint is published by the registered company Springer Nature Switzerland AG
The registered company address is: Gewerbestrasse 11, 6330 Cham, Switzerland

Preface

It would constitute a stunning progress in medicine if, in a few years, we contribute to engineering a predictive intelligence able to predict missing clinical data with high precision. Given the outburst of big and complex medical data with multiple modalities (e.g., structural magnetic resonance imaging (MRI) and resting function MRI (rsfMRI)) and multiple acquisition timepoints (e.g., longitudinal data), more intelligent predictive models are needed to improve diagnosis of a wide spectrum of diseases and disorders while leveraging minimal medical data. Basically, predictive intelligence in medicine (PRIME) aims to facilitate diagnosis at the earliest stage using minimal clinically non-invasive data. For instance, PRIME would constitute a breakthrough in early neurological disorder diagnosis as it would allow accurate early diagnosis using multimodal MRI data (e.g., diffusion and functional MRIs) and follow-up observations all predicted from only T1-weighted MRI acquired at the baseline timepoint.

Existing computer-aided diagnosis methods can be divided into two main categories: (1) analytical methods and (2) predictive methods. While analytical methods aim to efficiently analyze, represent, and interpret data (static or longitudinal), predictive methods leverage the data currently available to predict observations at later timepoints (i.e., forecasting the future) or predicting observations at earlier timepoints (i.e., predicting the past for missing data completion). For instance, a method which only focuses on classifying patients with mild cognitive impairment (MCI) and patients with Alzheimer's disease (AD) is an analytical method, while a method which predicts if a subject diagnosed with MCI will remain stable or convert to AD over time is a predictive method. Similar examples can be established for various neurodegenerative or neuropsychiatric disorders, degenerative arthritis, or in cancer studies, in which the disease/disorder develops over time.

Following the success of the first three editions of PRIME MICCAI in 2018, 2019, and 2020, the fourth edition of the workshop (PRIME MICCAI 2021) aimed to drive the field of 'high-precision predictive medicine', where late medical observations are predicted with high precision while providing explanation via machine and deep learning, and statistical-, mathematical- or physical-based models of healthy or disordered development and aging. Despite the terrific progress that analytical methods have made in the last twenty years in medical image segmentation, registration, or other related applications, efficient predictive intelligent models and methods are somewhat lagging behind. As such predictive intelligence develops and improves – and it is likely to do so exponentially in the coming years – this will have far-reaching consequences for the development of new treatment procedures and novel technologies. These predictive models will begin to shed light on one of the most complex healthcare and medical challenges we have ever encountered, and, in doing so, change our basic understanding of who we are.

What are the key challenges we aim to address?

The main aim of PRIME MICCAI is to propel the advent of predictive models in a broad sense, with particular application to medical data. To this end, the workshop papers of 8 to 12 pages in length describing new cutting-edge predictive models and methods that solve challenging problems in the medical field. We envision that the PRIME MICCAI workshop will become a nest for high-precision predictive medicine, one that is set to transform multiple fields of healthcare technologies in unprecedented ways. Topics of interests for the workshop include but are not limited to predictive methods dedicated to the following:

– Modeling and predicting disease development or evolution from a limited number of observations;
– Computer-aided prognostic methods (e.g., for brain diseases, prostate cancer, cervical cancer, dementia, acute disease, neurodevelopmental disorders);
– Forecasting disease or cancer progression over time;
– Predicting low-dimensional data (e.g., behavioral scores, clinical outcomes, age, gender);
– Predicting the evolution or development of high-dimensional data (e.g., shapes, graphs, images, patches, abstract features, learned features);
– Predicting high-resolution data from low-resolution data;
– Prediction methods using 2D, 2D+t, 3D, 3D+t, ND, and ND+t data;
– Predicting data of one image modality from a different modality (e.g., data synthesis);
– Predicting lesion evolution;
– Predicting missing data (e.g., data imputation or data completion problems);
– Predicting clinical outcomes from medical data (genomic, imaging data, etc).

Key Highlights

This year's workshop mediated ideas from both machine learning and mathematical/statistical/physical modeling research directions in the hope of providing a deeper understanding of the foundations of predictive intelligence developed for medicine, as well as where we currently stand and what we aspire to achieve through this field. PRIME MICCAI 2021 featured a single-track workshop with keynote speakers with deep expertise in high-precision predictive medicine using machine learning and other modeling approaches – which are believed to stand in opposing directions. The workshop was run virtually and keynote talks were streamed live this year due to the COVID-19 pandemic. Pre-recorded videos of accepted papers and keynote presentations were posted on the PRIME web page[1]. Eventually, this will increase the outreach of PRIME publications to a broader audience while steering a wide spectrum of MICCAI publications from being 'only analytical' to being 'jointly analytical and predictive.'

We received a total of 26 submissions. All papers underwent a rigorous double-blind review process, with at least three, and mostly four, members of the

---

[1] http://basira-lab.com/prime-miccai-2021/.

Program Committee reviewing each paper. The Program Committee was composed of 28 well-known research experts in the field. The selection of the papers was based on technical merit, significance of results, relevance, and clarity of presentation. Based on the reviewing scores and critiques, all but one submission was scored highly by reviewers, i.e., had an average score above the acceptance threshold.

Diversity and inclusion are important components of PRIME MICCAI, and this year the workshop strongly supported gender balance and geographic diversity within the Program Committee. The authors of the accepted PRIME papers were affiliated with institutions in four continents: Africa, Europe, America, and Asia. We also provided two PRIME scholarships to register the papers of talented minority students in low-middle income countries (both were from Africa). The eligibility criteria of the PRIME scholarship were included in the CMT submission system. We will strive to continue this initiative in the upcoming years and see a similar trend in other conferences and workshops. To promote research reproducibility, we included a reproducibility checklist to encourage authors to share their source code and data in support of open science. The authors of the best PRIME papers were also invited to submit an extended version of their work to a MELBA[2] special issue.

August 2021

Islem Rekik
Ehsan Adeli
Sang Hyun Park
Julia Schnabel

---

[2] https://www.melba-journal.org/.

Organization

Chairs

Islem Rekik	Istanbul Technical University, Turkey
Ehsan Adeli	Stanford University, USA
Sang Hyun Park	DGIST, South Korea
Julia Schnabel	Technical University of Munich and Helmholtz Center Munich, Germany

Program Committee

Ahmed Nebli	Université de Sousse, Tunisia
Alaa Bessadok	Université de Sousse, Tunisia
Changqing Zhang	Tianjin University, China
Dong Hye Ye	Marquette University, USA
Duygu Sarikaya	Gazi University, Turkey
Febrian Rachmadi	RIKEN, Japan
Gang Li	University of North Carolina at Chapel Hill, USA
Heung-Il Suk	Korea University, South Korea
Ilwoo Lyu	Ulsan National Institute of Science and Technology, South Korea
Islem Mhiri	Université de Sousse, Tunisia
Jaeil Kim	Kyungpook National University, South Korea
Le Lu	PAII Inc, USA
Lichi Zhang	Shanghai Jiao Tong University, China
Manhua Liu	Shanghai Jiao Tong University, China
Maria Deprez	King's College London, UK
Maria A. Zuluaga	EURECOM, France
Mayssa Soussia	University of North Carolina at Chapel Hill, USA
Minjeong Kim	University of North Carolina at Greensboro, USA
Pew-Thian Yap	University of North Carolina at Chapel Hill, USA
Qian Wang	Shanghai Jiao Tong University, China
Qingyu Zhao	Stanford University, USA
Robert Jenssen	UiT The Arctic University of Norway, Norway
Seung Yeon Shin	National Institutes of Health, USA
Ulas Bagci	Northwestern University, USA
Victor Gonzalez	Universidad de León, Spain
Xiaoxiao Li	Princeton University, USA
Yu Zhang	Stanford University, USA
Ziga Spiclin	University of Ljubljana, Slovenia

Contents

Low-Dose CT Denoising Using Pseudo-CT Image Pairs 1
Dongkyu Won, Euijin Jung, Sion An, Philip Chikontwe,
and Sang Hyun Park

A Few-Shot Learning Graph Multi-trajectory Evolution Network
for Forecasting Multimodal Baby Connectivity Development from
a Baseline Timepoint..................................... 11
Alaa Bessadok, Ahmed Nebli, Mohamed Ali Mahjoub, Gang Li,
Weili Lin, Dinggang Shen, and Islem Rekik

One Representative-Shot Learning Using a Population-Driven
Template with Application to Brain Connectivity Classification
and Evolution Prediction 25
Umut Guvercin, Mohammed Amine Gharsallaoui, and Islem Rekik

Mixing-AdaSIN: Constructing a De-biased Dataset Using Adaptive
Structural Instance Normalization and Texture Mixing................ 37
Myeongkyun Kang, Philip Chikontwe, Miguel Luna, Kyung Soo Hong,
June Hong Ahn, and Sang Hyun Park

Liver Tumor Localization and Characterization from Multi-phase MR
Volumes Using Key-Slice Prediction: A Physician-Inspired Approach 47
Bolin Lai, Yuhsuan Wu, Xiaoyu Bai, Xiao-Yun Zhou, Peng Wang,
Jinzheng Cai, Yuankai Huo, Lingyun Huang, Yong Xia, Jing Xiao,
Le Lu, Heping Hu, and Adam Harrison

Improving Tuberculosis Recognition on Bone-Suppressed Chest X-Rays
Guided by Task-Specific Features 59
Yunbi Liu, Genggeng Qin, Yun Liu, Mingxia Liu, and Wei Yang

Template-Based Inter-modality Super-Resolution of Brain Connectivity 70
Furkan Pala, Islem Mhiri, and Islem Rekik

Adversarial Bayesian Optimization for Quantifying Motion Artifact
Within MRI 83
Anastasia Butskova, Rain Juhl, Dženan Zukić, Aashish Chaudhary,
Kilian M. Pohl, and Qingyu Zhao

False Positive Suppression in Cervical Cell Screening via Attention-Guided
Semi-supervised Learning 93
Xiaping Du, Jiayu Huo, Yuanfang Qiao, Qian Wang, and Lichi Zhang

Investigating and Quantifying the Reproducibility of Graph Neural
Networks in Predictive Medicine.   104
   Mohammed Amine Gharsallaoui, Furkan Tornaci, and Islem Rekik

Self Supervised Contrastive Learning on Multiple Breast Modalities Boosts
Classification Performance   117
   Shaked Perek, Mika Amit, and Efrat Hexter

Self-guided Multi-attention Network for Periventricular
Leukomalacia Recognition   128
   Zhuochen Wang, Tingting Huang, Bin Xiao, Jiayu Huo, Sheng Wang,
   Haoxiang Jiang, Heng Liu, Fan Wu, Xiang Zhou, Zhong Xue, Jian Yang,
   and Qian Wang

Opportunistic Screening of Osteoporosis Using Plain Film Chest X-Ray   138
   Fakai Wang, Kang Zheng, Yirui Wang, Xiaoyun Zhou, Le Lu, Jing Xiao,
   Min Wu, Chang-Fu Kuo, and Shun Miao

Multi-task Deep Segmentation and Radiomics for Automatic Prognosis
in Head and Neck Cancer   147
   Vincent Andrearczyk, Pierre Fontaine, Valentin Oreiller, Joel Castelli,
   Mario Jreige, John O. Prior, and Adrien Depeursinge

Integrating Multimodal MRIs for Adult ADHD Identification
with Heterogeneous Graph Attention Convolutional Network   157
   Dongren Yao, Erkun Yang, Li Sun, Jing Sui, and Mingxia Liu

Probabilistic Deep Learning with Adversarial Training and Volume Interval
Estimation - Better Ways to Perform and Evaluate Predictive Models
for White Matter Hyperintensities Evolution   168
   Muhammad Febrian Rachmadi, Maria del C. Valdés-Hernández,
   Rizal Maulana, Joanna Wardlaw, Stephen Makin, and Henrik Skibbe

A Multi-scale Capsule Network for Improving Diagnostic Generalizability
in Breast Cancer Diagnosis Using Ultrasonography   181
   Chanho Kim, Won Hwa Kim, Hye Jung Kim, and Jaeil Kim

Prediction of Pathological Complete Response to Neoadjuvant
Chemotherapy Using Multi-scale Patch Learning with Mammography   192
   Ho Kyung Shin, Won Hwa Kim, Hye Jung Kim, Chanho Kim,
   and Jaeil Kim

The Pitfalls of Sample Selection: A Case Study on Lung
Nodule Classification.   201
   Vasileios Baltatzis, Kyriaki-Margarita Bintsi, Loïc Le Folgoc,
   Octavio E. Martinez Manzanera, Sam Ellis, Arjun Nair, Sujal Desai,
   Ben Glocker, and Julia A. Schnabel

Anatomical Structure-Aware Pulmonary Nodule Detection via Parallel
Multi-task RoI Head   212
    Haoyi Tao, Yuanfang Qiao, Lichi Zhang, Yiqiang Zhan, Zhong Xue,
    and Qian Wang

Towards Cancer Patients Classification Using Liquid Biopsy   221
    Sebastian Cygert, Franciszek Górski, Piotr Juszczyk,
    Sebastian Lewalski, Krzysztof Pastuszak, Andrzej Czyżewski,
    and Anna Supernat

Foreseeing Survival Through 'Fuzzy Intelligence': A Cognitively-Inspired
Incremental Learning Based *de novo* Model for Breast Cancer Prognosis
by Multi-Omics Data Fusion   231
    Aviral Chharia and Neeraj Kumar

Improving Across Dataset Brain Age Predictions Using Transfer Learning. . .   243
    Lara Dular Žiga Špiclin,
    and The Alzheimer's Disease Neuroimaging Initiative

Uncertainty-Based Dynamic Graph Neighborhoods
for Medical Segmentation   255
    Ufuk Demir, Atahan Ozer, Yusuf H. Sahin, and Gozde Unal

FLAT-Net: Longitudinal Brain Graph Evolution Prediction from a Few
Training Representative Templates   266
    Guris Özen, Ahmed Nebli, and Islem Rekik

Author Index   279

Low-Dose CT Denoising Using Pseudo-CT Image Pairs

Dongkyu Won, Euijin Jung, Sion An, Philip Chikontwe,
and Sang Hyun Park$^{(\boxtimes)}$

Department of Robotics Engineering, DGIST, Daegu, South Korea
{won548,euijin,sion_an,philipchicco,shpark13135}@dgist.ac.kr

Abstract. Recently, self-supervised learning methods able to perform image denoising without ground truth labels have been proposed. These methods create low-quality images by adding random or Gaussian noise to images and then train a model for denoising. Ideally, it would be beneficial if one can generate high-quality CT images with only a few training samples via self-supervision. However, the performance of CT denoising is generally limited due to the complexity of CT noise. To address this problem, we propose a novel self-supervised learning-based CT denoising method. In particular, we train pre-train CT denoising and noise models that can predict CT noise from Low-dose CT (LDCT) using available LDCT and Normal-dose CT (NDCT) pairs. For a given test LDCT, we generate Pseudo-LDCT and NDCT pairs using the pre-trained denoising and noise models and then update the parameters of the denoising model using these pairs to remove noise in the test LDCT. To make realistic Pseudo LDCT, we train multiple noise models from individual images and generate the noise using the ensemble of noise models. We evaluate our method on the 2016 AAPM Low-Dose CT Grand Challenge dataset. The proposed ensemble noise models can generate realistic CT noise, and thus our method significantly improves the denoising performance existing denoising models trained by supervised- and self-supervised learning.

1 Introduction

Computed Tomography (CT) is a widely used medical imaging modality to visualize patient anatomy using X-rays. However, X-rays are known to be harmful to the human body as radiation may cause genetic and cancerous diseases. To address this, Low-dose CT (LDCT) images are often taken by lowering the X-ray exposure to reduce the dose yet inevitably inducing higher noise with lower quality. Though this mitigates the potential side-effects to patients, LDCT images are insufficient for accurate medical diagnosis compared to high quality Normal-dose CT (NDCT) images. Thus, robust denoising techniques are required to enhance the quality of LDCT.

In literature, several studies have attempted to address CT denoising using filtering [5,7,10,14,15], dictionary learning [6,22], and deep learning methods [3,4,20,21], respectively. Broadly speaking, existing methods try to maximize

© Springer Nature Switzerland AG 2021
I. Rekik et al. (Eds.): PRIME 2021, LNCS 12928, pp. 1–10, 2021.
https://doi.org/10.1007/978-3-030-87602-9_1

some prior given well defined noise model measurements and also exploit correlations among pixels to regulate the denoised image. For data-driven approaches, human experts are often required to mathematically model pixel correlations based analysis of clean images; limiting applicability when data is scarce. Thus, deep learning methods [3,4,17,19–21,23] are attractive given the recent impressive results for CT denoising, especially the supervised learning methods.

Despite the continued mainstream adoption of deep learning based techniques for medical image analysis, performance is still hindered by the availability of large and rich amounts of annotated data. For example, achieving high quality denoising requires several paired LDCT and NDCT images for models to learn an optimal mapping. Indeed, several techniques [11,13] have opted to train a denoiser without ground-truth labels for natural images where assumptions about noise are modeled based on surrounding pixel values including masking strategies to enforce the model to learn data complexity. However, such assumptions may not hold for medical images.

In this paper, we propose a novel self-supervised learning framework for CT denoising using Pseudo-CT noise images. We aim to directly address the limited availability of LDCT/NDCT images by defining a self-supervised task on a pair of LDCT and Pseudo noise CT generated images that can model different types of unseen noise during training. Specifically, we first pre-train a CT denoiser that predicts NDCT from LDCT, and also pre-train a LDCT noise model tasked with generating a noise difference map i.e. NDCT-LDCT, which we term Pseudo-CT for simplicity. Second, we train a new denoiser initialized with weights of the pre-trained denoiser by iteratively optimizing a self-supervised task between its output and the Pseudo-CT. Through learning to generate different types of unseen noise by ensembling across noise predictions for self-supervision, our approach produces efficient and high-quality denoising models. The main contributions of this work are as follows: (1) we propose a novel self-supervised learning scheme for CT denoising, (2) we show that our strategy can create realistic Pseudo-LDCT by leveraging diverse noise generated from ensemble noise models and, (3) Empirical results confirm that our method improves the performance of several existing methods.

2   Related Work

In this section, we discuss recent advances for CT denoising in two aspects; supervised LDCT denoising [3,4,17,19–21,23] and self-supervised denoising methods [1,11–13].

LDCT Denoising. Here, we first highlight conventional CT denoising techniques such as filtering [5,7,10,14,15] and dictionary based learning [6,22]. These techniques generally employ informative filters or dictionary patches to achieve denoising. Though they show impressive results, performance is largely limited by difficulty in modeling complex CT noise distributions.

Recently, several deep learning-based methods [3,20,21] have emerged for CT denoising. For instance, Chen et al. [4] proposed a simple shallow convolutional neural network(CNN) with three layers to remove noise. Following, Yang et al. [21] proposed a deeper model with perceptual loss to better preserve structural information. A more recent work by Won et al. [20] adopted intra- and inter-convolutions to extract features in both low and high-frequency components of CT images. As for generative approaches, Nishio et al. [17] proposed a CNN-based auto-encoder using NDCT with noise as input and predicts denoised NDCT. Similarly, Wolterinkt et al. [19] proposed a generative adversarial network(GAN) based approach for denoising. The generator generates CT noise and then subtracts from LDCT images to generate NDCT images, with a discriminator used to classify the generated NDCT as real or fake. However, these methods rely on the availability of several LDCT and NDCT paired samples to be successful.

Self-supervised Denoising. To address the shortcomings of supervised methods, self-supervised techniques are very attractive for CT denoising. These methods assume paired data is unavailable for training and define self-supervised tasks to achieve denoising performance on par with supervised methods. Lehtinen et al. [13] proposed Noise2Noise; a novel strategy to predict a clean image only using noisy image pairs. Furthermore, Batson et al. [2] (Noise2Self) and Krull et al. [11] (Noise2Void) both proposed masking strategies for denoising using only noisy images. These works specifically addressed the blind-spot problem; a common phenomena where a networks produces degenerative predictions by learning the identity in limited pixel receptive fields. By hiding or replacing specific pixels in the images, models are able to remove pixel-wise independent noise showing improved performance. Recently, Lee et al. [12] proposed a self-supervised denoising strategy by updating denoiser with pseudo image pairs. They generated pseudo clean and noisy images by using a pre-trained model with random Gaussian noise and then showed the self-similar image pairs help improving the performance of the denoiser. This seems reasonable for natural images, yet is unclear if such approaches can be directly applied to the medical domain.

Nevertheless, Hasan et al. [8] made early attempts with Noise2Noise [13] for CT denoising. Here, several generators networks called hybrid-collaborative generators were proposed to predict NDCT from many LDCT pairs. Hendriksen et al. [9] employed Noise2Self for sinogram reconstruction by dividing sinograms into several groups and later predict a sub-sinogram from the sub-sinograms. Notably, the borrowed techniques are domain specific solutions for denoising in natural images. Consequently, this may degrade denoising performance in CT and they cannot accurately reflect noise properties observed in CT. We address this by defining a new self-supervised learning scheme with iterative updates for model training.

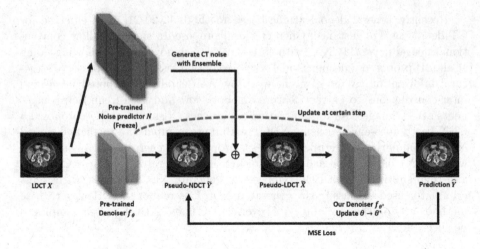

Fig. 1. Proposed self-supervised LDCT denoising method.

3   Method

Let us assume we already have a pre-trained LDCT denoiser f_θ trained with LDCT X and NDCT Y images i.e. f_θ can be trained with any learning method, supervised or self-supervised. Our goal is to denoise an unseen LDCT test image via self-supervised fine-tuning of f_θ with Pseudo noise CTs. To achieve this, we also pre-train a LDCT noise model N to generate Pseudo noisy LDCT $\tilde{X}$ using the difference map between LDCT and NDCT as supervison, with f_θ producing $\tilde{Y}$ i.e. Pseudo-NDCT. Following, to train an improved CT denoiser $f(\theta^\star)$ initialized the weights of f_θ, we simultaneously update f_θ and $f(\theta^\star)$ at different time steps using a self-supervised loss between the Pseudo-CT images and the predictions i.e. $\hat{Y} = f(\theta^\star)$. Our proposed method is shown in Fig. 1 with specific details expanded below.

Pre-training Noise Models. The proposed method requires a pre-trained LDCT noise model N to predict CT noise from X. Here, N is trained to predict noise map Z, which is the difference image between X and Y. In particular, we constructed pairs $\{X, Z\}$ to train N with each noise map obtained by subtracting X from Y. For this work, we train several N networks equal to the number of CT subjects in the training dataset used by f_θ i.e. we used 3 LDCT noise models. For all subsequent procedures, these models are fixed without parameter update.

Pseudo-CT Image Generation. To generate pseudo images for supervision, both f_θ and N are employed. Pseudo-CT images consist of a pair of $\tilde{X}$ and $\tilde{Y}$ and are generated as follows,

$$\tilde{Y} = f_\theta(X), \tag{1}$$

$$Z_m = N_m(X), \tag{2}$$

Algorithm 1. Parameter update procedure

Input: Low-Dose CT image X
Output: Denoised Image $\hat{Y}$
Pre-trained model: denoising model f_θ, noise models $\{N_1, \cdots, N_m\}$ where m
                is the number of training set
Hyperparameter: learning rate α, update period C

1 **for** $i \leftarrow 1$ to m **do**
2   Generate noise $Z_i = N_i(X)$

3 $count = 0$
4 $\theta^* \leftarrow \theta$
5 **while** *not done* **do**
6   Generate Pseudo-NDCT image $\tilde{Y} = f_\theta(X)$
7   Generate Pseudo-LDCT image $\tilde{X} = \tilde{Y} + Ensemble(Z)$
8   Parameter update $\theta^* \leftarrow \theta^* + \alpha \bigtriangledown_{\theta^*} \mathcal{L}(\tilde{Y}, f_{\theta^*}(\tilde{X}))$
9     $count \leftarrow count + 1$
10   **if** $count \equiv 0 \pmod{C}$ **then**
11        $\theta \leftarrow \theta^*$

12 Return $\hat{Y} = f_\theta(X)$

$$\tilde{X} = \tilde{Y} + Ensemble(Z), \tag{3}$$

where m is the index of N, with f_θ used to generate $\tilde{Y}$. As mentioned, f_θ is a pre-trained model tasked to predict X as Y. Once X is fed to f_θ, we obtain a clean version of X and denote this as Pseudo-NDCT $\tilde{Y}$. Also, the ensemble of N predictions is applied to generate $\tilde{X}$ by feeding X into N_m to predict noise Z_m. Specifically, to generate Z by ensembling; we randomly select pixels among Z_m for all required pixels. Repeating this process several times enables us to create diverse noise maps per input facilitating stable learning. Thus, by adding Z to $\tilde{Y}$ we can obtain the final Pseudo-LDCT image $\tilde{X}$. During training, only $\tilde{X}$ and $\tilde{Y}$ will be used for parameter updates.

Pseudo-CT Self-supervision. Given the generated Pseudo-CT images, improved denoising model $f(\theta^*)$ can be optimized via self-supervision. First, $f(\theta^*)$ is initialized with the parameters f_θ, with f_θ and N initially frozen when generating $\tilde{X}$ and $\tilde{Y}$ at each training step. Formally, $f(\theta^*)$ is optimized with the loss:

$$L(\tilde{Y}, \hat{Y}) = \frac{1}{K} \sum_{k=1}^{K} \|f_{\theta^*}(\tilde{X}) - \tilde{Y}\|_2^2. \tag{4}$$

where $\hat{Y}$ is prediction from $f(\theta^*)$ and K is the mini-batch size per step. In order to generate and provide high-quality Pseudo-LDCT images to $f(\theta^*)$, we update θ with θ^* after a fixed number of steps. This allows f_θ to provide continuously improved Pseudo-CT images to $f(\theta^*)$ and eventually improves CT denoising performance. Algorithm 1 provides the details of our method.

Table 1. Quantitative results on supervised and self-supervised method and various Pseudo-CT image generation.

Method	PSNR(std)	SSIM(std)
NDCT	46.21 ± 1.37	0.9773 ± 0.0072
N2V [11]	46.46 ± 1.23	0.9790 ± 0.0054
N2V+Ours w/o θ Update	49.22 ± 0.82	0.9893 ± 0.0027
N2V+Ours	**49.57 ± 0.82**	**0.9905 ± 0.0019**
N2N [13]	46.16 ± 0.93	0.9769 ± 0.0039
N2N+Ours w/o θ Update	46.69 ± 1.14	0.9794 ± 0.0055
N2N+Ours	**49.85 ± 0.86**	**0.9910 ± 0.0018**
N2C [3]	50.30 ± 0.93	0.9920 ± 0.0018
N2C+Ours w/o θ Update	50.47 ± 0.98	0.9922 ± 0.0018
N2C+Ours	**50.50 ± 0.98**	**0.9922 ± 0.0018**
N2C+Hist	48.65 ± 1.42	0.9872 ± 0.0045
N2C+Gaussian	49.94 ± 0.87	0.9911 ± 0.0019
N2C+Model+Hist	50.36 ± 0.95	0.9920 ± 0.0018

Implementation Details. For this work, we used REDCNN [3] as the pre-trained LDCT denoising model and our final CT denoiser, with U-Net [18] as the LDCT noise model. All methods employed CT patches of 64×64 in size. Data augmentation was used in all training and update steps. For data augmentation, random rescaling between [0.5, 2], and random flipping [Horizontal, Vertical] was used. An L1 loss was used to train the CT denoising and LDCT noise models, with Adam optimizer and a learning rate (LR) of 0.0001. A ReduceLROnPlateau scheduler was employed to adjust the LR by 50% of the initial value whenever the loss did not improve during training. All experiments were implemented on a Intel Xeon Gold 6132, 192GB RAM, NVIDIA TITAN Xp workstation with the PyTorch deep learning framework.

4   Experimental Results

Dataset and Experimental Settings. For evaluation, the 2016 AAPM Low-Dose CT Grand Challenge dataset [16] was used and split into 3 train and 7 test, respectively. It consists of abdominal LDCT and NDCT images obtained from 10 patients with image size 512 × 512. The voxel space of CT images is 0.5 mm × 0.5 mm with 3 mm slice thickness. To demonstrate that our method works with any existing learning method, we applied our method to N2C (Noise2Clean), N2N (Noise2Noise) [13], and N2V (Noise2Void) [11]. N2C is a supervised learning method that uses all available labeled paired data. N2N and N2V are self-supervised learning methods which use noisy pair images and masking schemes. Also, to demonstrate the effectiveness of our noise model, we compared LDCT denoising performance against existing noise generation techniques

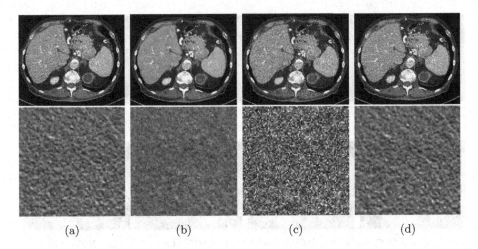

(a)   (b)   (c)   (d)

Fig. 2. Comparison of generated Pseudo-LDCT images and their noise. (a) LDCT, (b) Random noise histogram, (c) Gaussian noise, (d) Ours. The right-bottom patch indicates the noise lying on each image. The noise quality of (b) and (c) shows that random noise cannot represent (a). In contrast, (d) shows almost similar to (a)

i.e. Random Noise histogram (Hist), Gaussian noise (Gaussian), and Single noise model+Noise histogram (Model+Hist). Hist samples noise from the difference map based on the histogram between LDCT and NDCT images, whereas Gaussian samples the noise from a Gaussian distribution with zero mean and 0.02 standard deviation. Model+Hist is the combination of a single pre-trained LDCT noise model (using all subjects in the training dataset) and Hist. For evaluation, Peak Signal-to-Noise Ratio (PSNR) and Structural Similarity (SSIM) are reported.

Quantitative Results. Table 1 shows the average PSNR and SSIM scores for supervised and self-supervised methods. Among the compared methods, N2C showed the highest LDCT denoising performance without the proposed techniques; with more significant improvements for all methods when the our methods were applied. Though N2N and N2V report higher scores than NDCT, with similar trends on natural image denoising, they were not very effective when applied to CT. We believe training by merely adding random noise (e.g., Gaussian), often very different from noise observed in CT is not useful for denoising, leading to poor results. Indeed, our method improved every learning method's performance showing that the technique is model agnostic.

Effect of Noise Model. For parameter updates, it is crucial to generate high-quality Pseudo-LDCT images i.e. images similar to an actual LDCT image. If high-quality Pseudo-LDCT images are provided to the model for training, CT denoising performance can be improved. In contrast, using low-quality Pseudo-CT images may adversely affects the model learning, and consequently lead

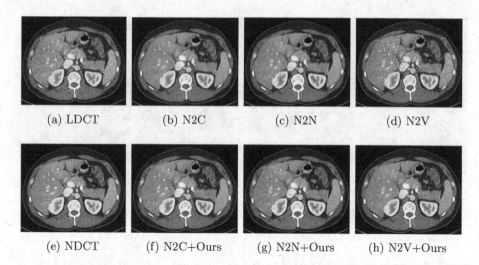

(a) LDCT   (b) N2C   (c) N2N   (d) N2V

(e) NDCT   (f) N2C+Ours   (g) N2N+Ours   (h) N2V+Ours

Fig. 3. Comparison of existing methods and Ours. (a) LDCT, (b) N2C, (c) N2N, (d) N2V, (e) NDCT, (f) N2C+Ours, (g) N2N+Ours, (h) N2V+Ours.

to decreased denoising performance. Table 1 also shows the average PSNR and SSIM in different Pseudo-CT image generation settings. Here, both Hist and Gaussian reported lower performance compared to N2C. Based on our observations, this serves to show that random noise employed in natural images is not useful for CT denoising. In contrast, when our noise model was combined with Hist i.e. Model+Hist, performance gains over N2C were noted. This implies that our noise model can generate reasonable noise similar to the actual CT noise, and also improves performance. Furthermore, our ensemble noise models without parameter update show improved results over Model+Hist without the need for additional random noise.

In Figs. 2 and 3, we show comparison results of Pseudo-LDCT images and their generated noise using various methods, as well as the predictions of our method for each. In Fig. 2, Pseudo-LDCT images and their noise with Hist and Gaussian highlight a huge discrepancy between the actual LDCT image and its noise. In the case of random noise that is shown to be independently distributed across the entire image, it is often useful for natural image denoising. Moreover, this types of noise does not accurately reflect the nature of CT images. Through the proposed method, we show it is possible to generate Pseudo-CT images that preserve overall CT image characteristics with high quality (Fig. 3).

5   Conclusions

In this paper, we proposed a novel self-supervised CT denoising method using Pseudo-CT images. We showed by using a noise model and generated Pseudo-CT images to define a self-supervised learning scheme, we can produce robust denoising model without relying on a large collection of labeled images. Also, we provide concrete evidence via extensive experimentation that our approach is model-agnostic and can improve CT denoising performance for existing approaches.

Acknowledgments. This work was supported by the grant of the medical device technology development program funded by the Ministry of Trade, Industry and Energy (MOTIE, Korea) (20006006), and the grant of the High-Potential Individuals Global Training Program (2019-0-01557) supervised by the IITP(Institute for Information & Communications Technology Planning & Evaluation).

References

1. Batson, J., Royer, L.: Noise2Self: blind denoising by self-supervision. In: International Conference on Machine Learning, pp. 524–533. PMLR (2019)
2. Batson, J., Royer, L.: Noise2Self: blind denoising by self-supervision. In: Chaudhuri, K., Salakhutdinov, R. (eds.) Proceedings of the 36th International Conference on Machine Learning, ICML 2019, Long Beach, California, USA, 9–15 June 2019, Proceedings of Machine Learning Research, vol. 97, pp. 524–533. PMLR (2019). http://proceedings.mlr.press/v97/batson19a.html
3. Chen, H., et al.: Low-dose CT with a residual encoder-decoder convolutional neural network. IEEE Trans. Med. Imaging **36**(12), 2524–2535 (2017)
4. Chen, H., et al.: Low-dose CT via convolutional neural network. Biomed. Opt. Express **8**(2), 679–694 (2017)
5. Chen, Y., et al.: Thoracic low-dose CT image processing using an artifact suppressed large-scale nonlocal means. Phys. Med. Biol. **57**(9), 2667 (2012)
6. Chen, Y., et al.: Improving abdomen tumor low-dose CT images using a fast dictionary learning based processing. Phys. Med. Biol. **58**(16), 5803 (2013)
7. Feruglio, P.F., Vinegoni, Č., Gros, J., Sbarbati, A., Weissleder, R.: Block matching 3D random noise filtering for absorption optical projection tomography. Phys. Med. Biol. **55**(18), 5401 (2010)
8. Hasan, A.M., Mohebbian, M.R., Wahid, K.A., Babyn, P.: Hybrid collaborative Noise2Noise denoiser for low-dose CT images. IEEE Trans. Radiat. Plasma Med. Sci. **5**, 235–244 (2020)
9. Hendriksen, A.A., Pelt, D.M., Batenburg, K.J.: Noise2Inverse: self-supervised deep convolutional denoising for tomography. IEEE Trans. Comput. Imaging **6**, 1320–1335 (2020)
10. Kang, D., et al.: Image denoising of low-radiation dose coronary CT angiography by an adaptive block-matching 3D algorithm. In: Medical Imaging 2013: Image Processing, vol. 8669, p. 86692G. International Society for Optics and Photonics (2013)
11. Krull, A., Buchholz, T.O., Jug, F.: Noise2Void-learning denoising from single noisy images. In: Proceedings of the IEEE/CVF Conference on Computer Vision and Pattern Recognition, pp. 2129–2137 (2019)
12. Lee, S., Lee, D., Cho, D., Kim, J., Kim, T.H.: Restore from restored: single image denoising with pseudo clean image (2020)
13. Lehtinen, J., et al.: Noise2Noise: learning image restoration without clean data. In: International Conference on Machine Learning, pp. 2965–2974. PMLR (2018)
14. Li, Z., et al.: Adaptive nonlocal means filtering based on local noise level for CT denoising. Med. Phys. **41**(1), 011908 (2014)

15. Ma, J., et al.: Low-dose computed tomography image restoration using previous normal-dose scan. Med. Phys. **38**(10), 5713–5731 (2011)

16. McCollough, C.: TU-FG-207A-04: overview of the low dose CT grand challenge. Med. Phys. **43**(6Part35), 3759–3760 (2016)

17. Nishio, M., et al.: Convolutional auto-encoder for image denoising of ultra-low-dose CT. Heliyon **3**(8), e00393 (2017)

18. Ronneberger, O., Fischer, P., Brox, T.: U-Net: convolutional networks for biomedical image segmentation. In: Navab, N., Hornegger, J., Wells, W.M., Frangi, A.F. (eds.) MICCAI 2015. LNCS, vol. 9351, pp. 234–241. Springer, Cham (2015). https://doi.org/10.1007/978-3-319-24574-4_28

19. Wolterink, J.M., Leiner, T., Viergever, M.A., Išgum, I.: Generative adversarial networks for noise reduction in low-dose CT. IEEE Trans. Med. Imaging **36**(12), 2536–2545 (2017)

20. Won, D.K., An, S., Park, S.H., Ye, D.H.: Low-dose CT denoising using octave convolution with high and low frequency bands. In: Rekik, I., Adeli, E., Park, S.H., Valdés Hernández, M.C. (eds.) PRIME 2020. LNCS, vol. 12329, pp. 68–78. Springer, Cham (2020). https://doi.org/10.1007/978-3-030-59354-4_7

21. Yang, Q., Yan, P., Kalra, M.K., Wang, G.: CT image denoising with perceptive deep neural networks. arXiv preprint arXiv:1702.07019 (2017)

22. Zhang, H., Zhang, L., Sun, Y., Zhang, J.: Projection domain denoising method based on dictionary learning for low-dose CT image reconstruction. J. Xray Sci. Technol. **23**(5), 567–578 (2015)

23. Zhong, A., Li, B., Luo, N., Xu, Y., Zhou, L., Zhen, X.: Image restoration for low-dose CT via transfer learning and residual network. IEEE Access **8**, 112078–112091 (2020)

A Few-Shot Learning Graph Multi-trajectory Evolution Network for Forecasting Multimodal Baby Connectivity Development from a Baseline Timepoint

Alaa Bessadok[1,2,3], Ahmed Nebli[1,4], Mohamed Ali Mahjoub[2,3],
Gang Li[5], Weili Lin[5], Dinggang Shen[6,7], and Islem Rekik[1(✉)]

[1] BASIRA Lab, Faculty of Computer and Informatics, Istanbul Technical University, Istanbul, Turkey
irekik@itu.edu.tr
[2] Higher Institute of Informatics and Communication Technologies, University of Sousse, Sousse 4011, Tunisia
[3] National Engineering School of Sousse, University of Sousse, LATIS- Laboratory of Advanced Technology and Intelligent Systems, Sousse 4023, Tunisia
[4] National School for Computer Science (ENSI), Mannouba, Tunisia
[5] Department of Radiology and BRIC, University of North Carolina at Chapel Hill, Chapel Hill, NC, USA
[6] School of Biomedical Engineering, ShanghaiTech University, Shanghai, China
[7] Department of Research and Development, United Imaging Intelligence Co., Ltd., Shanghai, China
http://basira-lab.com

Abstract. Charting the baby connectome evolution trajectory during the first year after birth plays a vital role in understanding dynamic connectivity development of baby brains. Such analysis requires acquisition of longitudinal connectomic datasets. However, both neonatal and postnatal scans are rarely acquired due to various difficulties. A small body of works has focused on predicting baby brain evolution trajectory from a neonatal brain connectome derived from a single modality. Although promising, large training datasets are essential to boost model learning and to generalize to a multi-trajectory prediction from different modalities (i.e., functional and morphological connectomes). Here, we unprecedentedly explore the question: *"Can we design a few-shot learning-based framework for predicting brain graph trajectories across different modalities?"* To this aim, we propose a Graph Multi-Trajectory Evolution Network (GmTE-Net), which adopts a teacher-student paradigm where the teacher network learns on pure neonatal brain graphs and the student network learns on simulated brain graphs given a set of different timepoints. To the best of our knowledge, this is the first teacher-student architecture tailored for brain graph multi-trajectory growth prediction that is based on few-shot learning and generalized to graph neural networks (GNNs). To boost the performance of the student network, we

I. Rekik et al. (Eds.): PRIME 2021, LNCS 12928, pp. 11–24, 2021.
https://doi.org/10.1007/978-3-030-87602-9_2

introduce a local topology-aware distillation loss that forces the predicted graph topology of the student network to be consistent with the teacher network. Experimental results demonstrate substantial performance gains over benchmark methods. Hence, our GmTE-Net can be leveraged to predict atypical brain connectivity trajectory evolution across various modalities. Our code is available at https://github.com/basiralab/GmTE-Net.

Keywords: Few-shot learning using graph neural network · Multimodal brain graph evolution prediction · Baby connectome development · Knowledge distillation network

1   Introduction

During the first year after birth, the baby brain undergoes the most critical and dynamic postnatal development in structure, function, and morphology [1]. Such dramatic changes are highly informative for future brain diseases that can be easily preventable if *predicted* at an early stage [2,3]. However, this problem is challenging due to the scarcity of longitudinal datasets. Thus, there is an increasing need to foresee longitudinal missing baby connectomic data given what already exists [4]. Network literature [5] shows that a typical brain connectome is defined as a graph of brain connectivities where nodes denote regions of interest (ROIs) and edges encode the pairwise connectivity strengths between ROI pairs. As such, [6] attempted to predict baby brain evolution trajectory from a single neonatal brain graph. The authors designed an ensemble of regressors for a sample selection task, thereby progressively predicting a follow-up baby brain connectome from a single timepoint. While effective, this method is limited in the following ways: (i) the learning model lacks preservation of topological properties of the original brain graphs, and (ii) the framework is not designed in an end-to-end learning manner which may lead to accumulated errors across the learning steps. To overcome these limitations, [7] proposed a model which learns an adversarial brain graph embedding using graph convolution networks (GCN) [8] and introduced a sample selection strategy guided by a population brain *template*. Another work also proposed a infinite sample selection strategy for brain network evolution prediction from baseline [9]. However, both models were dichotomized into subparts during the learning step. In another study, [10] proposed an end-to-end learning framework based on Generative Adversarial Network (GAN) that successfully predicts the follow-up brain graphs in a cascaded manner. However, such design choice exclusively predicts *uni-trajectory* evolution where brain graphs are derived from a single modality, limiting its generelizability to *multi-trajectory* graph evolution prediction where each trajectory represents a different brain modality (i.e., structural and functional connectomes) (Fig. 1).

A second major limitation is requiring many training brain graphs to boost model learning, reducing its learning capacity in a frugal setting where only *a few* training samples exist. Specifically, due to the high medical image acquisition costs, it is often challenging to acquire all modalities for each subject. Hence, it is mandatory to learn how to predict brain graphs from a small number of graphs effectively. This has initiated a new line of research, namely few-shot learning, which aims to bridge the gap between the breadth of deep learning algorithms and the necessity to use small data sets [11,12]. In other words, few-shot learning aims to extract the most relative information from each training data point without the need for acquiring a large number of instance.

In this context, several works have demonstrated the feasibility of learning given few samples [13,14]. However, none of these addressed the problem of multi-trajectory evolution prediction. To tackle these limitations, **we propose Graph Multi-Trajectory Evolution Network (GmTE-Net), the first few-shot learning framework designed for baby brain graph multi-trajectory evolution prediction using a teacher-student (TS) paradigm** [15]. Namely, TS is a set of two networks where an extensive network called teacher aims to transfer its knowledge to a smaller network called student. Intuitively, if the teacher network can efficiently transfer its knowledge to the student network, then we can affirm the high learning quality of the teacher, and we can condition the learning curve of the student. Such architecture is highly suitable for few-shot learning [16] thanks to its high ability to generalize to different data distributions and enforce robustness against perturbations.

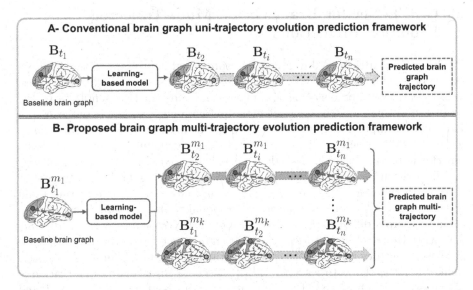

Fig. 1. Conventional brain graph uni-trajectory evolution prediction methods and the proposed brain graph multi-trajectory evolution prediction architecture.

Also, unlike [10], this architecture breaks free from the cascaded scheme of stacked networks enabling prediction to go "multi-trajectory". Our model aims to ensure the biological soundness of the predicted brain graphs by defining a *global topology loss* to train the teacher network. For the student network, we propose a novel *local topology-aware distillation loss* which enforces the student model to capture the potentially complex dynamics of the local graph structure generated by the teacher over time. This enables topology-aware knowledge transfer from the teacher, yielding a compact yet high-performance student model. Ultimately, we present the multi-level contributions of our work:

1. *On a conceptual level.* Our proposed GmTE-Net is the first teacher-student framework tailored for jointly predicting multiple trajectories of infant brain graphs from a single neonatal observation.
2. *On a methodological level.* GmTE-Net is a novel knowledge distillation approach where the student model benefits from the few-shot learning of the teacher model. We also propose a local topology-aware distillation loss to transfer the local graph knowledge from the teacher to the student. We further propose a global topology loss to optimize the learning of the teacher network.
3. *On clinical level.* GmTE-Net provides a great opportunity for charting the rapid and dynamic brain developmental trajectories of baby connectomes which help to spot early neurodevelopmental disorders e.g., autism.

2   GmTE-Net for Brain Graph Multi-Trajectory Evolution Prediction

Problem Definition. A brain can be represented as a graph $\mathcal{B} = \{\mathbf{R}, \mathbf{E}, \mathbf{v}\}$ where $\mathbf{R}$ is a set of nodes (i.e., brain ROIs) and $\mathbf{E}$ is a set of weighted edges. $\mathbf{v}$ is a feature vector denoting a compact and reduced representation of the brain connectivity matrix measuring the pairwise edge weight between nodes. Each baby sample s is represented by a set of brain graph *multi-trajectories* $\mathcal{T}^s = \{\mathcal{T}^{m_j}\}_{j=1}^{n_m}$ where $s \in \{1, \ldots, n_s\}$ and n_m is the number of modalities (e.g., functional, morphological). A single trajectory derived from the m_j-th modality can be written as $\mathcal{T}^{m_j} = \{\mathbf{v}_{t_i}^{m_j}\}_{i=1}^{n_t}$ where $\mathbf{v}_{t_i}^{m_j}$ denotes brain feature vector at the i-th timepoint t_i, where $i \in \{1, \ldots, n_t\}$. Given a baseline testing brain feature vector $\mathbf{v}_{t_1}^{m_1}$ derived from a single modality m_1, we aim to predict its evolution across different n_m trajectories: one *base* trajectory derived from m_1 and other $\{n_m - 1\}$ trajectories derived from different modalities. Note that for the base trajectory we only need to predict the evolving graph for $t \in \{2, \ldots, n_t\}$ unlike the other trajectories $\{\{\hat{\mathbf{v}}_{t_i}^{m_j}\}_{i=1}^{n_t}\}_{j=2}^{n_m}$ where the neonatal observation is also missing. We note that for simplicity, $k = n_m$ and $n = n_t$ in Fig. 1 and Fig. 2.

A- Multimodal Longitudinal Brain Graph Feature Extraction. The ultimate goal of this work is to predict a set of evolution trajectories given a single brain graph, each mapping a particular brain modality. Since a brain graph is encoded in a symmetric matrix composed of weighted edge values, we propose

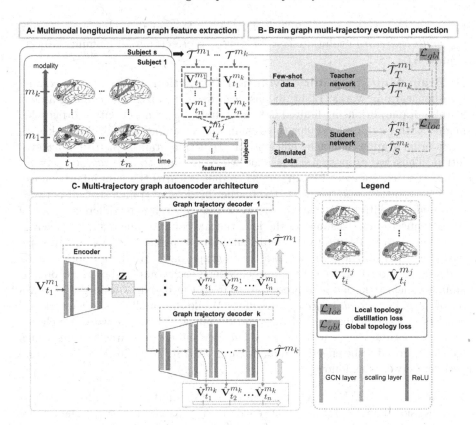

Fig. 2. *Pipeline of the proposed GmTE-Net framework for predicting multi-trajectory evolution of baby brain graphs from a single neonatal brain graph.* **(A) Multimodal longitudinal brain graph feature extraction.** Each subject is represented by n_m trajectories (i.e., $\mathcal{T}^{m_j}$), each representing a set of brain graphs derived from multiple modalities at a specific timepoint. Since each brain graph is encoded in a symmetric matrix, we vectorize the off-diagonal upper-triangular part and stack the resulted vectors in a matrix $\mathbf{V}_{t_i}^{m_j}$ for each brain modality. **(B) Brain multi-trajectory evolution prediction.** We first train a teacher network with a few-shot learning strategy. Given a feature matrix $\mathbf{V}_{t_1}^{m_1}$ representing a baseline timepoint, we aim to predict multi-trajectory evolution where each trajectory is a set of follow-up brain graphs of a specific modality (i.e., $\hat{\mathcal{T}}_T^{m_j}$). Second, we propose to train a student network on simulated data to boost the generalization capability in learning from any data distribution. We regularize the teacher network with a *global topology loss* and the student network with a *local topology-aware distillation loss* so that they capture both global and local node properties. **(C) Multi-trajectory graph autoencoder architecture.** Both teacher and student are GCN-based graph autoencoders encapsulating an encoder and a set of decoders each aiming to predict a graph trajectory for a specific modality.

to vectorize its off-diagonal upper-triangular part. We stack the resulting feature vectors of all subjects into a matrix $\mathbf{V}_{t_i}^{m_j}$ representing the brain features derived from the modality m_j at timepoint t_i. We repeat the same step for each timepoint to obtain a tensor $\mathcal{T}^{m_j} = \{\mathbf{V}_{t_i}^{m_j}\}_{i=1}^{n_t}$. For our training set, we obtain n_m *multi-trajectory* evolution $\mathcal{T} = \{\mathcal{T}^{m_j}\}_{j=1}^{n_m}$ representing all training subjects (Fig. 2–A).

B- Brain Multi-trajectory Evolution Prediction. Owing to the existing few-shot learning works [13,14], it has been demonstrated that training a deep learning framework using limited data is feasible, but it never outperforms same models with large data. Under this assumption, training a single network with a few brain graphs cannot be efficient for predicting brain *multi-trajectory* evolution. Thus, we propose a TS scheme where we first train a teacher network using a limited number of brain graphs so that it learns the *pure* infant connectivity patterns. Second, we freeze the teacher model and train a student network on *simulated* data such that the network can generalize to different types of brain connectomes. To simulate the longitudinal connectomic data, we generate a random dataset having the same distribution properties (i.e., mean and standard deviation) of the pure neonatal brain graph data. We design both teacher and student models as two graph autoencoders; each composed of an encoder E aiming to learn the latent representation from the baseline brain graph and a set of $\{D^{m_j}\}_{j=1}^{n_m}$ *graph trajectory* decoders. While the vanilla TS framework [15] consists in a larger teacher network and a shallow student network, we choose to design both networks with identical encoder and decoder architectures which has been proved to be effective in many tasks [17,18]. We define the encoder and decoder networks as follows:

1. ***The encoder.*** $E(\mathbf{V}_{t_i}^{m_j}, \mathbf{A})$ aims to map a set of input brain graphs into a low-dimensional space. To do so, our encoder takes as an input $\mathbf{V}_{t_i}^{m_j}$, the feature matrix stacking brain vectors of samples at the initial timepoint t_i for m_j and an adjacency matrix $\mathbf{A}$. We initialize the adjacency matrix as an identity one to (i) eliminate redundancy of edge weights and (ii) compensate for the vectorized edge features. The proposed encoder architecture is composed of two layers, each containing a GCN layer, followed by a Rectified Linear Unit (ReLU) and a dropout function.

2. ***The decoder.*** $D(\mathbf{Z}^l, \mathbf{A})$ aims to predict a modality-specific trajectory m_j for the follow-up brain graphs. To do so, our decoder takes as input the latent space $\mathbf{Z}^l$, and an identity matrix $\mathbf{I}$. Similarly to the encoder, we define the decoder as a two-layer GCN, each followed by a ReLU and a dropout functions. This elementary GCN-based architecture is cascaded n_t times where each predicts a brain feature matrix at a different timepoint. Specifically, the first GCN layer might play a scaling role in case of graph resolution (i.e., node size) difference between initial and target domains. In such scenario, it maps the base brain graph to the follow-up trajectory-specific resolutions. As so, our decoder is able to predict brain graphs even if there are resolution shifts across timepoints.

Our proposed autoencoder is capable of foreseeing multiple trajectories given a single brain graph modality acquired at the initial timepoint. *In essence, from a single shared low-dimensional space of baseline brain graphs it charts*

out the follow-up brain graphs for different trajectories and with potential reso-
lution shifts across trajectories. We define the propagation rule of our GCN-based
autoencoder as follows:

$$\mathbf{Z}^{(l)} = f_{ReLU}(\mathbf{F}^{(l)}, \mathbf{A}|\mathbf{W}^{(l)}) = ReLU(\widetilde{\mathbf{D}}^{-\frac{1}{2}}\widetilde{\mathbf{A}}\widetilde{\mathbf{D}}^{-\frac{1}{2}}\mathbf{F}^{(l)}\mathbf{W}^{(l)})$$

$\mathbf{Z}^{(l)}$ is the learned graph representation resulting from the layer l. For the
encoder, $\mathbf{F}^{(l)}$ is defined as $\mathbf{V}_{t_i}^{m_j}$ in the first layer and as the learned embeddings $\mathbf{Z}^l$ in the second layer. We define the output of the decoder for n_t layers
as the trajectory $\mathcal{T}^{m_j} = \{\mathbf{V}_{t_i}^{m_j}\}_{i=1}^{n_t}$. $\mathbf{W}^{(l)}$ is a filter denoting the graph convolutional weights in layer l. We define the graph convolution function as $f_{(.)}$ where
$\widetilde{\mathbf{A}} = \mathbf{A} + \mathbf{I}$ with $\mathbf{I}$ being an identity matrix, and $\widetilde{\mathbf{D}}_{ii} = \sum_j \widetilde{\mathbf{A}}_{ij}$ is a diagonal
matrix (Fig. 2–C).

Despite the effective graph representation learning of GCN, it might fail in
preserving both global and local graph topological structures. To enforce the
teacher model to capture the high-order representation of input brain graphs,
we introduce a *global topology loss* $\mathcal{L}_{glob}^{m_j}$ defined as the average mean absolute
error (MAE) between the ground truth brain features $\mathbf{V}_{t_i}^{m_j}$ and the predicted
ones $\hat{\mathbf{V}}_{t_i}^{m_j}$ over timepoints. This loss aims to learn the global graph structure
given few examples per modality, thereby synthesizing biologically sound brain
graph *multi-trajectory* predictions. We define the teacher loss function and our
global topology loss as follows:

$$\mathcal{L}_{Teacher} = mean(\sum_{j=1}^{n_m} \mathcal{L}_{glob}^{m_j}); \quad \mathcal{L}_{glob}^{m_j} = mean(\sum_{i=1}^{n_t} \ell_{MAE}(\mathbf{V}_{t_i}^{m_j}, \hat{\mathbf{V}}_{t_i}^{m_j}))$$

Following the teacher training, we freeze it in order to train the student model.
To generalize the student network to be distribution agnostic, we train it on a
large set of simulated data. To this end, it is difficult to assume node property
preservation for the predicted brain graphs with respect to the ground-truth
training. Interestingly, brain graphs have unique topological properties for functional, structural and morphological connectivities [19] that should be preserved
when synthesizing the brain graphs of a specific trajectory [20,21]. Thus, we
introduce a *local topology-aware distillation loss* that aims to force the student
network to mimic the local topology of the predicted graphs from the teacher
model. As so, we are training the student using the local topological knowl-
edge distilled from the teacher network. We denote our loss by $\mathcal{L}_{loc}^{m_j}$ defined as
the MAE between the centrality scores of the ground truth and the predicted
graphs. We choose the closeness centrality measure since it is mostly used in
graph theory [22]. It quantifies the closeness of a node to all other nodes [23]
which is defined by the following formula: $C(r^a) = \frac{r-1}{\sum_{r^a \neq r^b} p_{r^a r^b}}$, where r denotes
the number of nodes and $p_{r^a r^b}$ the length of the shortest path between nodes
r^a and r^b. So, we first create a centrality vector for each brain graph where its
elements represent the computed centrality scores for the graph nodes. Then, we
vertically stack the resulting vectors and create a centrality matrices $\mathbf{C}$ and $\hat{\mathbf{C}}$

for ground-truth and predicted brain graphs, respectively. Therefore, we define the student loss function and the *local topology-aware distillation loss* as follows:

$$\mathcal{L}_{Student} = mean(\sum_{j=1}^{n_m} \mathcal{L}_{loc}^{m_j}); \quad \mathcal{L}_{loc}^{m_j} = mean(\sum_{i=1}^{n_t} \ell_{MAE}(\mathbf{C}_{t_i}^{m_j}, \hat{\mathbf{C}}_{t_i}^{m_j}))$$

In that way, the student network is not only preserving the local topology of the brain graphs but also their global topology yielding an accurate student model.

3   Results and Discussion

Dataset. We evaluated our framework on 11 baby brain graphs of developing infant subjects each has 4 serial t_2-w MRI and resting-state fMRI (rsfMRI) scans acquired at 1, 3, 6, and 9 months of age, respectively. We create for each subject two brain graphs derived from two different brain parcellations: a functional brain graph of size 116×116 using AAL template [24] and a morphological brain graph of size 35×35 using Desikan-Killiany Atlas [25]. Out of these 11 subjects, we have four complete (i.e., have all observations and modalities across all timepoints). Using a single Tesla V100 GPU (NVIDIA GeForce GTX TITAN with 32 GB memory), we train our teacher network using these four complete subjects and test both teacher and student networks using the whole dataset.

Comparison Methods and Evaluation. Due to the lack of existing frameworks that aim to predict the multi-trajectory evolution of brain graphs, we benchmark our GmTE-Net against its two ablated versions and two variants, each trained with three different loss functions computed between the predicted and ground truth graphs. Among the three loss functions we have, $\mathcal{L}_a$, $\mathcal{L}_b$, and $\mathcal{L}^\star$, denoting the MAE loss, the Wasserstein distance, and our proposed loss optimizing the training of both teacher and student networks. We detail our benchmarking methods as follows:

1. *Ablated (augmented + real data).* Here, we use a single decoder with 4 stacked layers to predict the multi-trajectory brain evolution where we used both real and simulated data to train both teacher and student networks.
2. *Ablated (few-shot learning).* Here, we use the same architecture, but we train the teacher on a few-shots from our dataset, and we train the student on the simulated brain graphs.
3. *GmTE-Net and GmTE-Net*.* In these versions, we train our proposed GmTE-Net with and without few-shot learning strategy, respectively.

To evaluate our proposed GmTE-Net, we compute for each timepoint the MAE and the centrality scores (i.e., eigenvector (EC) and PageRank (PC) centralities metrics) between the ground truth and the predicted graphs. Mainly, we evaluate both teacher and student networks. Then, we report the results for each timepoint in Table 3 and Table 4 and we report the average prediction results across four timepoints in Table 1 and Table 2. Specifically, Table 1

Table 1. Morphological connectomic data-based comparison of our GmTE-Net with its two ablated versions in terms of architecture (first two rows) and its two variants (last two rows). We highlight in bold, blue, green and purple the best performances got for each method when using $\mathcal{L}_a$, $\mathcal{L}_b$ and $\mathcal{L}^*$ and loss functions, respectively.

Models		Student evaluation			Teacher evaluation		
		MAE(graph)	MAE(EC)	MAE(PC)	MAE(graph)	MAE(EC)	MAE(PC)
Ablated (augmented + real data)	$\mathcal{L}_a$	0.40821	0.02948	0.00393	0.34529	0.02818	0.00371
	$\mathcal{L}_b$	0.48441	0.0407	0.00569	0.48505	0.04715	0.00641
	$\mathcal{L}^*$	0.48436	0.04074	0.00546	0.35087	0.0284	0.00377
Ablated (few-shot learning)	$\mathcal{L}_a$	0.49334	0.08182	0.01016	0.3481	**0.0123**	0.00177
	$\mathcal{L}_b$	0.48429	0.03993	0.00572	0.48358	0.04024	0.00566
	$\mathcal{L}^*$	0.48429	0.03993	0.00573	0.34932	**0.01225**	**0.00176**
GmTE-Net	$\mathcal{L}_a$	0.40856	0.03032	0.00404	**0.31938**	0.02851	0.00376
	$\mathcal{L}_b$	0.48444	0.03948	**0.00555**	0.48316	0.04779	0.00666
	$\mathcal{L}^*$	0.48413	0.04666	0.00625	0.33871	0.02845	0.00376
GmTE-Net*	$\mathcal{L}_a$	**0.40391**	**0.01589**	**0.00217**	0.32215	0.01242	**0.00174**
	$\mathcal{L}_b$	**0.48369**	**0.03875**	0.00565	0.48398	0.03997	0.00522
	$\mathcal{L}^*$	**0.48374**	**0.03846**	**0.00545**	**0.3145**	0.01314	0.00189

Table 2. Functional connectomic data-based comparison of our GmTE-Net with its two ablated versions in terms of architecture (first two rows) and its two variants (last two rows). We highlight in bold, blue, green and purple the best performances got for each method when using $\mathcal{L}_a$, $\mathcal{L}_b$ and $\mathcal{L}^*$ and loss functions, respectively.

Models		Student evaluation			Teacher evaluation		
		MAE(graph)	MAE(EC)	MAE(PC)	MAE(graph)	MAE(EC)	MAE(PC)
Ablated (augmented + real data)	$\mathcal{L}_a$	0.11913	**0.00972**	**0.0008**	0.13245	**0.00959**	**0.00078**
	$\mathcal{L}_b$	0.10437	0.01332	**0.00108**	**0.10438**	0.01356	0.00113
	$\mathcal{L}^*$	0.48436	0.04074	0.00546	0.35087	0.0284	0.00377
Ablated (few-shot learning)	$\mathcal{L}_a$	0.10347	0.04136	0.00251	0.0987	0.01375	0.00119
	$\mathcal{L}_b$	0.10442	0.01454	0.00117	0.10442	0.01319	0.00106
	$\mathcal{L}^*$	0.10442	0.01454	0.00118	0.09867	0.014	0.00121
GmTE-Net	$\mathcal{L}_a$	0.11572	0.00989	0.00081	0.12914	0.00969	0.00079
	$\mathcal{L}_b$	**0.10434**	**0.01318**	0.00109	0.10441	0.01374	0.00112
	$\mathcal{L}^*$	**0.10436**	**0.01234**	**0.00103**	0.11921	**0.00956**	**0.00078**
GmTE-Net*	$\mathcal{L}_a$	**0.10065**	0.01432	0.00124	**0.09826**	0.01446	0.00125
	$\mathcal{L}_b$	0.10439	0.01378	0.00111	0.10443	0.01428	0.00116
	$\mathcal{L}^*$	0.10439	0.01368	0.00111	**0.09814**	0.01426	0.00125

shows the significant outperformance of GmTE-Net* over the baseline methods when tested on morphological brain graphs. This demonstrates the breadth of our few-shot learning strategy in accurately training the teacher network. Thus, we can confirm the high-quality knowledge transferred from the teacher to the student network. Our GmTE-Net* trained with the proposed global and local loss functions significantly outperformed its variants when using other losses ($p < 0.05$ using two-tailed paired t-test), which highlights the impact of our proposed topology-aware distillation loss in accurately learning the topology of the brain graphs predicted from the teacher network. While GmTE-Net*

Table 3. *Morphological data-based comparison of our GmTE-Net with its two ablated versions in terms of architecture (first two rows) and its two variants (last two rows). The best performances for each timepoint are highlighted in blue, green and purple when using $\mathcal{L}_a$, $\mathcal{L}_b$ and $\mathcal{L}^\star$ and loss functions, respectively. We further highlight in bold the best performance across four timepoints (i.e., average).*

Models	Losses		Student evaluation				Teacher evaluation		
		Timepoints	MAE(graph)	MAE(EC)	MAE(PC)	Timepoints	MAE(graph)	MAE(EC)	MAE(PC)
Ablated (augmented + real data)	$\mathcal{L}_a$	t_0	0.335965	0.036643	0.00511	t_0	0.291466	0.03059	0.004119
		t_1	0.331479	0.026421	0.003494	t_1	0.280166	0.026346	0.003473
		t_2	0.481598	0.028848	0.003642	t_2	0.403836	0.028906	0.003648
		t_3	0.483778	0.02601	0.003484	t_3	0.405698	0.026874	0.0036
		Average	0.408205	0.029481	0.003932	Average	0.345292	0.028179	0.00371
	$\mathcal{L}_b$	t_0	0.34096	0.038585	0.004985	t_0	0.340525	0.040181	0.005548
		t_1	0.371163	0.04101	0.005817	t_1	0.372369	0.049025	0.0069
		t_2	0.571771	0.04229	0.006097	t_2	0.572566	0.051062	0.006595
		t_3	0.653756	0.040905	0.005877	t_3	0.654722	0.048351	0.006607
		Average	0.484413	0.040697	0.005694	Average	0.485045	0.047155	0.006412
	$\mathcal{L}^\star$	t_0	0.341511	0.043517	0.006172	t_0	0.294021	0.031239	0.004282
		t_1	0.371031	0.035268	0.00465	t_1	0.284265	0.026825	0.003616
		t_2	0.571445	0.043749	0.005747	t_2	0.411524	0.029098	0.003664
		t_3	0.653458	0.040413	0.005281	t_3	0.413663	0.026434	0.003509
		Average	0.484361	0.040737	0.005462	Average	0.350868	0.028399	0.003768
Ablated (few-shot learning)	$\mathcal{L}_a$	t_0	0.351988	0.130276	0.015552	t_0	0.311441	0.02595	0.003645
		t_1	0.379058	0.045972	0.006921	t_1	0.292633	0.007219	0.001158
		t_2	0.580456	0.088472	0.010039	t_2	0.414301	0.007849	0.001055
		t_3	0.661841	0.06257	0.008138	t_3	0.37403	0.008166	0.001235
		Average	0.493336	0.081823	0.010163	Average	0.348101	**0.012296**	0.001773
	$\mathcal{L}_b$	t_0	0.341153	0.045592	0.006457	t_0	0.339081	0.046843	0.006548
		t_1	0.371633	0.036616	0.005368	t_1	0.371176	0.038149	0.005246
		t_2	0.571151	0.038609	0.00547	t_2	0.570984	0.038375	0.005369
		t_3	0.653206	0.038898	0.005605	t_3	0.653075	0.037581	0.00549
		Average	0.484286	0.039929	0.005725	Average	0.483579	0.040237	0.005663
	$\mathcal{L}^\star$	t_1	0.341153	0.045592	0.006457	t_1	0.311496	0.025645	0.003565
		t_2	0.371633	0.036616	0.005368	t_2	0.293318	0.006916	0.001131
		t_3	0.571151	0.038609	0.00547	t_3	0.415802	0.007973	0.001086
		t_4	0.653206	0.038898	0.005605	t_4	0.37667	0.008465	0.001269
		Average	0.484286	0.039929	0.005725	Average	0.349322	0.01225	0.001763
GmTE-Net	$\mathcal{L}_a$	t_0	0.33616	0.039172	0.005343	t_0	0.264755	0.031116	0.00421
		t_1	0.329668	0.026591	0.003621	t_1	0.257026	0.026932	0.003614
		t_2	0.480684	0.028641	0.003606	t_2	0.371957	0.028914	0.003633
		t_3	0.487717	0.026864	0.003599	t_3	0.383764	0.02707	0.003589
		Average	0.408557	0.030317	0.004042	Average	**0.319375**	0.028508	0.003761
	$\mathcal{L}_b$	t_0	0.341053	0.034078	0.005061	t_0	0.338462	0.044651	0.0064
		t_1	0.371283	0.04419	0.006003	t_1	0.370603	0.045793	0.006445
		t_2	0.571659	0.040527	0.005615	t_2	0.570774	0.052012	0.007025
		t_3	0.653747	0.039143	0.005525	t_3	0.652816	0.048719	0.006787
		Average	0.484435	0.039484	**0.005551**	Average	0.483164	0.047794	0.006664
	$\mathcal{L}^\star$	t_0	0.341785	0.041216	0.006017	t_0	0.283261	0.032299	0.004392
		t_1	0.370725	0.049689	0.006571	t_1	0.273054	0.025628	0.003464
		t_2	0.571063	0.050196	0.006391	t_2	0.396226	0.029045	0.003637
		t_3	0.652935	0.045524	0.006245	t_3	0.402294	0.026811	0.003542
		Average	0.484127	0.046656	0.006245	Average	0.338709	0.028446	0.003759
GmTE-Net*	$\mathcal{L}_a$	t_0	0.339821	0.036816	0.004804	t_0	0.308983	0.026169	0.003618
		t_1	0.337102	0.009204	0.001357	t_1	0.279894	0.006663	0.000973
		t_2	0.482349	0.008957	0.001198	t_2	0.386167	0.008259	0.001078
		t_3	0.456356	0.008567	0.001304	t_3	0.313551	0.008602	0.001276
		Average	**0.403907**	**0.015886**	**0.002166**	Average	0.322149	0.012423	**0.001736**
	$\mathcal{L}_b$	t_0	0.341082	0.034719	0.004827	t_0	0.340471	0.043005	0.006099
		t_1	0.369858	0.037938	0.00565	t_1	0.370992	0.038875	0.005325
		t_2	0.570924	0.041427	0.006112	t_2	0.571186	0.038807	0.004706
		t_3	0.652887	0.040915	0.006021	t_3	0.653264	0.039177	0.004769
		Average	**0.483687**	**0.03875**	0.005653	Average	0.483978	0.039966	0.005225
	$\mathcal{L}^\star$	t_0	0.339521	0.0389	0.005933	t_0	0.306318	0.029756	0.004309
		t_1	0.371142	0.044609	0.006036	t_1	0.274529	0.00688	0.001014
		t_2	0.571028	0.034936	0.004895	t_2	0.377451	0.007728	0.001058
		t_3	0.653257	0.03541	0.004927	t_3	0.299714	0.00821	0.00117
		Average	**0.483737**	**0.038464**	**0.005448**	Average	**0.314503**	**0.013143**	**0.001888**

Table 4. *Functional data-based comparison of our GmTE-Net with its two ablated versions in terms of architecture (first two rows) and its two variants (last two rows).* The best performances for each timepoint are highlighted in blue, green and purple when using $\mathcal{L}_a$, $\mathcal{L}_b$ and $\mathcal{L}^*$ and loss functions, respectively. We further highlight in bold the best performance across four timepoints (i.e., average).

Models	Losses	Timepoints	Student evaluation MAE(graph)	MAE(EC)	MAE(PC)	Timepoints	Teacher evaluation MAE(graph)	MAE(EC)	MAE(PC)
Ablated (augmented + real data)	L1	t_0	0.121212	0.011029	0.000903	t_0	0.125928	0.010554	0.000862
		t_1	0.104294	0.00949	0.000765	t_1	0.117978	0.009507	0.000762
		t_2	0.122563	0.009394	0.000751	t_2	0.137761	0.009428	0.000751
		t_3	0.128451	0.008972	0.00076	t_3	0.148127	0.008881	0.000763
		Average	0.11913	**0.009721**	**0.000795**	Average	0.132449	**0.009592**	**0.000784**
	L2	t_0	0.120395	0.014975	0.001222	t_0	0.120375	0.014974	0.001208
		t_1	0.095587	0.013357	0.001074	t_1	0.095627	0.012812	0.001073
		t_2	0.109044	0.01304	0.001025	t_2	0.109065	0.013409	0.001109
		t_3	0.09246	0.011906	0.000988	t_3	0.092451	0.013031	0.001111
		Average	0.104371	0.01332	**0.001077**	Average	**0.104379**	0.013557	0.001126
	L*	t_0	0.341511	0.043517	0.006172	t_0	0.294021	0.031239	0.004282
		t_1	0.371031	0.035268	0.00465	t_1	0.284265	0.026825	0.003616
		t_2	0.571445	0.043749	0.005747	t_2	0.411524	0.029098	0.003664
		t_3	0.653458	0.040413	0.005281	t_3	0.413663	0.026434	0.003509
		Average	0.484361	0.040737	0.005462	Average	0.350868	0.028399	0.003768
Ablated (few-shot learning)	L1	t_0	0.119765	0.0859	0.0039	t_0	0.114709	0.01055	0.000919
		t_1	0.09452	0.029714	0.002193	t_1	0.090791	0.014592	0.001337
		t_2	0.108232	0.026141	0.002032	t_2	0.099395	0.014356	0.001176
		t_3	0.091351	0.023682	0.001901	t_3	0.089889	0.015502	0.001309
		Average	0.103467	0.041359	0.002507	Average	0.098696	0.01375	0.001185
	L2	t_0	0.12038	0.014879	0.001205	t_0	0.120424	0.014109	0.001185
		t_1	0.095688	0.015012	0.001188	t_1	0.095698	0.013133	0.001057
		t_2	0.109088	0.014142	0.00113	t_2	0.109068	0.005663	0.001031
		t_3	0.092526	0.014113	0.001176	t_3	0.092475	0.012375	0.000986
		Average	0.104421	0.014536	0.001175	Average	0.104416	0.013191	0.001065
	L*	t_1	0.12038	0.014879	0.001205	t_1	0.11471	0.010551	0.000923
		t_2	0.095688	0.015012	0.001188	t_2	0.090882	0.014529	0.001345
		t_3	0.109088	0.014142	0.00113	t_3	0.099082	0.015019	0.00124
		t_4	0.092526	0.014113	0.001176	t_4	0.090015	0.015918	0.001343
		Average	0.104421	0.014536	0.001175	Average	0.098672	0.014004	0.001213
GmTE-Net	L1	t_0	0.121237	0.011289	0.000926	t_0	0.12924	0.010334	0.000852
		t_1	0.103771	0.009661	0.000776	t_1	0.121497	0.009674	0.000778
		t_2	0.120055	0.009583	0.000761	t_2	0.134557	0.009574	0.000765
		t_3	0.117831	0.009018	0.000768	t_3	0.131255	0.009197	0.000772
		Average	0.115724	0.009888	0.000808	Average	0.129137	0.009695	0.000792
	L2	t_0	0.120372	0.014684	0.001222	t_0	0.120417	0.013327	0.001078
		t_1	0.095583	0.012329	0.001014	t_1	0.095659	0.013698	0.001119
		t_2	0.109018	0.012985	0.001066	t_2	0.109085	0.013917	0.001122
		t_3	0.092406	0.012718	0.001051	t_3	0.092496	0.014031	0.001163
		Average	**0.104345**	**0.013179**	0.001088	Average	0.104414	0.013743	0.00112
	L*	t_0	0.120275	0.014645	0.001188	t_0	0.126361	0.01046	0.000858
		t_1	0.095648	0.011079	0.000926	t_1	0.116775	0.009398	0.000751
		t_2	0.109045	0.011617	0.000975	t_2	0.126008	0.009335	0.000738
		t_3	0.092465	0.012018	0.001037	t_3	0.107681	0.009034	0.000765
		Average	**0.104358**	**0.01234**	**0.001032**	Average	0.119206	**0.009557**	**0.000778**
GmTE-Net*	L1	t_0	0.11815	0.008892	0.000783	t_0	0.113811	0.010551	0.000924
		t_1	0.092031	0.015175	0.001381	t_1	0.090502	0.015233	0.001388
		t_2	0.102706	0.015379	0.001279	t_2	0.09887	0.01507	0.001244
		t_3	0.089694	0.017842	0.001528	t_3	0.089863	0.016986	0.001444
		Average	**0.100645**	0.014322	0.001243	Average	**0.098261**	0.01446	0.00125
	L2	t_0	0.120361	0.015588	0.001248	t_0	0.120448	0.013349	0.001091
		t_1	0.095634	0.013961	0.001142	t_1	0.095685	0.014943	0.001208
		t_2	0.109066	0.013336	0.001057	t_2	0.10907	0.014648	0.00119
		t_3	0.09248	0.012231	0.001012	t_3	0.092502	0.014172	0.001166
		Average	0.104385	0.013779	0.001115	Average	0.104426	0.014278	0.001164
	L*	t_0	0.120442	0.015209	0.001224	t_0	0.113638	0.010754	0.000949
		t_1	0.095632	0.013632	0.001099	t_1	0.090207	0.014794	0.001395
		t_2	0.10903	0.013166	0.001063	t_2	0.098658	0.01509	0.001258
		t_3	0.092448	0.012727	0.001063	t_3	0.090071	0.016392	0.00139
		Average	0.104388	0.013684	0.001112	Average	**0.098143**	0.014257	0.001248

outperformed other comparison methods when evaluated on the morphological dataset, it ranked second-best in MAE(graph), MAE(EC), and MAE (PC), following GmTE-Net when evaluated on functional brain graphs as displayed in Table 2. This shows that shifting the size of GCN layers to predict the brain graphs with a different resolution from the base one is not sufficient for getting an accurate prediction. Thereby, as a future research direction, we will include a domain alignment component in our architecture which will boost the prediction of brain graphs with different resolutions [26,27].

4   Conclusion

We proposed GmTE-Net, the first teacher-student framework tailored for predicting baby brain graph multi-trajectory evolution over time from a single neonatal observation. Our key contributions consist in: (i) designing a graph autoencoder able to chart out multiple trajectories with shifting graph resolutions from a single learned shared embedding of a testing neonatal brain graph, (ii) circumventing the learning on a few brain graphs such that we train the teacher network solely on a few shots and the student network on simulated brain graphs, and (iii) introducing a local topology-aware distillation loss to force the student model to preserve the node structure of the predicted graphs from teacher network. For the first time, we take the baby connectomics field one step further into foreseeing realistic baby brain graphs that may help to better elucidate how infant brain disorders develop. As a future direction, we will incorporate a domain alignment module into each decoder in our framework so that we reduce the domain fracture resulting in the distribution difference across brain modalities [27].

Acknowledgements. This work was funded by generous grants from the European H2020 Marie Sklodowska-Curie action (grant no. 101003403, http://basira-lab.com/normnets/) to I.R. and the Scientific and Technological Research Council of Turkey to I.R. under the TUBITAK 2232 Fellowship for Outstanding Researchers (no. 118C288, http://basira-lab.com/reprime/). However, all scientific contributions made in this project are owned and approved solely by the authors. A.B is supported by the same TUBITAK 2232 Fellowship.

References

1. Zhang, H., Shen, D., Lin, W.: Resting-state functional MRI studies on infant brains: a decade of gap-filling efforts. Neuroimage **185**, 664–684 (2019)
2. Rekik, I., Li, G., Lin, W., Shen, D.: Predicting infant cortical surface development using a 4d varifold-based learning framework and local topography-based shape morphing. Med. Image Anal. **28**, 1–12 (2016)
3. Rekik, I., Li, G., Yap, P.T., Chen, G., Lin, W., Shen, D.: Joint prediction of longitudinal development of cortical surfaces and white matter fibers from neonatal mri. Neuroimage **152**, 411–424 (2017)

4. Bessadok, A., Mahjoub, M.A., Rekik, I.: Graph neural networks in network neuroscience. arXiv preprint arXiv:2106.03535 (2021)
5. Fornito, A., Zalesky, A., Breakspear, M.: The connectomics of brain disorders. Nat. Rev. Neurosci. **16**, 159–172 (2015)
6. Ghribi, O., Li, G., Lin, W., Shen, D., Rekik, I.: Multi-regression based supervised sample selection for predicting baby connectome evolution trajectory from neonatal timepoint. Med. Image Anal. **68**, 101853 (2021)
7. Goktas, A.S., Bessadok, A., Rekik, I.: Residual embedding similarity-based network selection for predicting brain network evolution trajectory from a single observation. arXiv preprint arXiv:2009.11110 (2020)
8. Kipf, T.N., Welling, M.: Semi-supervised classification with graph convolutional networks. arXiv preprint arXiv:1609.02907 (2016)
9. Ezzine, B.E., Rekik, I.: Learning-guided infinite network atlas selection for predicting longitudinal brain network evolution from a single observation. In: Shen, D., et al. (eds.) MICCAI 2019. LNCS, vol. 11765, pp. 796–805. Springer, Cham (2019). https://doi.org/10.1007/978-3-030-32245-8_88
10. Nebli, A., Kaplan, U.A., Rekik, I.: Deep evographnet architecture for time-dependent brain graph data synthesis from a single timepoint. In: Rekik, I., Adeli, E., Park, S.H., Valdés Hernández, M.C. (eds.) PRIME 2020. LNCS, vol. 12329, pp. 144–155. Springer, Cham (2020). https://doi.org/10.1007/978-3-030-59354-4_14
11. Tian, Y., Maicas, G., Pu, L.Z.C.T., Singh, R., Verjans, J.W., Carneiro, G.: Few-shot anomaly detection for polyp frames from colonoscopy. In: Martel, A.L., et al. (eds.) MICCAI 2020. LNCS, vol. 12266, pp. 274–284. Springer, Cham (2020). https://doi.org/10.1007/978-3-030-59725-2_27
12. Li, A., Luo, T., Lu, Z., Xiang, T., Wang, L.: Large-scale few-shot learning: knowledge transfer with class hierarchy. In: Proceedings of the IEEE/CVF Conference on Computer Vision and Pattern Recognition, pp. 7212–7220 (2019)
13. Li, X., Yu, L., Jin, Y., Fu, C.-W., Xing, L., Heng, P.-A.: Difficulty-aware meta-learning for rare disease diagnosis. In: Martel, A.L., et al. (eds.) MICCAI 2020. LNCS, vol. 12261, pp. 357–366. Springer, Cham (2020). https://doi.org/10.1007/978-3-030-59710-8_35
14. Yuan, P., et al.: Few is enough: task-augmented active meta-learning for brain cell classification. In: Martel, A.L., et al. (eds.) MICCAI 2020. LNCS, vol. 12261, pp. 367–377. Springer, Cham (2020). https://doi.org/10.1007/978-3-030-59710-8_36
15. Hinton, G., Vinyals, O., Dean, J.: Distilling the knowledge in a neural network. arXiv preprint arXiv:1503.02531 (2015)
16. Rajasegaran, J., Khan, S., Hayat, M., Khan, F.S., Shah, M.: Self-supervised knowledge distillation for few-shot learning. arXiv preprint arXiv:2006.09785 (2020)
17. Hu, M., et al.: Knowledge distillation from multi-modal to mono-modal segmentation networks. In: Martel, A.L., et al. (eds.) MICCAI 2020. LNCS, vol. 12261, pp. 772–781. Springer, Cham (2020). https://doi.org/10.1007/978-3-030-59710-8_75
18. Zhou, Y., Chen, H., Lin, H., Heng, P.-A.: Deep semi-supervised knowledge distillation for overlapping cervical cell instance segmentation. In: Martel, A.L., et al. (eds.) MICCAI 2020. LNCS, vol. 12261, pp. 521–531. Springer, Cham (2020). https://doi.org/10.1007/978-3-030-59710-8_51
19. Bassett, D.S., Sporns, O.: Network neuroscience. Nat. Neurosci. **20**, 353–364 (2017)
20. Liu, J., et al.: Complex brain network analysis and its applications to brain disorders: a survey. Complexity 2017 (2017)
21. Joyce, K.E., Laurienti, P.J., Burdette, J.H., Hayasaka, S.: A new measure of centrality for brain networks. PloS one **5**, e12200 (2010)

22. Fornito, A., Zalesky, A., Bullmore, E.: Fundamentals of Brain Network Analysis. Academic Press, Cambridge (2016)
23. Freeman, L.C.: A set of measures of centrality based on betweenness. Sociometry, pp. 35–41 (1977)
24. Tzourio-Mazoyer, N., et al.: Automated anatomical labeling of activations in spm using a macroscopic anatomical parcellation of the mni mri single-subject brain. Neuroimage **15**, 273–289 (2002)
25. Fischl, B., et al.: Sequence-independent segmentation of magnetic resonance images. Neuroimage **23**, S69–S84 (2004)
26. Pilanci, M., Vural, E.: Domain adaptation on graphs by learning aligned graph bases. IEEE Trans. Knowl. Data Eng. (2020). IEEE
27. Redko, I., Morvant, E., Habrard, A., Sebban, M., Bennani, Y.: A survey on domain adaptation theory. arXiv preprint arXiv:2004.11829 (2020)

One Representative-Shot Learning Using a Population-Driven Template with Application to Brain Connectivity Classification and Evolution Prediction

Umut Guvercin, Mohammed Amine Gharsallaoui, and Islem Rekik$^{(\boxtimes)}$ ⓘ

BASIRA Lab, Faculty of Computer and Informatics, Istanbul Technical University, Istanbul, Turkey
irekik@itu.edu.tr
http://basira-lab.com

Abstract. Few-shot learning presents a challenging paradigm for training discriminative models on a few training samples representing the target classes to discriminate. However, classification methods based on deep learning are ill-suited for such learning as they need large amounts of training data –let alone *one-shot* learning. Recently, graph neural networks (GNNs) have been introduced to the field of network neuroscience, where the brain connectivity is encoded in a graph. However, with scarce neuroimaging datasets particularly for rare diseases and low-resource clinical facilities, such data-devouring architectures might fail in learning the target task. In this paper, we take a very different approach in training GNNs, where we aim to *learn with one sample and achieve the best performance* –a formidable challenge to tackle. Specifically, we present the first one-shot paradigm where a GNN is trained on a *single population-driven template* –namely a connectional brain template (CBT). A CBT is a compact representation of a population of brain graphs capturing the unique connectivity patterns shared across individuals. It is analogous to brain image atlases for neuroimaging datasets. Using a *one-representative* CBT as a training sample, we alleviate the training load of GNN models while boosting their performance across a variety of classification and regression tasks. We demonstrate that our method significantly outperformed benchmark one-shot learning methods with downstream classification and time-dependent brain graph data forecasting tasks while competing with the "train on all" conventional training strategy. Our source code can be found at https://github.com/basiralab/one-representative-shot-learning.

Keywords: One-shot learning · Graph neural networks · Connectional brain templates · Time-dependent graph evolution prediction · Brain disorder classification

1 Introduction

Deep learning models have achieved remarkable results across different medical imaging learning tasks such as segmentation, registration and classification

I. Rekik et al. (Eds.): PRIME 2021, LNCS 12928, pp. 25–36, 2021.
https://doi.org/10.1007/978-3-030-87602-9_3

[1,2]. Despite their ability to extract meaningful and powerful representations from labelled data, they might fail to operate in a frugal setting where the number of the samples to train on is very limited. Besides, training such data-devouring architectures is computationally expensive and might not work on scarce neuroimaging datasets particularly for rare diseases [3] and in countries with low-resource clinical facilities [4] –limiting their generalizibility and diagnostic impact. Thus, deep networks seem less powerful when the goal is to learn a mapping classification or regression function on the fly, from only a few samples. Such problem is usually remedied by the few-shot learning (FSL) paradigm [5–7], where only few labeled samples are used to learn a new concept that can generalize to unseen distributions of testing samples. While training with lesser amount of labelled data reduces the cost in terms of computational power and required time, it also overcomes the data scarcity issue. Several works developed novel ways of leveraging FSL in medical image-based learning tasks. For instance, [8] presented a learning-based method that is trained with few examples while leveraging data augmentation and unlabeled image data to enhance model generalizability. [9] also leveraged unlabeled data for segmenting 3D multi modal medical images. Their implementation of semi-supervised approach with generative adversarial networks obviated the need of pre-trained model. [10] used the meta-train data from common diseases for rare disease diagnosis and tackled the low-data regime problem while leveraging meta-learning. However, such works resort to data augmentation strategies or generative models to better estimate the unseen distributions of the classes to discriminate. However, generating *real* and *biologically sound* data samples, particularly in the context where the ultimate goal is disease diagnosis and biomarker discovery, becomes a far cry from learning on a few samples *alone*.

In this paper, we take a very different approach and ask whether we can **learn with one sample and achieve the best performance without any augmentation** –a formidable challenge to tackle. Our approach aims to train FSL models on a single sample that is *well-representative* of a particular class to learn–demonstrating that *one is enough* if it captures well the unique traits of the given class. Our one-representative shot model does not require costly optimization and is efficient in one go. Deep learning methods have proven their efficiency in many applications. However, these methods are less successful when dealing with non-Euclidian data such as graphs and manifolds [11]. To circumvent this limitation, many studies have proposed graph neural networks (GNNs), which extend the efficiency of deep learning to a broader range of data structures such as graphs [12]. Particularly, GNNs have been widely used to tackle many problems in network neuroscience as detailed in a recent review paper [13]. So far, we also note that there is a limited number of works where FSL was adopted to train GNNs. For example, [14] designed a GNN [15] architecture trained on a few shots. It aims to learn a complete graph network where nodes store features and class information. [16] proposes attentive GNN model for few-shot learning that can mitigate the over-fitting and over-smoothing issues in deep GNN models. However, to the best of our knowledge there is no study that focuses on training GNN models with *a single shot* and where each sample is encoded as a whole graph. Recent adversarial GNN models have been tailored for brain graph evolution prediction from a single timepoint such as EvoGraphNet [17] or brain graph classification [18]. However, such architectures cross-pollinating the field of network neuroscience where the brain is represented as a graph encoding the pairwise interactions between anatomical regions of interest (ROIs), are still data-hungry and needy for high computational facilities.

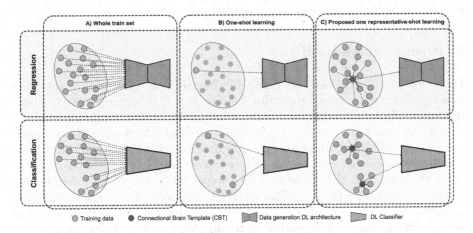

Fig. 1. *Proposed one representative-shot learning compared to the conventional 'train on all' strategy and one-shot learning.* **A) Whole train set.** The learned (regressor or classifier) is trained on the whole set. **B) One-shot learning.** The learner is trained with only on one sample. **C) Proposed one representative-shot learning.** The regressor is trained with the representative sample (i.e., CBT) of the whole population. The classifier is trained with the CBTs of each class.

In order to train a GNN with a single sample, one faces the diversity of selecting samples and creating new one(s) or possibly both. Since the lack of data increases the importance of the quality and diversity of the training set, randomly selecting samples is neither viable nor stable. Clearly, learning from a single exemplar is only possible given sufficient prior knowledge on the learning domain. The main challenge is then to find a single sample that captures the heterogeneous distribution of the class to learn. Here we propose to leverage the emerging concept of a connectional brain template [19–21], where a population of brain graphs is encoded in a compact connectivity matrix capturing the unique connectivity patterns shared across individuals of a given population. This can be regarded as network 'atlas' by analogy to brain image template such as the Montreal Neurological Institute (MNI) template [22]. We propose to use such a compact and universal sample (i.e., CBT) to train GNN models for brain graph classification and time-dependent generation tasks. In opposition to the state-of-the-art methods which are data-hungry and needy for high computational facilities, our proposed approach can learn from one sample using the

paradigm-shift we are creating through the **one-representation training** of deep learning models. The principle introduced is highly innovative and fully agnostic to the deployed methods. Inspired from the CBT concept, this pioneering work can drive the whole field of geometric deep learning in a completely new direction related to the optimization of the amount of data needed to train a GNN model for a specific task. In addition to the computational gain, we achieved striking results compared to other benchmark one-shot learning methods and a competitive performance to the "train on all" conventional training strategy across different connectomic datasets and learning tasks.

2   Proposed Method

Overview. In this section, we detail the foundational principles of the proposed representative one-shot learning method (Fig. 1) in comparison to training strategies. The first benchmark is resourceful (Fig. 1-A) as it consumes all available training data. In this case, it is generally assumed that the available training samples are well distributed across possible data domain. The second benchmark is the conventional one-shot learning where we only train our method on one sample of the data (Fig. 1-B). When we dive deeper into concrete implementations of such strategies, we face a peculiar challenge: the deep model performance can be highly sensitive to training and test set distribution shifts via local or global perturbations. This can be remedied by training on a single sample encapsulating sufficient prior knowledge on the learning domain, thereby allowing generalizability to unseen samples. Our proposed one representative-shot learning method (Fig. 1-C) supports such hypothesis where we investigate the potential of one shot (i.e., a CBT) in training well different learning models. Here, we focus on two different target learning tasks to evaluate our method. For regression, we use a CBT sample to represent the whole population; for classification, we use one class-specific CBT sample to represent each class.

The CBT as One Representative Shot. A well-representative connectional brain template is centered (i.e., with minimal distance to all samples of its parent population) and discriminative (i.e., can easily distinguished from a different CBT generated from a different class). Let $\mathcal{D} = \{\mathbf{X}_1, \ldots, \mathbf{X}_n\}$ denote a population of brain connectivity matrices, where $\mathbf{X}_s \in \mathbb{R}^{r \times r}$ encodes the interaction weights between r ROIs for a subject s. Here we leverage the state-of-the-art deep graph normalizer (DGN) network [21] to learn an integral and holistic connectional template of the input population $\mathcal{D}$. DGN takes two or more weighted (possibly unweighted) graphs and maps them onto an output population center graph (i.e., connectional template). This learning task is fundamentally rooted in mapping connectivity patterns onto a high-dimensional vector representation for each node in each graph in the given population, namely a node feature vector. During the mapping process, the unique domain-specific topological properties of the population graphs are preserved thanks to a topology-constrained normalization loss function which penalizes the deviation from the ground-truth population topology. Next, we derive the CBT edges from the pairwise relationship of node embeddings. The final CBT encoded in a matrix $\mathbf{C} \in \mathbb{R}^{r \times r}$ is learned by minimizing the Frobenius distance to a random set S of individuals to avoid overfitting as follows:

$$\underset{\mathbf{C}}{\operatorname{argmin}} \sum_{s \in S} \|\mathbf{C} - \mathbf{X}_s\|_F$$

Regression Learning Task (Time-Series Brain Graph Evolution Prediction). Foreseeing the brain evolution [23–25] as a complex highly interconnected system is crucial for mapping dynamic interactions between different brain ROIs for different neurological states in the dataset. In fact, atypical alterations in the brain connectome evolution trajectory might indicate the emergence of neurological disorders such as Alzheimer's disease (AD) or dementia [26,27]. Hence, learning to predict longitudinal brain dysconnectivity from a baseline timepoint helps to diagnose such disorders at an early stage. Such regression task aims at learning how to generate real brain graphs at follow-up timepoints from a single baseline graph acquired at timepoint t_1. Recently, EvoGraphNet [17] was proposed as the first GNN for predicting the evolution of brain graphs at $t_i > t_1$ by cascading a set of time-dependent graph generative adversarial networks (gGANs), where the predicted graph at the current timepoint t_i is passed on to train the next gGAN in the cascade, thereby generating the brain graph at timepoint t_{i+1}. To optimize the learning process of generation, EvographNet minimizes the adversarial loss between a pair of generator G_i and descriminator D_i at each time point t_i. EvoGraphNet architecture assumes sparse changes in the brain connectivity structure over time and includes $l1$ loss for that reason. Besides, it also introcuces $\mathcal{L}_{KL}$, a loss based on Kullback-Leibler (KL) divergence [28], to enforce the alignment between the predicted and ground-truth domains. The full loss is defined as follows:

$$\mathcal{L}_{FUll} = \sum_{i=1}^{T} (\lambda_1 \mathcal{L}_{adv}(G_i, D_i) + \frac{\lambda_2}{n} \sum_{tr=1}^{n} \mathcal{L}_{l1}(G_i, t_r) + \frac{\lambda_3}{n} \sum_{tr=1}^{n} \mathcal{L}_{KL}(t_i, t_r)) \quad (1)$$

where λ_1, λ_2 and λ_3 are hyperparameters to moderate the importance of each component. For our one-representative shot learning, to train EvoGraphNet architecture with a single sample trajectory, we learn time-dependent CBTs from the available training samples at each timepoint $\{t_i\}_{i=1}^{T}$, independently. Each gGAN in the cascade learns how to evolve the population CBT $\mathbf{C}^{t_i}$ from timepoint t_i into the CBT $\mathbf{C}^{t_{i+1}}$ at the follow-up timepoint t_{i+1}.

Diagnostic Learning Task (Brain Graph Classification). The graph classification task aims to predict the label of a graph instance. This task is performed heavily on chemical and social domains [29]. In brain graph classification, instances consist of brain connectomes. Graph attention networks (GAT) [30] are one of the most powerful convolution-style neural networks that operate on graph-structured data. It uses masked self-attentional layers. Its performance does not depend on knowing the full graph structure and it can operate on different number of neighbours while assigning different importance to each node. Considering its performance and efficiency and our sample structure, we choose GAT to perform our experiments with brain graph classification task into different neurological states. In order to integrate one representative-shot learning approach with the brain graph classification task, we use one CBT per class. This means that every class is represented by one sample that is derived from its whole class population using DGN [21]. Then GAT model is then trained on *only two samples (i.e., class-specific templates)* in a supervised manner. Next, the trained method is fed with testing samples and according to the predicted possibility values, testing samples are labelled.

Given that GAT is based on self-attention computations, we denote the shared attentional mechanism as $a : \mathbb{R}^{F'} \times \mathbb{R}^{F'} \to \mathbb{R}$ where F and F' are the dimensions of the node features in the current and the following layers, respectively. For each node i in the graph, we compute the attention coefficients of its direct neighbors:

$$e_{ij} = a(\mathbf{W}h_i, \mathbf{W}h_j)$$

where $\mathbf{W} \in \mathbb{R}^{F \times F'}$ is a learnt weight matrix and $h_i, h_j \in \mathbb{R}^F$ denote the feature vectors of the nodes i and j, respectively. The coefficients are then normalized using a softmax function:

$$\alpha_{ij} = softmax_j(e_{ij}) = \frac{exp(e_{ij})}{\sum_{k \in \mathcal{N}_i} exp(e_{ik})}$$

where $\mathcal{N}_i$ is the set of neighbors of node i in the graph. Finally, the new node features are calculated using a non-linearity function:

$$\sigma : h_i' = \sigma\left(\sum_{j \in \mathcal{N}_i} \alpha_{ij} \mathbf{W}h_j\right)$$

3   Results and Discussion

3.1   Regression Learning Task

Longitudinal Dataset. We used 113 subjects from the OASIS-2[1] longitudinal dataset [31]. This set consists of a longitudinal collection of 150 subjects with a ranging age from 60 to 96. Each subject has 3 visits (i.e., timepoints), separated by at least one year. Each subject is represented by one cortical morphological brain network derived from T1-w MRI. Specifically, we use Desikan-Killiany cortical template to partition the cortical surface into 35 ROIs, each hemisphere (left and right). Next, by computing the absolute difference in the maximum principal curvature between pairs of ROIs we define the morphological weight between them.

---

[1] https://www.oasis-brains.org/.

Table 1. Mean absolute error (MAE) between ground-truth and predicted brain graphs at follow-up timepoints by our proposed method and comparison methods including standard deviations across different folds.

Method	t_1		t_2	
	Left Hemisphere	Right Hemisphere	Left Hemisphere	Right Hemisphere
EvoGraphNet [17] (all)	0.053 ± 0.0069	0.070 ± 0.0221	0.069 ± 0.0149	$\mathbf{0.161 \pm 0.1208}$
Random one-shot	0.098 ± 0.0051	0.126 ± 0.0268	0.201 ± 0.0168	0.289 ± 0.1454
Linear average one-shot	0.078 ± 0.0199	0.064 ± 0.0090	0.183 ± 0.0466	0.212 ± 0.1599
CBT one-shot (ours)	$\mathbf{0.047 \pm 0.0031}$	$\mathbf{0.053 \pm 0.0035}$	$\mathbf{0.056 \pm 0.0017}$	0.214 ± 0.0759

Parameters. In order to find the CBT, we used 5-fold cross-validation strategy. We set the random training sample set S size to 10, learning rate to 0.0005 and trained 100 epochs with early stopping criteria. For EvoGraphNet [17], we use adaptive learning rates for the generators decreasing from 0.01 to 0.1 while we fixed the discriminator learning rate to 0.0002. The λ parameters for the loss functions are set to $\lambda_1 = 2$, $\lambda_2 = 2$, and $\lambda_3 = 0.001$, respectively. We used the AdamW [32] optimizer with 0.5 and 0.999 moment estimates. We trained our model over 300 epochs using a single Tesla V100 GPU (NVIDIA GeForce GTX TITAN with 32 GB memory). These hyperparameters were selected empirically using grid search.

Evaluation and Comparison Methods. We compared our one-representative shot learning method against EvoGraphNet [17] (trained on all samples) and two other one-shot learning strategies, namely (i) linear average one-shot and (ii) random one-shot. In the linear average one-shot learning, we created the CBT by simply linearly averaging the population samples at each timepoint, independently. For the random one-shot, we picked a single random sample to train on at baseline timepoint. We repeated the random sampling 20 times and report the average results. All methods used EvoGraphNet architecture with the set parameters. The first training strategy of EvoGraphNet presents an upper bound as it is trained with all samples. In Table 1, we report the mean absolute error (MAE) between the ground-truth and predicted graphs.

Our method clearly outperformed both linear average and random one-shot learning methods. Remarkably, it also outperformed the EvoGraphNet except for the right hemispheric data at t_2. Although EvoGraphNet (all) presents an upper bound in the expected performance, our method, while being trained on single sample, outperformed it in three out of four cases. The standard deviation is also higher in other methods especially in random one-shot learning. Comparing the CBT one-shot against the linear average one demonstrates that linear averaging does not produce a highly-representative template of the parent population, hence the low performance. This also implicitly shows the ability of DGN in learning well representative and centered CBTs that can be leveraged in downstream learning tasks. Our CBT one-shot method also achieved the highest stability against train and test distribution shifts via cross-validation by lowering the MAE standard deviation across folds. This clearly indicates that our single population-representative shot is enough to train a cascade of gGAN architectures without resorting to expensive augmentation and prolonged batch-based training. Overall, our method provided significant results in terms of accuracy and stability, and we even achieved a better performance than when training on all samples.

3.2 Diagnostic Learning Task

Dataset. We evaluated our CBT one-shot learning framework on a brain dementia dataset consisting of 77 subjects (41 subjects diagnosed with Alzheimer's diseases (AD) and 36 with Late Mild Cognitive Impairment (LMCI)) from the Alzheimer's Disease Neuroimaging Initiative (ADNI) database GO public dataset [33]. Each subject is represented by a morphological network derived from the maximum principle curvature of the cortical surface.

Parameters. For CBT estimation, we used the same DGN parameter setting as in the previous section for the brain graph evolution prediction task. For the main architecture of GAT[30], we used a dropout rate of 0.6 and removed weak connections by thresholding each brain connectivity matrix using its mean value. We set the initial learning rate to 0.0001 and the weight decay to 0.0005. For Leaky ReLU, we set the alpha to 0.2. These hyperparameters were chosen empirically.

Evaluation and Comparison Methods. We also used 5-fold cross-validation strategy for classification evaluation. We compared our method against GAT and both linear average and random one-shot learning strategies. All methods used GAT for brain graph classification. The classification results on both right and left brain graph datasets are detailed in Table 2 and Table 3, respectively. In addition to regression, we evaluated our CBT one-shot training strategy using a classification GNN model (here GAT) to further demonstrate the *generalizability* of our method *across various learning tasks*. Using two different datasets, we ranked *first* for the left hemisphere dataset and *second best* for the right hemisphere in terms of classification accuracy. Both Tables 2 and 3 indicate that the specificity scores are low. However, our method brought way higher sensitivity which is more important for clinical applications as we need to avoid false negatives. We note that the brain changes are very subtle between LMCI and AD patients, hence the low classification accuracy across all methods given the difficulty of this task. In fact, the distribution of samples drawn from both classes are highly overlapping. Many studies have pointed out the difficulty of the discrimination task between LMCI and AD patients [34]. In fact, this task is considered as a hard problem in the AD literature which is least addressed relatively [35]. We also notice that for the right hemisphere we had a highly competitive performance with the upper-bound method (GAT trained on all). Again, the best classification performance across both hemispheres was achieved using our CBT one-shot strategy. Our reproducible conclusive results across different learning tasks challenges the assumption of deep learning models being 'data-hungry' and prone to failure when learning from a few samples. We have shown that it is

Table 2. AD/LMCI brain classification results by the proposed method and benchmarks for the right hemisphere dataset. 4 different metrics are used to compare: accuracy, sensitivity, specificity, AUC. Results are the average of the 5 folds. **Bold**: best. Underlined: second best.

Method	Accuracy	Sensitivity	Specificity	AUC
GAT [30] (all)	**0.52**	0.44	<u>0.6</u>	**0.52**
Linear average one-shot	0.49	0.24	**0.78**	<u>0.51</u>
Random one-shot	0.5	<u>0.5</u>	0.5	0.5
CBT one-shot (ours)	<u>0.51</u>	**0.69**	0.3	0.49

Table 3. AD/LMCI brain classification results by the proposed method and benchmarks for the left hemisphere dataset. 4 different metrics are used to compare: accuracy, sensitivity, specificity, AUC. Results are the average of the 5 folds. **Bold**: best. Underlined: second best.

Method	Accuracy	Sensitivity	Specificity	AUC
GAT [30] (all)	<u>0.51</u>	0.36	<u>0.68</u>	<u>0.52</u>
Linear average one-shot	0.42	0.18	**0.7**	0.44
Random one-shot	0.5	<u>0.49</u>	0.51	0.5
CBT one-shot (ours)	**0.54**	**0.67**	0.4	**0.53**

possible to *learn with one sample and achieve the best performance without any augmentation* if the sample is well-representative of the given population –recall that random and averaging cases did not perform well. Possible extensions of this novel line of research include the debunking of the learning behavior of GNN models in resourceful and frugal training conditions and eventually laying the theoretical principles of training such heavy networks on *templates*.

4   Conclusion

In this paper, we showed that different GNN architectures trained using a single representative training graph can outperform models trained on large datasets across different tasks. Specifically, our method uses population-driven templates (CBTs) for training. We have showed that our model generated promising results in two different tasks: brain graph evolution prediction and classification. The concept of training on a CBT has the potential to be generalized to other GNN models. Although many theoretical investigations have yet to be taken, leveraging a representative CBT for training such models may become a useful component in alleviating the costly batch-based training of GNNs models. The CBT one-shot learning opens a new chapter in enhancing the GNNs performance while potentially reducing their computational time by capturing the commonalities across brain connectivities populations via the template representation.

Acknowledgements. This work was funded by generous grants from the European H2020 Marie Sklodowska-Curie action (grant no. 101003403, http://basira-lab.com/ normnets/) to I.R. and the Scientific and Technological Research Council of Turkey to I.R. under the TUBITAK 2232 Fellowship for Outstanding Researchers (no. 118C288, http://basira-lab.com/reprime/). However, all scientific contributions made in this project are owned and approved solely by the authors. M.A.G is supported by the same TUBITAK 2232 Fellowship.

References

1. Shen, D., Wu, G., Suk, H.I.: Deep learning in medical image analysis. Annu. Rev. Biomed. Eng. **19**, 221–248 (2017)
2. Yi, X., Walia, E., Babyn, P.: Generative adversarial network in medical imaging: a review. Med. Image Anal. **58**, 101552 (2019)
3. Schaefer, J., Lehne, M., Schepers, J., Prasser, F., Thun, S.: The use of machine learning in rare diseases: a scoping review. Orphanet J. Rare Dis. **15**, 1–10 (2020)
4. Piette, J.D., et al.: Impacts of e-health on the outcomes of care in low-and middle-income countries: where do we go from here? Bull. World Health Organ. **90**, 365–372 (2012)
5. Kadam, S., Vaidya, V.: Review and analysis of zero, one and few shot learning approaches. In: Abraham, A., Cherukuri, A.K., Melin, P., Gandhi, N. (eds.) ISDA 2018 2018. AISC, vol. 940, pp. 100–112. Springer, Cham (2020). https://doi.org/ 10.1007/978-3-030-16657-1_10
6. Sun, Q., Liu, Y., Chua, T.S., Schiele, B.: Meta-transfer learning for few-shot learning. In: Proceedings of the IEEE/CVF Conference on Computer Vision and Pattern Recognition, pp. 403–412 (2019)
7. Li, X., Sun, Z., Xue, J.H., Ma, Z.: A concise review of recent few-shot meta-learning methods. arXiv preprint arXiv:2005.10953 (2020)
8. Zhao, A., Balakrishnan, G., Durand, F., Guttag, J.V., Dalca, A.V.: Data augmentation using learned transformations for one-shot medical image segmentation. In: Proceedings of the IEEE/CVF Conference on Computer Vision and Pattern Recognition, pp. 8543–8553 (2019)
9. Mondal, A.K., Dolz, J., Desrosiers, C.: Few-shot 3D multi-modal medical image segmentation using generative adversarial learning. arXiv preprint arXiv:1810.12241 (2018)
10. Li, X., Yu, L., Jin, Y., Fu, C.-W., Xing, L., Heng, P.-A.: Difficulty-aware meta-learning for rare disease diagnosis. In: Martel, A.L., et al. (eds.) MICCAI 2020. LNCS, vol. 12261, pp. 357–366. Springer, Cham (2020). https://doi.org/10.1007/ 978-3-030-59710-8_35
11. Bronstein, M.M., Bruna, J., LeCun, Y., Szlam, A., Vandergheynst, P.: Geometric deep learning: going beyond Euclidean data. IEEE Signal Process. Mag. **34**, 18–42 (2017)
12. Kipf, T.N., Welling, M.: Semi-supervised classification with graph convolutional networks. arXiv preprint arXiv:1609.02907 (2016)
13. Bessadok, A., Mahjoub, M.A., Rekik, I.: Graph neural networks in network neuroscience. arXiv preprint arXiv:2106.03535 (2021)
14. Garcia, V., Bruna, J.: Few-shot learning with graph neural networks. arXiv preprint arXiv:1711.04043 (2017)

15. Gori, M., Monfardini, G., Scarselli, F.: A new model for learning in graph domains. In: Proceedings of the 2005 IEEE International Joint Conference on Neural Networks, vol. 2, pp. 729–734. IEEE (2005)
16. Cheng, H., Zhou, J.T., Tay, W.P., Wen, B.: Attentive graph neural networks for few-shot learning. arXiv preprint arXiv:2007.06878 (2020)
17. Nebli, A., Kaplan, U.A., Rekik, I.: Deep EvoGraphNet architecture for time-dependent brain graph data synthesis from a single timepoint. In: Rekik, I., Adeli, E., Park, S.H., Valdés Hernández, M.C. (eds.) PRIME 2020. LNCS, vol. 12329, pp. 144–155. Springer, Cham (2020). https://doi.org/10.1007/978-3-030-59354-4_14
18. Bi, X., Liu, Z., He, Y., Zhao, X., Sun, Y., Liu, H.: GNEA: a graph neural network with ELM aggregator for brain network classification. Complexity **2020** (2020)
19. Rekik, I., Li, G., Lin, W., Shen, D.: Estimation of brain network atlases using diffusive-shrinking graphs: application to developing brains. In: Niethammer, M., et al. (eds.) IPMI 2017. LNCS, vol. 10265, pp. 385–397. Springer, Cham (2017). https://doi.org/10.1007/978-3-319-59050-9_31
20. Dhifallah, S., Rekik, I., Initiative, A.D.N., et al.: Estimation of connectional brain templates using selective multi-view network normalization. Med. Image Anal. **59**, 101567 (2020)
21. Gurbuz, M.B., Rekik, I.: Deep graph normalizer: a geometric deep learning approach for estimating connectional brain templates. In: Martel, A.L., et al. (eds.) MICCAI 2020. LNCS, vol. 12267, pp. 155–165. Springer, Cham (2020). https://doi.org/10.1007/978-3-030-59728-3_16
22. Brett, M., Christoff, K., Cusack, R., Lancaster, J., et al.: Using the Talairach atlas with the MNI template. Neuroimage **13**, 85–85 (2001)
23. Gilmore, J.H., et al.: Longitudinal development of cortical and subcortical gray matter from birth to 2 years. Cereb. Cortex **22**, 2478–2485 (2012)
24. Mills, K.L., et al.: Structural brain development between childhood and adulthood: convergence across four longitudinal samples. Neuroimage **141**, 273–281 (2016)
25. Meng, Y., Li, G., Lin, W., Gilmore, J.H., Shen, D.: Spatial distribution and longitudinal development of deep cortical sulcal landmarks in infants. Neuroimage **100**, 206–218 (2014)
26. Lohmeyer, J.L., Alpinar-Sencan, Z., Schicktanz, S.: Attitudes towards prediction and early diagnosis of late-onset dementia: a comparison of tested persons and family caregivers. Aging Mental Health **25**, 1–12 (2020)
27. Ezzine, B.E., Rekik, I.: Learning-guided infinite network atlas selection for predicting longitudinal brain network evolution from a single observation. In: Shen, D., et al. (eds.) MICCAI 2019. LNCS, vol. 11765, pp. 796–805. Springer, Cham (2019). https://doi.org/10.1007/978-3-030-32245-8_88
28. Kullback, S., Leibler, R.A.: On information and sufficiency. Ann. Math. Stat. **22**, 79–86 (1951)
29. Errica, F., Podda, M., Bacciu, D., Micheli, A.: A fair comparison of graph neural networks for graph classification. arXiv preprint arXiv:1912.09893 (2019)
30. Veličković, P., Cucurull, G., Casanova, A., Romero, A., Lio, P., Bengio, Y.: Graph attention networks. arXiv preprint arXiv:1710.10903 (2017)
31. Marcus, D.S., Fotenos, A.F., Csernansky, J.G., Morris, J.C., Buckner, R.L.: Open access series of imaging studies: longitudinal MRI data in nondemented and demented older adults. J. Cogn. Neurosci. **22**, 2677–2684 (2010)
32. Loshchilov, I., Hutter, F.: Fixing weight decay regularization in adam (2018)
33. Mueller, S.G., et al.: The Alzheimer's disease neuroimaging initiative. Neuroimaging Clin. **15**, 869–877 (2005)

34. Goryawala, M., Zhou, Q., Barker, W., Loewenstein, D.A., Duara, R., Adjouadi, M.: Inclusion of neuropsychological scores in atrophy models improves diagnostic classification of Alzheimer's disease and mild cognitive impairment. Comput. Intell. Neurosci. **2015** (2015)
35. Mahjoub, I., Mahjoub, M.A., Rekik, I.: Brain multiplexes reveal morphological connectional biomarkers fingerprinting late brain dementia states. Sci. Rep. **8**, 1–14 (2018)

Mixing-AdaSIN: Constructing a De-biased Dataset Using Adaptive Structural Instance Normalization and Texture Mixing

Myeongkyun Kang[1], Philip Chikontwe[1], Miguel Luna[1], Kyung Soo Hong[2], June Hong Ahn[2(✉)], and Sang Hyun Park[1(✉)]

[1] Robotics Engineering, Daegu Gyeongbuk Institute of Science and Technology (DGIST), Daegu, Korea
{mkkang,shpark13135}@dgist.ac.kr
[2] Division of Pulmonology and Allergy, Department of Internal Medicine, Regional Center for Respiratory Diseases, Yeungnam University Medical Center, College of Medicine, Yeungnam University, Daegu, Korea
fireajh@yu.ac.kr

Abstract. Following the pandemic outbreak, several works have proposed to diagnose COVID-19 with deep learning in computed tomography (CT); reporting performance on-par with experts. However, models trained/tested on the same in-distribution data may rely on the inherent data biases for successful prediction, failing to generalize on out-of-distribution samples or CT with different scanning protocols. Early attempts have partly addressed bias-mitigation and generalization through augmentation or re-sampling, but are still limited by collection costs and the difficulty of quantifying bias in medical images. In this work, we propose Mixing-AdaSIN; a bias mitigation method that uses a generative model to generate de-biased images by *mixing* texture information between different labeled CT scans with semantically similar features. Here, we use Adaptive Structural Instance Normalization (AdaSIN) to enhance de-biasing generation quality and guarantee structural consistency. Following, a classifier trained with the generated images learns to correctly predict the label without bias and generalizes better. To demonstrate the efficacy of our method, we construct a biased COVID-19 vs. bacterial pneumonia dataset based on CT protocols and compare with existing state-of-the-art de-biasing methods. Our experiments show that classifiers trained with de-biased generated images report improved in-distribution performance and generalization on an external COVID-19 dataset.

1 Introduction

Recently, several methods have been proposed to improve COVID-19 patient diagnosis or treatment planning using CT scans [12,16,24,25]. The most methods are often trained and evaluated on single source data; producing models that

© Springer Nature Switzerland AG 2021
I. Rekik et al. (Eds.): PRIME 2021, LNCS 12928, pp. 37–46, 2021.
https://doi.org/10.1007/978-3-030-87602-9_4

exploit underlying biases in the data for better predictions. Yet, they fail to generalize when the bias shifts in external data reporting lower performance. This has critical implications, especially in medical imaging where biases are hard to define or accurately quantify. To address this, extensive data augmentation [5] or re-sampling is often employed; though is still limited by collection (multi-institute data) and how to express the bias when it is unknown. Thus, there is a need for methods that mitigate bias in part or fully towards improved model performance and better generalization.

In general, models trained on biased data achieve high accuracy and despite their capacity lack the motivation to learn the complexity of the intended task. For instance, a model trained on our in-house biased dataset with COVID-19 and bacterial pneumonia reported 97.18% (f1-score) on the validation set, yet degrades to 33.55% when evaluated on an unbiased test-set. Here, we believe bias may originate from varied CT protocols based on exam purpose, scanners, and contrast delivery requirements [3]. Though contrast CT is a standard protocol, it is challenging for practitioners to meet requirements during the pandemic due to extra processes such as contrast agent injection and disinfection [11, 21]. Further, protocols may also vary for other pneumonia that exhibit similar imaging characteristics with COVID-19 in CT. Consequently, we believe biased datasets are often constructed unexpectedly and sometimes unavoidably due to the aforementioned factors.

Among the existing techniques proposed to remove model dependence on bias, augmentation is a de-facto technique for medical images; with other methods pre-defining the bias the trained model should be independent of. This assumes bias is easily defined, but one has to take extra care in the medical setup where such assumptions do not hold. To address this, we propose to construct a de-biased dataset where spurious features based on texture information become uninformative for accurate prediction. A key motivation is that accurate prediction of COVID-19 from other pneumonia's is dependent on the CT protocols related to texture features and contrast. Thus, we propose to generate COVID-19 CTs with bacterial pneumonia protocol characteristics and vice versa for bacterial pneumonia with COVID-19, respectively.

Specifically, we propose Mixing-AdaSIN; a generative model based bias removal framework that leverages Adaptive Structural Instance Normalization (AdaSIN) and texture *mixing* to generate de-biased images used to train classifiers robust to the original bias. For image generation, we employ two main components: (a) texture *mixing*, which enables realistic image generation, and (b) AdaSIN, which guarantees structural consistency and prevents bias retainment in the input image via modifying the distribution of the structure feature. To prevent incorrect image generation, we first pre-train a contrastive encoder [8] to learn key CT features and later use it to search similar image pairs for the texture *mixing* step in the proposed generative framework. For evaluation, we construct biased train/validation sets based on the CT protocol and an unbiased test set from the non-overlapping CT protocols of the train/validation sets, respectively. The proposed method reports high bias mitigation performance

(66.97% to 80.97%) and shows improved generalization performance when verified on an external dataset. The main contributions of this work are summarized as follows:

– We propose a generative model that can sufficiently mitigate the bias present in the training dataset using AdaSIN and texture *mixing.*
– We show that the use of contrastive learning for texture transfer pair selection prevents incorrect image generation.
– We constructed a biased COVID-19 vs. bacterial pneumonia dataset to verify bias mitigation performance. Our approach not only enabled improvements for standard classification models such as ResNet18 but also current state-of-the-art COVID-19 classification models.
– We also demonstrate the generalization performance of our classifier trained with the de-biased data on an external public dataset.

2   Related Works

CT Based COVID-19 Classification. Several methods have been proposed to address this task since the inception of the pandemic [12, 16, 24, 25]. For instance, Li et al. [16] encode CT slices using a 2D CNN and aggregate slice predictions via max-pooling to obtain patient-level diagnosis. Wang et al. [24] proposed COVID-Net, an approach that utilizes the long-range connectivity among slices to increase diagnostic performance. Later, Wang et al. [25] further improve COVID-Net by employing batch normalization and contrastive learning to make the model robust to multi-center training. However, these models did not address the bias in the dataset and thus may fail to generalize on other datasets.

Bias Mitigation. To mitigate bias, Alvi et al. [1] employed a bias classifier and a confusion loss to regularize the extracted features to make them indistinguishable from a bias classifier. Kim et al. [15] proposed to mitigate bias through mutual information minimization and the use of a gradient reversal layer. Though these methods can mitigate distinct biases such as color, they fail to mitigate bias in the medical domain since the bias from CT protocols is subtle and hard to distinguish even for humans [2].

Another line of work is the augmentation based models that utilize techniques such as arbitrary style transfer. Geirhos et al. [5] proposed shape-ResNet, an approach that finetunes a model pre-trained on AdaIN [10] generated images. Li et al. [18] proposed a multitask learning approach that enables accurate prediction by either using shape, texture, or both types of features. Though successful, a key drawback is the heavy reliance on artistic image generation techniques that may be detrimental for medical images and subsequent diagnoses. To address this, our approach is able to capture subtle differences in the image to generate consistent texture updated images. Thus, classifiers trained on the generated images can avoid subtle bias information.

Generative Texture Transfer. Methods [10] and [6] both proposed to generate texture updated images based on arbitrary style transfer, with adaptive

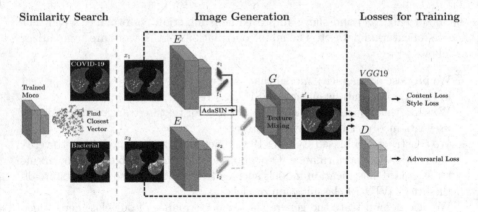

Fig. 1. Diagram of the proposed method.

and conditional instance normalization employed to transfer texture information. CycleGAN [27] is another popular method for texture transfer that uses the idea of consistency across several model outputs. However, these techniques not only change the texture but also induce structural distortion to the outputs which may introduce new forms of bias. Recently, Park et al. [20] achieved better results by using an autoencoder with texture swapping as well as a co-occurrence patch discriminator to capture high-level detail. In this method, the discriminator model may often change the original structural characteristics which is undesirable for medical images. Since our main objective is to maintain the structural information, we avoid techniques such as cycle consistency and patch discriminators that often produce structurally distorted images.

3 Methods

Given a chest CT dataset $\mathcal{D} = \{X^1, ..., X^n\}$ where X^i is a set of 2D CT slices $X^i = \{x_1^i, ..., x_m^i\}$ each with its label $y^i \in \{0, 1\}$ denoting bacterial pneumonia and COVID-19 samples, respectively. Our goal is to generate a dataset $\mathcal{D}' = \{X^{1\prime}, ..., X^{n\prime}\}$ where each $X^{i\prime}$ is a set of texture updated CT slices that may contain the bias information of the other label CT protocol. To achieve this, we first pre-train a contrastive encoder using $\mathcal{D}$ to learn representative slice features and then use it for slice similarity search i.e. $x_1^i, x_2^i \sim \mathcal{D}, y^i \neq y^j$ with semantically similar image structures. Second, to generate $x_1^{i\prime}$ we feed searched pairs x_1^i, x_2^j to an encoder network E that outputs structural and texture features used as input for a generator network G. Here, AdaSIN and texture *mixing* is employed on the respective features for improved generation. Lastly, following standard practice for adversarial-based methods, we also employ a discriminator network D to improve the quality of generation. To enforce style and content similarity, a pre-trained VGG19 [23] is used to optimize content loss between x_1' and x_1, and a style loss between x_1' and x_2, respectively. Through this process, we can generate images to construct $\mathcal{D}'$ and then combine $\mathcal{D}$ and $\mathcal{D}'$ to train a classifier that will learn to ignore bias information. The overall framework is presented in Fig. 1 with specific details of each step categorized below.

Slice Similarity Search. As opposed to arbitrarily sampling pairs for texture transfer in our generative framework, we employ similarity based sampling of image pairs with similar structural features between the classes. Here, we pre-train a momentum contrastive network (MoCo) [8] using the training data and find the closest slice based on the $L1$ distance. This is crucial for generation since using arbitrary image pairs for texture transfer can produce artificial images without any clinical significance. Following, we construct image pairs for the entire dataset for image generation.

Image Generation with AdaSIN and Texture *Mixing*. To generate a de-biased and texture mixed image, an encoder network E takes as input the sampled image pairs to produce texture and structure features that are first modified via AdaSIN before being feed to G. To retain texture and structure, both are required as inputs for G. Specifically, the structure feature is a feature with spatial dimensions pooled at an earlier stage of E, whereas the texture feature is an embedding vector obtained following 1×1 convolution in the later stages of E. In addition, we believe that directly employing these features for generation can still retain the inherent bias contained in the features, thus we propose to modify the structural features via AdaIN [10]. To achieve this, we use the mean($\mu(\cdot)$) and standard deviation ($\sigma(\cdot)$) of the features before passing to G. Formally,

$$AdaSIN(s_1, s_2) = \sigma(s_2)\left(\frac{s_1 - \mu(s_1)}{\sigma(s_1)}\right) + \mu(s_2), \qquad (1)$$

where s_1 and s_2 denotes the extracted structure features of the input image pairs i.e. $x_1, x_2 \sim \mathcal{D}$. Next, G takes both features for an image generation and texture transfer. Texture transfer is achieved via convolution weight modulation and demodulation introduced in [14]. Herein, texture information is delivered to each convolutional layer in G except for the last layer.

To train the entire framework, we follow the arbitrary style transfer training procedure described in [10]. A VGG19 pre-trained model is used to extract features from the input pairs which is then used in the style L_{style} and content $L_{content}$ losses, respectively. The style loss minimizing the mean squared error (MSE) between the generated output and the texture image, whereas the content loss minimizing the MSE between generated output and the structure image [17]. Further, we use an adversarial loss to improve the image generation quality via $L_{GAN}(G, D) = -\mathbb{E}_{x_1, x_2 \sim \mathcal{D}}[D(G(AdaSIN(s_1, s_2), t_2))]$, with regularization (non-saturation loss) omitted for simplicity [7,14]. The final loss function is defined as:

$$L_{total} = L_{content} + \lambda L_{style} + L_{GAN}. \qquad (2)$$

Implementation Details. A Mask-cascade-RCNN-ResNeSt-200 [26] with deformable convolution neural network (DCN) [4] was employed to extract the

Table 1. The statistics and bias information (CT protocol) of the train/validation/test dataset.

Split	COVID	CT protocol	Bacterial	CT protocol
Train	42	Non-Contrast (kVp: 120).	60	Contrast (kVp: 100)
Val	10	Non-Contrast (kVp: 120).	14	Contrast (kVp: 100)
Test	21	Various CT protocols from different scanners: Contrast (18), etc. (3).	23	Various CT protocols from different scanners: Non-Contrast (18), etc. (5)

lung and lesion regions in the CT scans to mask out the non-lung regions and excluding slices that do not contain a lesion. The contrastive slice encoder was trained for 200 epochs whereas the generative model was trained with a batch size of 8 for 40,000 steps with Adam optimizer and learning rate of 0.002. Further, the discriminator D follows Karras et al.'s [13,14] implementation and we empirically set $\lambda = 10$.

4   Experiments and Results

Experiments Settings. For evaluation, we constructed an in-house biased dataset for training/validation and an unbiased testing dataset. First, we extracted the CT protocol per scan using metadata in the DICOM files and create splits COVID-19 and bacterial pneumonia based on the protocol. The dataset and bias information are shown in Table 1.

To evaluate the effect of bias mitigation using the generated images for classification, a pre-trained ResNet18 and recent COVID-19 classification models [9,16,24,25] i.e. COVNet, COVID-Net, Contrastive-COVIDNet were compared. The models were trained for 100 epochs with a batch size 64 and a learning rate 0.001. We also applied random crops, horizontal flips and intensity augmentations i.e. (brightness and contrast) as a baseline augmentation technique. Performance comparison of our approach against recent state-of-the-art non-generation based bias mitigation methods [1,15] applied for natural image classification is also reported. To verify the effectiveness of the proposed method, we include comparisons against a commonly used arbitrary style transfer model i.e. AdaIN, and the current state-of-the-art generation method i.e. swapping-autoencoder [10,20]. For a fair comparison of the generation based methods, the same texture pairs were utilized for a generation. Also, training and validation were performed three times for all methods with final average performance reported.

Results on Internal Dataset. In Table 2, we present the evaluation results on the biased COVID-19 vs. bacterial pneumonia dataset. Initially, the model shows high f1-score on the validation dataset i.e. 97.18%, yet significantly drops

Table 2. The test results (f1-score) on biased COVID-19 vs. bacterial pneumonia dataset. The results of validation dataset are written as Val. The model without intensity augmentation is denoted w/o aug.

	ResNet18 [9]	COVNet [16]	COVID-Net [24]	Contrastive COVID-Net [25]
Val w/o aug	97.18	89.33	78.30	93.31
Base	66.97	63.41	41.89	53.37
Base w/o aug	33.55	59.52	43.99	35.64
LNTL [15]	31.87	–	–	–
Blind eye [1]	32.72	–	–	–
AdaIN [10]	75.71	66.35	51.35	65.76
Swap [20]	76.84	67.64	56.28	62.60
Mixing-AdaSIN	**80.97**	74.61	61.57	66.16

to 33.55% on the unbiased test set. This shows that the classifier makes predictions based on the bias information. The results of the learning-based models i.e. Learning not to learn, and Blind eye show no considerable performance improvements and highlight the failure to mitigate the bias, especially for medical domain images as reported in [2]. Further, these methods were proposed to remove bias that is distinctly recognizable in the image such as color. Capturing and mitigating the subtle bias difference in the medical image is considerably harder for such techniques.

Among generation based methods, the proposed method reports the best performance. Even though AdaIN can transfer texture well, the quality of the generated image is extremely low. Consequently, this inhibits classifier training as shown by the limited performance improvements. Though swapping-autoencoder updates the texture with high quality image generation results, two major drawbacks were noted: (i) the generated image still retains bias, and (ii) it distorts key representative characteristics by artificial translation of lesions i.e. changing the lesion of COVID-19 to appear as bacterial pneumonia. Such phenomena may be due to the direct usage of structure features, and use of the co-occurrence

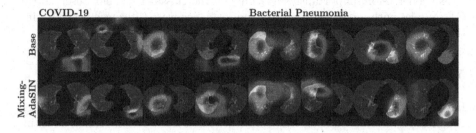

Fig. 2. Grad-CAM [22] visualizations of the ResNet18 [9] classifier. Herein, Grad-CAM of the base classifier pointed the normal lung area. On the other hand, the classifier trained with Mixing-AdaSIN pointed the lesion correctly. Hence, the model with debiasing can be more generalized.

Table 3. The test results (accuracy) of the ResNet18 [9] on the external COVID-19 dataset.

	Base	Base w/o aug	AdaIN [10]	Swap [20]	Mixing-AdaSIN
Accuracy	24.11	34.04	50.35	76.60	77.30

discriminator which leads to structural information deformation. On the other hand, our model employs two modules i.e. boosting a high-quality image generation via texture *mixing*; and minimizing the bias artifact transfer through an AdaSIN. We consider these techniques as instrumental in mitigating bias. In addition, the classifier with a de-biased method employed a more generalized feature as shown in Fig. 2. The Grad-CAM [22] of the trained classifier pointed to the lesion more correctly, thus a better generalization performance is expected.

Our proposed method can be easily applied to existing CT based COVID-19 classification models. In particular, COVNet reported 74.61% f1-score which represents successful mitigation of bias artifacts. However, COVID-Net and Contrastive COVID-Net showed relatively low accuracy, mainly due to slight differences in training details and architectures. Also, due to the long-range connectivity in the models, reliance on bias information is heavily induced during training.

Results on External Dataset. To verify the generalization efficacy of our trained classifier on external data, we employ the publicly available MosMed dataset [19]. It consists of 1110 CT scans from COVID-19 positive patients. However, the dataset contains CT scans that are not consistent with pneumonia observed in our original dataset. Thus, we selected scans of severe and critical patients only to evaluate the trained models. In addition, since we trained three classifiers from an internal experiment, we tested each classifier three times and final average performance reported. In Table 3, results are fairly consistent with improvements shown on the internal dataset evaluation. Our model shows a significant improvement over the baseline with +1% gain over swapping-autoencoder. More importantly, even though the classifier has not observed the CT samples with a different protocol, performance was still consistent verifying the utility of the proposed de-biasing technique.

5   Conclusion

In this work, we have proposed a novel methodology to train a COVID-19 vs. bacterial pneumonia classifier that is robust to bias information present on training data. We constructed an in-house biased training dataset in conjunction with an unbiased testing dataset and proved that our method allowed the classifier to learn the appropriate features to correctly predict the labels without considering bias information and achieved better generalization. All of this was possible thanks to an adequate image generation design that relies on two major components: (a) texture *mixing*, which enables realistic image generation, and (b) AdaSIN, which prevents bias flow from the input to the output image in the generation stage, while maintaining structural consistency. We proved the benefits of our pipeline by achieving the best bias mitigation performance when compared to other related methods in both our in-house dataset as well as in an external dataset. Considering that biases can be easily included when constructing datasets, we hope that our findings help to improve performance in various medical tasks.

Acknowledgment. This research was funded by the National Research Foundation of Korea (NRF) grant funded by the Korean Government (MSIT) (No. 201 9R1C1C1008727).

References

1. Alvi, M., Zisserman, A., Nellåker, C.: Turning a blind eye: explicit removal of biases and variation from deep neural network embeddings. In: Proceedings of the European Conference on Computer Vision Workshops (2018)
2. Bissoto, A., Valle, E., Avila, S.: Debiasing skin lesion datasets and models? Not so fast. In: Proceedings of the IEEE/CVF Conference on Computer Vision and Pattern Recognition Workshops, pp. 740–741 (2020)
3. Caschera, L., Lazzara, A., Piergallini, L., Ricci, D., Tuscano, B., Vanzulli, A.: Contrast agents in diagnostic imaging: present and future. Pharmacol. Res. **110**, 65–75 (2016)
4. Dai, J., et al.: Deformable convolutional networks. In: Proceedings of the IEEE International Conference on Computer Vision, pp. 764–773 (2017)
5. Geirhos, R., Rubisch, P., Michaelis, C., Bethge, M., Wichmann, F.A., Brendel, W.: ImageNet-trained CNNs are biased towards texture; increasing shape bias improves accuracy and robustness. In: International Conference on Learning Representations (2019)
6. Ghiasi, G., Lee, H., Kudlur, M., Dumoulin, V., Shlens, J.: Exploring the structure of a real-time, arbitrary neural artistic stylization network. In: British Machine Vision Conference (2017)
7. Gulrajani, I., Ahmed, F., Arjovsky, M., Dumoulin, V., Courville, A.: Improved training of Wasserstein GANs. In: Advances in Neural Information Processing Systems, pp. 5769–5779. Curran Associates Inc. (2017)
8. He, K., Fan, H., Wu, Y., Xie, S., Girshick, R.: Momentum contrast for unsupervised visual representation learning. In: Proceedings of the IEEE/CVF Conference on Computer Vision and Pattern Recognition, pp. 9729–9738 (2020)
9. He, K., Zhang, X., Ren, S., Sun, J.: Deep residual learning for image recognition. In: Proceedings of the IEEE Conference on Computer Vision and Pattern Recognition, pp. 770–778 (2016)
10. Huang, X., Belongie, S.: Arbitrary style transfer in real-time with adaptive instance normalization. In: Proceedings of the IEEE International Conference on Computer Vision, pp. 1501–1510 (2017)
11. Kalra, M.K., Homayounieh, F., Arru, C., Holmberg, O., Vassileva, J.: Chest CT practice and protocols for Covid-19 from radiation dose management perspective. Eur. Radiol. **30**, 1–7 (2020)

12. Kang, M., et al.: Quantitative assessment of chest CT patterns in Covid-19 and bacterial pneumonia patients: a deep learning perspective. J. Korean Med. Sci. **36**(5) (2021)

13. Karras, T., Aila, T., Laine, S., Lehtinen, J.: Progressive growing of GANs for improved quality, stability, and variation. In: International Conference on Learning Representations (2018)

14. Karras, T., Laine, S., Aittala, M., Hellsten, J., Lehtinen, J., Aila, T.: Analyzing and improving the image quality of StyleGAN. In: Proceedings of the IEEE/CVF Conference on Computer Vision and Pattern Recognition, pp. 8110–8119 (2020)

15. Kim, B., Kim, H., Kim, K., Kim, S., Kim, J.: Learning not to learn: training deep neural networks with biased data. In: Proceedings of the IEEE/CVF Conference on Computer Vision and Pattern Recognition, pp. 9012–9020 (2019)

16. Li, L., et al.: Using artificial intelligence to detect Covid-19 and community-acquired pneumonia based on pulmonary CT: evaluation of the diagnostic accuracy. Radiology **296**(2), E65–E71 (2020)

17. Li, Y., Wang, N., Liu, J., Hou, X.: Demystifying neural style transfer. In: Proceedings of the 26th International Joint Conference on Artificial Intelligence, IJCAI 2017, pp. 2230–2236. AAAI Press (2017)

18. Li, Y., et al.: Shape-texture debiased neural network training. In: International Conference on Learning Representations (2021)

19. Morozov, S.P., et al.: MosMedData: data set of 1110 chest CT scans performed during the Covid-19 epidemic. Digit. Diagn. **1**(1), 49–59 (2020)

20. Park, T., et al.: Swapping autoencoder for deep image manipulation. In: Advances in Neural Information Processing Systems, vol. 33, pp. 7198–7211. Curran Associates, Inc. (2020)

21. Pontone, G., et al.: Role of computed tomography in Covid-19. J. Cardiovasc. Comput. Tomogr. (2020)

22. Selvaraju, R.R., Cogswell, M., Das, A., Vedantam, R., Parikh, D., Batra, D.: Grad-CAM: visual explanations from deep networks via gradient-based localization. In: Proceedings of the IEEE International Conference on Computer Vision, pp. 618–626 (2017)

23. Simonyan, K., Zisserman, A.: Very deep convolutional networks for large-scale image recognition. In: International Conference on Learning Representations (2014)

24. Wang, L., Lin, Z.Q., Wong, A.: Covid-Net: a tailored deep convolutional neural network design for detection of Covid-19 cases from chest X-ray images. Sci. Rep. **10**(1), 1–12 (2020)

25. Wang, Z., Liu, Q., Dou, Q.: Contrastive cross-site learning with redesigned net for Covid-19 CT classification. IEEE J. Biomed. Health Inform. **24**(10), 2806–2813 (2020)

26. Zhang, H., et al.: ResNeSt: split-attention networks. arXiv preprint arXiv:2004.08955 (2020)

27. Zhu, J.Y., Park, T., Isola, P., Efros, A.A.: Unpaired image-to-image translation using cycle-consistent adversarial networks. In: Proceedings of the IEEE International Conference on Computer Vision, pp. 2223–2232 (2017)

Liver Tumor Localization and Characterization from Multi-phase MR Volumes Using Key-Slice Prediction: A Physician-Inspired Approach

Bolin Lai[1], Yuhsuan Wu[1], Xiaoyu Bai[1,2], Xiao-Yun Zhou[3], Peng Wang[5],
Jinzheng Cai[3], Yuankai Huo[4], Lingyun Huang[1], Yong Xia[2], Jing Xiao[1],
Le Lu[3], Heping Hu[5], and Adam Harrison[3(✉)]

[1] Ping An Technology, Shanghai, China
[2] Northwestern Polytechnical University, Xian, China
[3] PAII Inc., Bethesda, MD, USA
[4] Vanderbilt University, Nashville, TN, USA
[5] Eastern Hepatobiliary Surgery Hospital, Shanghai, China

Abstract. Using radiological scans to identify liver tumors is crucial for proper patient treatment. This is highly challenging, as top radiologists only achieve F1 scores of roughly 80% (hepatocellular carcinoma (HCC) vs. others) with only moderate inter-rater agreement, even when using multi-phase magnetic resonance (MR) imagery. Thus, there is great impetus for computer-aided diagnosis (CAD) solutions. A critical challenge is to *robustly* parse a 3D MR volume to localize diagnosable regions of interest (ROI), especially for edge cases. In this paper, we break down this problem using key-slice prediction (KSP), which emulates physician workflows by predicting the slice a physician would choose as "key" and then localizing the corresponding key ROIs. To achieve robustness, the KSP also uses curve-parsing and detection confidence re-weighting. We evaluate our approach on the largest multi-phase MR liver lesion test dataset to date (430 *biopsy-confirmed* patients). Experiments demonstrate that our KSP can localize diagnosable ROIs with high reliability: 87% patients have an average 3D overlap of ≥40% with the ground truth compared to only 79% using the best tested detector. When coupled with a classifier, we achieve an HCC vs. others F1 score of 0.801, providing a fully-automated CAD performance comparable to top human physicians.

Keywords: Liver · Tumor localization · Tumor characterization

1 Introduction

Liver cancer is the fifth/eighth most common malignancy in men/women worldwide [4]. During treatment planning, non-invasive diagnostic imaging is preferred, as invasive procedures, *i.e.*, biopsies or surgeries, can lead to hemmorages,

---

B. Lai, Y. Wu and X. Bai—Equal contribution.
This work was done when X. Bai was an intern at Ping An Technology.

© Springer Nature Switzerland AG 2021
I. Rekik et al. (Eds.): PRIME 2021, LNCS 12928, pp. 47–58, 2021.
https://doi.org/10.1007/978-3-030-87602-9_5

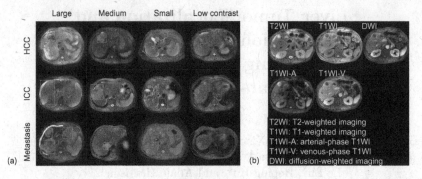

Fig. 1. (a) HCC, ICC and metastasis lesion examples on T2WIs, including large, medium, small and low contrast tumors, all rendered on their key slices. The third-from-left metastasis shows an example of a lesion cluster. (b) Different MR sequences of the same patient.

infections, and even death [13]. Multi-phase magnetic resonance (MR) imagery is considered the most informative radiological option [20], with T2-weighted imaging (T2WI) able to reveal tumor edges and aggressiveness [2]. Manual lesion differentiation is workload-heavy and ideally is executed by highly experienced radiologists that are not always available in every medical center. Studies on human reader performance, which focus on differentiating hepatocellular carcinoma (HCC) from other types, report low specificities [2] and moderate inter-rater agreement [9]. Thus, there is a need for computer-aided diagnosis (CAD) solutions, which is the topic of our work. Unlike other approaches, we propose a physician-inspired workflow that predicts key slices that would be chosen by clinicians. This allows us to achieve greater reliability and robustness.

A major motivation for CAD is addressing challenging cases that would otherwise be biopsied or even incorrectly operated on. For instance, a 2006 retrospective study discovered that pre-operative imaging misinterpreted 20% of its liver transplant patients as having HCC [10]. While several CAD approaches have been reported, many do not focus on histopathologically-confirmed studies [1,7,25,29], which are the cases most requiring CAD intervention. Prior CAD studies, except for Zhen *et al.* [32], also only focus on computed tomography (CT), despite the greater promise of MR. Most importantly, apart from two studies [6,18], CAD works typically assume a manually drawn region of interest (ROI) is available. In doing so, they elide the major challenge of parsing a medical volume to determine diagnosable ROIs. Without this capability, manual intervention remains necessary and the system also remains susceptible to inter-user variations. The most obvious localization strategy, *e.g.*, that of [18], would follow computer vision practices and directly applies a detector. However, detectors aim to find *all* lesions and their *entire* 3D extent in a study, whereas the needs for liver lesion characterization are distinct: *reliably* localize one or more key diagnosable ROI(s). This different goal warrants its own study, which we investigate.

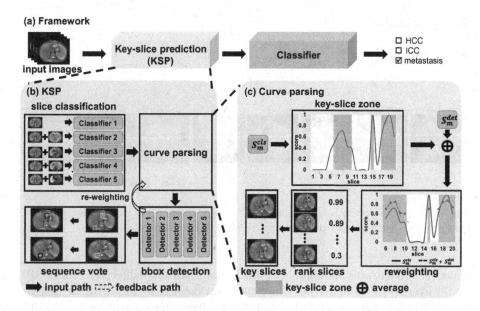

Fig. 2. The proposed (a) framework is composed of (b) KSP and lesion characterization. KSP consists of slice classification, detection and (c) curve parsing.

In this work, we develop a robust and fully-automated CAD system to differentiate malignant liver tumors into the HCC, ICC and metastasis subtypes. Figure 1(a) depicts these three types. To localize key diagnosable ROIs, we use a physician-inspired approach that departs from standard detection frameworks seen in computer vision and used elsewhere [18]. Instead, we propose key-slice prediction (KSP), which breaks down the parsing problem similarly to clinical practice, *i.e.*, first robustly identifying and ranking key slices in the volume to predict which are key slices and, from each of these, regressing a single diagnosable ROI. This follows, at least in spirit, protocols like the ubiquitous response evaluation criteria in solid tumors (RECIST) [8]. In concrete terms KSP comprises multi-sequence classification, detection, and curve parsing. Once localized, each ROI is classified using a standard classifier.

We test our approach on 430 multi-phase MR studies (2150 scans), *which is the largest test cohort studied for liver lesion CAD to date*. Moreover, all of our patient studies are histopathologically confirmed, well-representing the challenging cases requiring CAD. Using our KSP framework, we achieve very high reliability, with 87% of our predicted ROIs overlapping with the ground truth by $\geq 40\%$, outperforming the best detector alternative (only 79% with an overlap $\geq 40\%$).

2 Methods

2.1 Overview

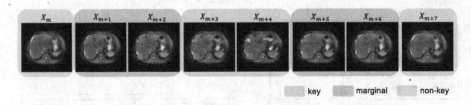

Fig. 3. The definition of key, marginal and non-key slices. Eight consecutive slices are demonstrated with ground truth bboxes showing the location of tumors.

Figure 2(a) illustrates our approach, which comprises a key-slice predictor (KSP) and a liver lesion classifier. As illustrated and defined by Fig. 1(b), we assume we are given a dataset of MR volumes with five sequences/phases. Formally, assuming N studies, we define our dataset as $\mathcal{D} = \{\mathcal{X}_i, \mathcal{B}_i, y_i\}_{i=1}^{N}$, where $\mathcal{X}_i = \{X_{i,j}\}_{j=1}^{5}$ is the MR sequences and y_i is a study-level lesion-type label. Lesions are either (1) annotated by 2D bounding boxes (bboxes) using RECIST-style marks [8] or (2) when they are too numerous to be individually annotated, a bbox over each cluster is provided. See Fig. 1(a) for an illustration of the two types. Given the extreme care and multiple readers needed for lesion masks [3], bbox labels are much more practical to generate. We use m to represents individual slices, e.g., $X_{i,j,m}$, which also selects any corresponding bboxes, $\mathcal{B}_{i,m}$. From the bboxes, we can also define slices as being "key", "marginal", and "non-key", see Fig. 3 for examples. For cases with a single tumor in the liver, "marginal" slices are those within a buffer of one slice from the beginning or end of a lesion. Then slices with tumors between two "marginal" regions are defined as "key" slices and the remaining slices are defined as "non-key" slices. For cases with more than one tumor, we merge the designations of each individual lesion together to create a slice-wise label. Under this protocol, a "key" slice is any slice that captures one or more individual lesion "key" slices. In the remaining slices, a "marginal" slice is any slice that captures one or more individual lesion "marginal" slices. Finally, for those that are not defined as "key" or "marginal", they are treated as "non-key" slices. We will drop the i when appropriate.

2.2 Key-Slice Prediction

Because any popular classifier can be used for lesion classification, our methodological focus is on the KSP. Illustrated in Fig. 2(b), KSP decomposes localization into the simpler problem of key-slice ranking followed by key ROI regression.

Slice ranking identifies whether each MR slice is a key slice or not. Any state-of-the-art classifier can be used, trained on "key" and "non-key" slices, with "marginal" ones ignored. But care must be taken to handle multi-sequence MR data. In short, the MR sequences or phases where lesions are visible vary, somewhat unpredictably, from lesion to lesion. Early fusion (EF) models, *i.e.*, inputting a five-channel slice, can be susceptible to overfitting to the specific sequence behavior seen in the training set for individual instances. Examining each MR sequence more independently mitigates this risk. Thus, we perform late fusion (LF). More specifically, because T2WI is the most informative sequence for liver tumors [2], we use T2WI as an anchor (T2-anchor) and pair each of the remaining sequences with it (and also one T2WI-only sequence), training a separate model for each. Unlike standard LF, the T2-anchor LF approach indeed boosts the performance over EF (see our ablation study). Under the T2-anchor LF approach, we obtain confidence scores for each of the five T2-anchor sequences, j, and for each slice, m: $s_{j,m}^{\mathrm{cls}}$. We average the confidence score across all sequences to compute a slice-wise classification confidence:

$$s_m^{\mathrm{cls}} = \frac{1}{5} \sum_{j=1}^{5} s_{j,m}^{\mathrm{cls}}. \tag{1}$$

Our decomposition strategy means that selected key slices should contain at least one lesion. We take advantage of this prior knowledge to regress a single key ROI from each prospective key slice. To do this, we train any state-of-the-art detector on each T2-anchor sequence, producing a set of bbox confidence values and locations for each slice and sequence: $\mathcal{S}_{j,m}^{\mathrm{det}}$ and $\hat{\mathcal{B}}_{j,m}$, respectively, and we group the outputs across all sequences together: $\mathcal{S}_m^{\mathrm{det}}, \hat{\mathcal{B}}_m = \{\mathcal{S}_{j,m}^{\mathrm{det}}, \hat{\mathcal{B}}_{j,m}\}_{j=1}^5$. To produce single ROI, we use a voting scheme where for every possible pixel location we sum up the detection confidences of any bbox, $s_{m,k}^{\mathrm{det}} \in \hat{\mathcal{B}}_m$, that overlaps with it. We then choose the pixel location with the highest detection confidence sum. For all the bboxes that overlap the chosen pixel location, we take their mean location and size as the final slice-wise ROI. Importantly, we filter out low-confidence ROIs using a threshold t, which is determined by examining the overlap of resulting slice-wise ROIs with ground truth bboxes in validation. We choose the t that provides the best empirical cumulative distribution function of overlaps, and thus the best balance between false negatives and false positives.

The final step is to rank slices and their corresponding ROIs, as illustrated in Fig. 2(c). We start by producing a confidence curve across all slices using s_m^{cls}. From this curve we identify peaks, which ideally should each correspond to the presence of a true lesion. Each peak defines a key-slice zone, which is the adjoining region where confidence values are within 1/2 of the "peak". Only key slices in key-slice zones will be ranked and selected, and we only admit slices that contain at least one bbox with a confidence score $>t$.

Since good performance relies on selecting the correct slices, we use detection to build in redundancy and to better rank slices in the key-slice zone. Specifically, from all bbox confidences in a slice, $\mathcal{S}_m^{\mathrm{det}}$, we compute a slice-wise confidence by choosing the maximum bbox confidence. We bias these confidences toward larger bboxes, based on the assumption that they are more diagnosable:

$$s_m^{\text{det}} = \max\left(\left\{(a_{m,k} + s_{m,k}^{\text{det}})/2\right\}_{k=1}^{\hat{K}_m}\right), \tag{2}$$

where k indexes the predicted bboxes and confidences in $\mathcal{S}_m^{\text{det}}$ and $\hat{\mathcal{B}}_m$ and $a_{m,k} \in (0,1]$ is the normalized bbox area across all slices. Next, we combine classification and detection scores:

$$s_m^{\text{cls+det}} = (s_m^{\text{cls}} + s_m^{\text{det}})/2. \tag{3}$$

We rank slices using (3) and select the top $T\%$ of slices. We choose T by examining the distribution of *within-study* precision and recalls across all validation studies. We choose the T giving an *across-study* average recall ≥ 0.5 and a first quartile (Q1) precision of ≥ 0.6 (see Results). This strategy is applied for all evaluated detectors. The corresponding slice-wise ROIs comprise the key predicted ROIs.

3   Results

Setup. We collected 430 multi-phase multi-sequence MR studies (2150 volumes) from the Eastern Hepatobiliary Surgery Hospital. The selection criteria were any patient who had surgical reaction or biopsy in the period between 2006 and 2019 where T1WI, T2WI, T1WI-V, T1WI-A, and DWI sequences are available. Lesion distribution was 207, 113, and 110 patients with HCC, ICC, and metastasis, respectively. The data was then split patient-wise using five-fold cross validation, with 70%, 10%, and 20% used for training, validation, and testing, respectively. Data splitting was executed on HCC, ICC and metastasis independently to avoid imbalanced distributions. RECIST marks were labeled on each slice, under the supervision of a hepatic physician with >10 years experience. From there, a bbox was generated. For clusters of lesions too numerous to individually mark, a bbox over each cluster was drawn, as shown in Fig. 1(a).

Implementation Details. For all experiments we augmented the data by random rotations and gamma intensity transforms. We resampled all MR volumes and aligned them using the DEEDS algorithm [16]. All volumes were preprocessed by clipping within the 0.1% and 99.9% percentile values.

As key-slice classifier we used a DenseNet121 [17] backbone. Following [28], we add an additional 1×1 convolutional layer before global pooling and use log-sum-exp (LSE) pooling, finding it outperforms the standard average pooling. Three adjacent slices are inputted to provide some 3D context. The batch size is set as 20, an Adam optimizer [19] with initial learning rate as 1×10^{-4} is used. The learning rate is decayed by 0.01 after every 1000 iterations.

As detectors we evaluated three KSP options: CenterNet has been used to achieve state-of-the-art results in DeepLesion [5]; ATSS achieves the best performance on COCO [31]; and 3DCE is a powerful detector specifically designed for

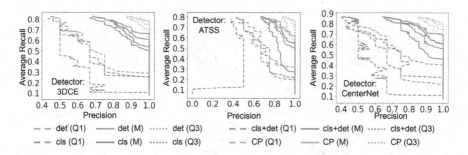

Fig. 4. PR curves of three key slice selection strategies. For each choice of top $T\%$ of slices selected for each study, we graph the corresponding average recall (across all patients) and the first quartile (Q1), median (M), and third quartile (Q3) precisions.

lesion localization [27]. To avoid overly tuning hyper-parameters, the network structure and loss use the same settings as the original papers. The batch size and learning rate are set as 30 and 1×10^{-4}, respectively, which follows a linear learning rate rule [12]. Each model is trained for 50 epochs. Random scaling and cropping was added as data augmentation for all tested detectors. All detectors were trained using the T2-anchor LF approach.

As for lesion characterization, we test radiomics [23], three standard classifiers, ResNet101 [15], DenseNet121 [17] and ResNeXt101 [26], as well as two texture based classifiers, DeepTEN [30] and SaDT [18]. In radiomics, the support vector machine (SVM) classifier is implemented with extracted features from manually localized tumors, including first order statistics, neighboring gray level dependence method (NGLDM) [21], gray level size zone matrix (GLSZM) [22], gray level run length matrix (GLRLM) [11] and gray-level co-occurrence matrix (GLCM) [14]. As for deep-learning-based classifiers, the batch size and learning rate are set as 8 and 1×10^{-4}, respectively. Networks are trained based on ground truth bounding boxes (bboxes) for 70 epochs. Random rotation, scaling and cropping are adopted as augmentation. Besides, ground truth bboxes are randomly shifted and resized to simulate the imperfect localization.

Key-Slice Prediction. We measure the impact of detection-based reweighting and curve parsing (CP). To do this, we rank key slices based on a) directly using detection output, *i.e.*, s_m^{det}, b) directly using classification output, *i.e.*, s_m^{cls}, c) using classification and detection confidences, *i.e.*, Eq.(3), and d) including CP in key-slice selection. As metrics, we select the top $T\%$ of ranked slices across all studies. For each choice of T, we calculate the within-study precision and recall, giving us a distribution of precision and recalls across studies. Thus, we graph the corresponding average recall (across patients) along with the median, first quartile (Q1) and third quartile (Q3) precision, providing typical, lower-, and upper-bound performances. In Fig. 4, detection-based re-weighting significantly outperforms detection and classification, boosting the Q1 and median precision, respectively. CP provides additional boosts in precision and recall, with notable

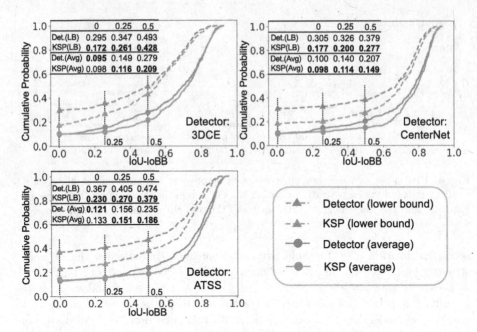

Fig. 5. Empirical CDF curves detection overlaps of three detectors and their KSP counterparts on the test set. Solid curves represent the average ROI overlap across selected slices for each patient, while dashed curves show the lower-bound (LB) overlap. Tables show exact percentages of patients with an average or LB overlap = 0, ≤0.25, and ≤0.50, where lower percentages indicate better performance.

boosts in lower-bound performance (robustness). To choose T we select the value corresponding to an average recall ≥0.5 and a Q1 precision ≥0.6, which balances between finding all slices while keeping good precision. For CenterNet, ATSS, and 3DCE this corresponds to keeping 48%, 54%, and 50% of the top slices, respectively.

Localization. Unlike standard detection setups, for CAD we are not interested in free response operating characteristics with arbitrary overlap cutoffs. Instead, we are only interested in whether we can select high-quality ROIs. Thus, we measure the overlap of selected ROIs against any ground truth bbox using the intersection over union (IoU). When ground truth bboxes are drawn over lesion clusters, we use the intersection over bounding box (IoBB) [24] as an IoU proxy. For each patient, we examine the average overlap across all selected ROIs and also the worst case, *i.e.*, lower bound (LB) overlap. We then directly observe the empirical cumulative distribution function (CDF) of these overlaps *across all patients*. We evaluate whether the KSP can enhance the performance of the three tested detectors, which would otherwise directly output key ROIs according to their bbox confidence scores.

Table 1. Lesion characterization performance. Radiomics is implemented based on the manual localization and SaDT [18] cannot be used without ROIs. Results when using DenseNet121 with ground truth bboxes are reported as the upper bound. The bold numbers indicate the best performance under each metric except the upper bound.

Methods	Accuracy	mean F1	F1(HCC)	F1(ICC)	F1(Meta.)
Radiomics [23]	58.65 ± 4.51	55.07 ± 6.34	71.37 ± 4.05	39.50 ± 10.94	54.36 ± 9.26
ResNet101 [15]	59.54 ± 3.53	50.03 ± 3.55	76.85 ± 4.55	12.74 ± 8.58	60.51 ± 4.55
DenseNet121 [17]	61.68 ± 5.43	51.24 ± 3.73	75.91 ± 10.37	17.54 ± 14.51	50.15 ± 7.57
ResNeXt101 [26]	60.51 ± 5.32	54.96 ± 5.60	78.15 ± 6.01	27.62 ± 11.89	59.12 ± 6.79
DeepTEN [30]	53.97 ± 3.38	54.74 ± 3.12	65.03 ± 6.60	41.39 ± 5.63	57.80 ± 6.40
KSP+ResNet101	64.91 ± 6.42	61.32 ± 5.91	77.00 ± 6.09	45.73 ± 6.72	61.26 ± 6.44
KSP+DenseNet121	**69.62 ± 3.13**	**66.49 ± 2.78**	80.12 ± 3.54	**55.34 ± 4.88**	**64.02 ± 6.81**
KSP+ResNeXt101	67.26 ± 4.18	62.82 ± 3.98	**80.33 ± 3.54**	45.86 ± 4.15	62.27 ± 7.98
KSP+DeepTEN	67.20 ± 2.79	64.14 ± 3.95	77.07 ± 3.26	51.61 ± 11.88	63.74 ± 8.48
KSP+SaDT	67.26 ± 3.91	63.25±3.63	79.69 ± 3.62	49.26 ± 8.51	60.80 ± 8.11
Upper bound	70.68 ± 2.97	68.02 ± 3.83	78.68 ± 3.61	57.59 ± 7.41	67.79 ± 7.78

From Fig. 5, the improvements provided by KSP are apparent on all detectors. When examining the mean overlap for each patient, the percentage of patients with low overlap (≤25% IoU-IoBB) is decreased by 0.5%~3.3%. *Much more significantly*, the LB performance indicates that KSP results in roughly 13% fewer patients with zero overlap and 12% fewer patients with low overlap. Thus, the KSP better ensures that no poor ROIs get selected and passed on to classification. It should be noted that the LB metrics directly measure robustness, which is the main motivating reason for the KSP framework. Hence, the corresponding LB improvements validate the KSP approach of hierarchically decomposing the problem into key-slice prediction and ROI regression.

Characterization. Finally, for the overall lesion characterization performance, we measure patient-wise accuracy, one-vs-all and mean F1 score(s) of the three tumor types, with emphasis on HCC-vs-others given its prominence in clinical work [2]. According to Fig. 5, CenterNet with KSP surpasses 3DCE and ATSS in average and LB overlap, respectively. Therefore, we choose it as our KSP detector and train and test various classifiers on its ROIs. Patient-wise diagnoses are produced by averaging classifications from detected ROIs weighted by confidence. As demonstrated in Table 1, compared with using classifiers alone, KSP significantly improves accuracy (+5%~+8%), mean F1 (+8%~+15%) and HCC F1 scores (+0.15%~+12%). DenseNet121 [26] performs best, garnering an HCC vs. others F1 score of 0.801, which is comparable to reported physician performance (0.791) [2]. In addition, we also produced an upper bound by testing DenseNet121 on oracle tumor locations, and there is only a marginal gap between it and our best results—1% in accuracy and 1.5% in mean F1 score. This further validates the effectiveness of KSP, suggesting that performance bottlenecks may now be due to classifier limitations, which we leave for future work.

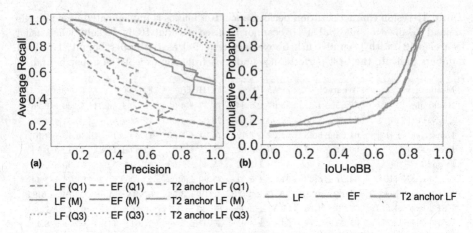

Fig. 6. (a) The comparison of early fusion (EF), standard late fusion (LF), and T2-anchor LF on key-slice prediction. Average recall (across all patients) and the first quartile (Q1), median (M), and third quartile (Q3) precisions are graphed. (b) The CDF curves of overlap of detections vs. ground truth bounding boxes when using the three fusion approaches.

Ablation Study. In Fig. 6(a), we measure three fusion approaches, including standard late fusion (LF), early fusion (EF), and also T2-anchor LF on key-slice classification. As metrics shows, when keeping the same average recall, T2-anchor LF outperforms both LF and EF in the first quartile (Q1), median (M) and third quartile (Q3) precision, showing its superiority in predicting key slices. In Fig. 6(b), the cumulative probability curves of CenterNet [33] with three fusion methods are demonstrated. To evaluate the detection performance independently, the curves show results directly from the detector without key-slice prediction. The curve of T2-anchor LF is lower than the other two especially when IoU-IoBB is smaller than 0.5. This indicates T2-anchor LF surpasses LF and EF by *decreasing* the ratio of bboxes of low overlaps with ground truth.

4   Discussion and Conclusion

As the medical image field progresses toward more clinically viable CAD, research efforts will likely increasingly focus on ensuring true robustness. We contribute toward this goal for liver lesion characterization. Specifically, we articulate a physician-inspired decompositional approach toward ROI localization that breaks down the complex problem into key-slice prediction and then ROI regression. Using our proposed framework, our KSP realization can achieve very high robustness, with 87% of its ROIs having an overlap of $\geq 40\%$. Overall, our fully automated CAD solution can achieve an HCC-vs-others F1 score of 80.1%. Importantly, this performance is reported on histopathologically-confirmed cases, which selects for the most challenging cases requiring CAD intervention. Even so, this matches reported clinical performances of 79.1% [2], despite such studies including both radiologically and histopathologically-confirmed cases. Given the challenging nature of liver lesion characterization, our proposed CAD system represents a step forward toward more clinically practical solutions.

References

1. Adcock, A., Rubin, D., Carlsson, G.: Classification of hepatic lesions using the matching metric. Comput. Vis. Image Underst. **121**, 36–42 (2014)
2. Aubé, C., et al.: EASL and AASLD recommendations for the diagnosis of HCC to the test of daily practice. Liver Int. **37**(10), 1515–1525 (2017)
3. Bilic, P., et al.: The liver tumor segmentation benchmark (LiTS) (2019)
4. Bosch, F.X., Ribes, J., Díaz, M., Cléries, R.: Primary liver cancer: worldwide incidence and trends. Gastroenterology **127**(5), S5–S16 (2004)
5. Cai, J., et al.: Lesion harvester: iteratively mining unlabeled lesions and hard-negative examples at scale. TMI (2020, accepted)
6. Chen, X., et al.: A cascade attention network for liver lesion classification in weakly-labeled multi-phase CT images. In: Wang, Q., et al. (eds.) DART/MIL3ID -2019. LNCS, vol. 11795, pp. 129–138. Springer, Cham (2019). https://doi.org/10.1007/978-3-030-33391-1_15
7. Diamant, I., et al.: Improved patch-based automated liver lesion classification by separate analysis of the interior and boundary regions. JBHI **20**(6), 1585–1594 (2015)
8. Eisenhauer, E., Therasse, P., Bogaerts, J., et al.: New response evaluation criteria in solid tumours: revised RECIST guideline (v1.1). EJC **45**(2), 228–247 (2009)
9. Fowler, K.J., et al.: Interreader reliability of LI-RADS version 2014 algorithm and imaging features for diagnosis of hepatocellular carcinoma: a large international multireader study. Radiology **286**(1), 173–185 (2018)
10. Freeman, R.B., et al.: Optimizing staging for hepatocellular carcinoma before liver transplantation: a retrospective analysis of the UNOS/OPTN database. Liver Transpl. **12**(10), 1504–1511 (2006)
11. Galloway, M.M.: Texture analysis using gray level run lengths. Comput. Graph. Image Process. **4**(2), 172–179 (1975)
12. Goyal, P., et al.: Accurate, large minibatch SGD: training ImageNet in 1 hour. arXiv abs/1706.02677 (2017)
13. Grant, A., Neuberger, J.: Guidelines on the use of liver biopsy in clinical practice. Gut **45**(Suppl. 4), IV1–IV11 (1999)
14. Haralick, R.M., Shanmugam, K., Dinstein, I.H.: Textural features for image classification. IEEE Trans. Syst. Man Cybern. 610–621 (1973)
15. He, K., Zhang, X., Ren, S., Sun, J.: Deep residual learning for image recognition. In: CVPR, pp. 770–778 (2016)
16. Heinrich, M.P., Jenkinson, M., Brady, M., Schnabel, J.A.: MRF-based deformable registration and ventilation estimation of lung CT. TMI **32**(7), 1239–1248 (2013)
17. Huang, G., Liu, Z., Van Der Maaten, L., Weinberger, K.Q.: Densely connected convolutional networks. In: CVPR, pp. 4700–4708 (2017)
18. Huo, Y., et al.: Harvesting, detecting, and characterizing liver lesions from large-scale multi-phase CT data via deep dynamic texture learning. arXiv:2006.15691 (2020)
19. Kingma, D.P., Ba, J.: Adam: a method for stochastic optimization. arXiv preprint arXiv:1412.6980 (2014)

20. Oliva, M.R., Saini, S.: Liver cancer imaging: role of CT, MRI, US and PET. Cancer Imaging 4(Spec No A), S42 (2004)
21. Sun, C., Wee, W.G.: Neighboring gray level dependence matrix for texture classification. Comput. Vis. Graph. Image Process. **23**(3), 341–352 (1983)
22. Thibault, G., et al.: Shape and texture indexes application to cell nuclei classification. Int. J. Pattern Recognit Artif Intell. **27**(01), 1357002 (2013)
23. Van Griethuysen, J.J., et al.: Computational radiomics system to decode the radiographic phenotype. Can. Res. **77**(21), e104–e107 (2017)
24. Wang, X., Peng, Y., Lu, L., Lu, Z., Bagheri, M., Summers, R.M.: ChestX-ray8: hospital-scale chest X-ray database and benchmarks on weakly-supervised classification and localization of common thorax diseases. In: CVPR, pp. 2097–2106 (2017)
25. Wu, J., Liu, A., Cui, J., Chen, A., Song, Q., Xie, L.: Radiomics-based classification of hepatocellular carcinoma and hepatic haemangioma on precontrast magnetic resonance images. BMC Med. Imaging **19**(1), 23 (2019)
26. Xie, S., Girshick, R., Dollár, P., Tu, Z., He, K.: Aggregated residual transformations for deep neural networks. In: CVPR, pp. 1492–1500 (2017)
27. Yan, K., Bagheri, M., Summers, R.M.: 3D context enhanced region-based convolutional neural network for end-to-end lesion detection. In: Frangi, A.F., Schnabel, J.A., Davatzikos, C., Alberola-López, C., Fichtinger, G. (eds.) MICCAI 2018. LNCS, vol. 11070, pp. 511–519. Springer, Cham (2018). https://doi.org/10.1007/978-3-030-00928-1_58
28. Yan, K., Lu, L., Summers, R.M.: Unsupervised body part regression via spatially self-ordering convolutional neural networks. In: ISBI, pp. 1022–1025. IEEE (2018)
29. Yang, W., Lu, Z., Yu, M., Huang, M., Feng, Q., Chen, W.: Content-based retrieval of focal liver lesions using bag-of-visual-words representations of single-and multiphase contrast-enhanced ct images. JDI **25**(6), 708–719 (2012)
30. Zhang, H., Xue, J., Dana, K.: Deep ten: texture encoding network. In: CVPR, pp. 708–717 (2017)
31. Zhang, S., Chi, C., Yao, Y., Lei, Z., Li, S.Z.: Bridging the gap between anchor-based and anchor-free detection via adaptive training sample selection. In: CVPR, pp. 9759–9768 (2020)
32. Zhen, S., et al.: Deep learning for accurate diagnosis of liver tumor based on magnetic resonance imaging and clinical data. Front. Oncol. **10**, 680 (2020)
33. Zhou, X., Wang, D., Krähenbühl, P.: Objects as points. arXiv:1904.07850 (2019)

Improving Tuberculosis Recognition on Bone-Suppressed Chest X-Rays Guided by Task-Specific Features

Yunbi Liu[1], Genggeng Qin[4], Yun Liu[5], Mingxia Liu[3](✉), and Wei Yang[1,2](✉)

[1] School of Biomedical Engineering, Southern Medical University,
Guangzhou 510515, China
[2] Pazhou Lab, Guangzhou 510335, China
[3] Department of Radiology and BRIC, University of North Carolina at Chapel Hill,
Chapel Hill, NC 27599, USA
mxliu@med.unc.edu
[4] Nanfang Hospital, Southern Medical University, Guangzhou, China
[5] Nankai University, Tianjin, China

Abstract. Early tuberculosis (TB) screening through chest X-rays (CXRs) is essential for timely detection and treatment of the disease. Previous studies have shown that texture abnormalities in CXRs can be enhanced by bone suppression techniques, which may potentially improve TB diagnosis performance. However, existing TB datasets with CXRs usually lack bone-suppressed images, making it difficult to take advantage of bone suppression for TB recognition. Also, existing bone suppression models are usually trained on a relatively small dual-energy subtraction (DES) dataset with CXRs, without considering the image specificity of TB patients. To this end, we propose a bone-suppressed CXR-based tuberculosis recognition (BCTR) framework, where diagnosis-specific deep features are extracted and used to guide a bone suppression (BS) model to generate bone-suppressed CXRs for TB diagnosis. Specifically, the BCTR consists of a *classification model* for TB diagnosis and an *image synthesis model* for bone suppression of CXRs. The classification model is first trained on original CXRs from multiple TB datasets such as the large-scale TBX11K. The image synthesis model is trained on a DES dataset to produce bone-suppressed CXRs. Considering the heterogeneity of CXR images from the TB datasets and the DES dataset, we proposed to extract *multi-scale task-specific features* from the trained classification model and transfer them (via channel-wise addition) to the corresponding layers in the image synthesis model to explicitly guide the bone suppression process. With the bone-suppressed CXRs as input, the classification model is further trained for multi-class TB diagnosis. Experimental results on five TB databases and a DES dataset suggest that our BCTR outperforms previous state-of-the-arts in automated tuberculosis diagnosis.

Keywords: Tuberculosis recognition · Chest X-ray · Bone suppression

© Springer Nature Switzerland AG 2021
I. Rekik et al. (Eds.): PRIME 2021, LNCS 12928, pp. 59–69, 2021.
https://doi.org/10.1007/978-3-030-87602-9_6

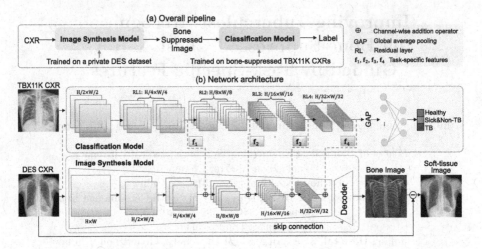

Fig. 1. Illustration of the proposed bone-suppressed CXR-based tuberculosis recognition (BCTR) framework, including two components: a classification model trained on the public TBX11K chest X-ray dataset for TB diagnosis, and an image synthesis model trained on a DES dataset for bone suppression of CXRs. To capture the disease specificity in CXRs, we extract multi-scale task-specific features from the classification model and transfer them (via channel-wise addition) to the corresponding layers in the image synthesis model to explicitly guide the bone suppression process.

1   Introduction

Tuberculosis (TB) is a potentially severe infectious disease that mainly affects the lungs. Early screening through chest X-rays (CXRs) is essential for timely detection and treatment of the disease [1–3]. Due to various manifestations of TB and overlapped anatomical structures over the lung areas in CXRs, it is often challenging to correctly interpret CXRs for TB recognition.

Learning-based methods can distinguish varied manifestations of tuberculosis based on CXRs, and accumulating evidence supports that bone suppression helps boost recognition performance of lung diseases [4–8]. Maduskar et al. [5] compared two machine-learning models to detect textural abnormalities in CXRs of TB suspects, with one trained and tested on original CXRs and another trained and tested on bone-suppressed CXRs. They found that bone suppression of CXRs could help improve automated detection of textural abnormalities in CXRs. However, existing TB datasets with CXRs usually lack bone-suppressed images, making it difficult to take advantage of bone suppression for TB recognition. Besides, existing bone suppression models are usually trained on a relatively small dual-energy subtraction (DES) dataset with fewer variety of CXRs [9–12], thus ignoring the specificity of tuberculosis in CXRs from other datasets. Recent studies have shown that modeling the disease-associated image specificity helps improve disease detection performance [13,14]. It is intuitively

Table 1. The information of five tuberculosis datasets with CXRs.

Dateset	Category#	Healthy	Sick&Non-TB	TB	Subject #	Label Type
TBX11K [15]	3	5,000	5,000	1,200	11,200	Image-level label & Bounding box annotations
MC [16]	2	80	–	58	138	Image-level label
SH [16]	2	326	–	336	662	Image-level label
DA [17]	2	78	–	78	156	Image-level label
DB [17]	2	75	–	75	150	Image-level label

reasonable to employ bone-suppressed CXRs to improve TB diagnosis by explicitly modeling the image specificity of tuberculosis.

In this paper, we develop a bone-suppressed CXR-based tuberculosis recognition (**BCTR**) framework, with its overall pipeline shown in Fig. 1(a). The BCTR consists a classification model for TB diagnosis and an image synthesis model for bone suppression of CXRs (see Fig. 1(b)). To explicitly capture the TB specificity in CXRs, we proposed a feature transfer learning strategy. That is, we extract multi-scale diagnosis-oriented (i.e., task-specific) CXR features from the trained classification model, and transfer them to the corresponding layers in the image synthesis model to guide the bone suppression process. Experiments on five TB datasets and a DES dataset suggest the superiority of our BCTR over previous state-of-the-arts in CXR-based tuberculosis recognition.

2   Materials and Method

2.1   Data and Image Pre-processing

Tuberculosis Datasets. Five public TB datasets are used in this work, including (1) TBX11K [15], (2) Montgomery County chest X-ray set (MC) [16], (3) Shenzhen chest X-ray set (SH) [16], (4) DA [17], and (5) DB [17]. The TBX11K is a large-scale and newly initiated dataset, consisting of 11,200 X-rays from three categories (i.e., 5,000 healthy cases, 5,000 sick but non-TB cases, and 1,200 cases with manifestations of TB). Each X-ray scan is corresponding to a specific subject in TBX11K. The TBX11K dataset has not only image-level labels for TB diagnosis but also bounding box annotations for TB detection. The MC, SH, DA and DB datasets contain only two categories (i.e., TB and non-TB). We used a total of 12,278 CXRs from these five TB chest X-ray datasets in the experiments. The information of these five TB dataset is summarized in Table 1.

DES Dataset. The private DES dataset contains 504 subjects, and each subject is represented by a triplet, i.e., a CXR, a DES bone image, and a DES soft-tissue image. These images were acquired by using a digital radiography (DR) machine with a two-exposure DES unit (Discovery XR 656, GE Healthcare) at a local hospital. Each image was stored in the DICOM format with a 14-bit depth. The size of each image ranges from $1,800 \times 1,800$ to $2,022 \times 2,022$ pixels, and the spatial resolution ranges from $0.1931 \times 0.1931 \, \text{mm}^2$ to $0.1943 \times 0.1943 \, \text{mm}^2$.

Image Pre-processing. All X-ray images from five public TB datasets are resized to 512×512 with the intensity range of $[0, 255]$. For the private DES dataset, we first normalize each conventional CXR (denoted as $\mathbf{I}$) in the DES dataset according to their corresponding window width (ww) and window level (wl), satisfying $\mathbf{I} = \frac{2\mathbf{I}-(2ww-wl)}{2wl} \times 255$. Accordingly, each bone image $\mathbf{B}$ is scaled to $\mathbf{B} = \mathbf{B}/wl \times 255$. Besides, we automatically segment the lung areas in CXRs, and manual labelled the motion artifact areas in bone images to generate a weighted mask $\mathbf{M}$ for each CXR image. This helps mitigate the effect of the areas outside lung and motion artifacts. In a mask $\mathbf{M}$, the areas outside lung are set to 0.25, the motion artifact areas are set to 0, and the remaining valid areas are set to 1.

2.2 Proposed Method

Overall Pipeline. The proposed BCTR framework aims to classify each bone-suppressed CXR image into three categories, *i.e.*, healthy, sick but non-TB and TB. As shown in Fig. 1(a), the TB recognition pipeline through BCTR consists of two sequential steps. In the *first* step, the to-be-tested CXR is fed into a image synthesis model (trained on a DES dataset) to generate a bone-suppressed image. In the *second* step, the bone-suppressed CXR is fed into a multi-class classification model (trained on the TBX11K dataset and four small TB datasets) to predict whether it belongs to the category of healthy, sick but non-TB or TB. Two unique strategies are developed in the BCTR to improve TB recognition, including a bone suppression strategy and a task-specific feature transfer scheme.

Network Architecture. As shown in Fig. 1(b), the proposed network consists of two models, i.e., (1) a classification network for TB diagnosis and (2) an image synthesis network (guided by diagnosis-oriented features generated from the classification model) for bone suppression of the original CXR images.

(1) Classification Network. We employ the ResNet18 [18] as the classification network because it is a shallow network and is easy to be trained from scratch. The loss function of the classification network is a cross-entropy loss to calculate the difference between the estimated multi-class probability scores and the true category label, which is formulated as follows

$$\mathcal{L}_p(\mathbf{q}) = -\sum_{i=1}^{N} y_i \log(p(i)), \tag{1}$$

where N is the number of classes ($N = 3$ in this work), and p_i and y_i is the estimated and real probabilities of the input $\mathbf{q}$ belonging to the class i, respectively. This network also produces multi-scale CXR features, i.e., $\mathbf{f}_1$, $\mathbf{f}_2$, $\mathbf{f}_3$ and $\mathbf{f}_4$ that are diagnosis-oriented.

We train this network twice. The first training is based on original CXRs from TBX11K and four small TB datasets, to provide diagnosis-oriented CXR features. The second is trained on the bone-suppressed CXRs generated by the image synthesis network from scratch for TB recognition.

(2) Image Synthesis Network. We develop an image synthesis network for bone suppression guided by diagnosis-oriented CXR features. As shown in Fig. 1(b), given a single CXR $\mathbf{I}$ as input, this network outputs a synthesized bone image $\mathbf{B}^*$ (corresponding to the ground-truth bone image $\mathbf{B}$). Based on $\mathbf{I}$ and $\mathbf{B}^*$, one can obtain the bone-suppressed/soft-tissue image. A UNet architecture is used in our image synthesis network, which consists of an encoder part, a decoder part and several skip connections to bridge the information between the encoder and the decoder [19]. There are six convolutional layers in the encoder part, with a max pooling operation followed by each convolutonal layer. The channel numbers of the six convolutional layers are 16, 32, 64, 128, 256 and 512, respectively. All the convolution kernels are 3×3 with the stride of 1. The decoder remains symmetric with the encoder, containing five deconvolutional layers and an output convolutional layer. Five skip connections are used to connect the corresponding layers between the encoder and the decoder.

Multi-scale Feature Transfer. To generate diagnosis-oriented images, we propose to feed CXRs from the DES dataset to the trained classification network to generate multi-scale features (i.e., $\mathbf{f}_1$, $\mathbf{f}_2$, $\mathbf{f}_3$ and $\mathbf{f}_4$). Then, a multi-scale feature transfer learning strategy is developed to use these features to train the image synthesis network for bone suppression of CXRs. The channel number of the last four convolutional layers in the encoder part of the image synthesis network is consistent with that of the output feature maps from the four residual layers of the classification network. In this way, these diagnosis-oriented CXR features can be explicitly transferred to the image synthesis network (via channel-wise addition) to guide the image synthesis process.

When training the image synthesis network, we aim to minimize the l_1 distance between the synthesized bone image $G(\mathbf{x})$ and the corresponding ground-truth bone image $\mathbf{b}$ in both intensity and gradient domains, to encourage $G(\mathbf{x})$ to be pixel-wise close to $\mathbf{b}$. This loss is mathematically formulated as follows

$$
\begin{aligned}
\mathcal{L}(\mathbf{x}, \mathbf{f}_1, \mathbf{f}_2, \mathbf{f}_3, \mathbf{f}_4) &= E_{\mathbf{x}\sim p(\mathbf{X}), \mathbf{b}\sim p(\mathbf{B})}\|G(\mathbf{x}) - \mathbf{b}\|_1 \\
&+ E_{\mathbf{x}\sim p(\mathbf{X}), \mathbf{b}\sim p(\mathbf{B})}\| \nabla_{1,G(\mathbf{x})} - \nabla_{1,\mathbf{b}} \|_1 \\
&+ E_{\mathbf{x}\sim p(\mathbf{X}), \mathbf{b}\sim p(\mathbf{B})}\| \nabla_{2,G(\mathbf{x})} - \nabla_{2,\mathbf{b}} \|_1,
\end{aligned}
\tag{2}
$$

where $\mathbf{x}$ is the input CXR, $\nabla_{1,G(\mathbf{x})}$ and $\nabla_{2,G(\mathbf{x})}$ are the gradients of the predicted bone image in horizontal and vertical directions, respectively, and $\nabla_{1,\mathbf{b}}$ and $\nabla_{2,\mathbf{b}}$ are the gradients of the ground-truth bone image. Besides, we focus on computing the losses within the lung areas and mask the regions with motion artifacts out by using the mask $\mathbf{M}$. Accordingly, the loss function of the proposed image synthesis network is finally formulated as follows

$$
\mathcal{L}_{syn} = \mathcal{L}(\mathbf{x}, \mathbf{f}_1, \mathbf{f}_2, \mathbf{f}_3, \mathbf{f}_4) \otimes \mathbf{M}.
\tag{3}
$$

Table 2. Recognition results (%) of different methods on the test data from five TB datasets. The term "Pretrained?" indicates whether a model(except BCTR-3) is pretrained on ImageNet. In the BCTR-3, the classification network pretrained on CXRs is finetuned on the bone-suppressed images produced by the image synthesis network. "BS CXR" indicates bone-suppressed CXR.

Method	Pretrained?	Input	ACC	AUC	SEN	SPE	AP	AR
SSD [20]	Yes	CXR	84.7	93.0	78.1	89.4	82.1	83.8
RetinaNet [21]	Yes	CXR	87.4	91.8	81.6	89.8	84.8	86.8
Faster R-CNN [22]	Yes	CXR	89.7	93.6	**91.2**	89.9	87.7	**90.5**
FCOS [23]	Yes	CXR	88.9	92.4	87.3	89.9	86.6	89.2
SSD [20]	No	CXR	88.2	93.8	88.4	89.5	86.0	88.6
RetinaNet [21]	No	CXR	79.0	87.4	60.0	90.7	75.9	75.8
Faster R-CNN [22]	No	CXR	81.3	89.7	72.5	87.3	78.5	79.9
DenseNet121 [24]	No	CXR	85.7	94.5	82.7	91.5	84.4	86.0
BCTR-1	No	CXR	89.3	94.7	70.6	95.6	87.4	86.4
BCTR-2	No	BS CXR	89.4	95.5	69.6	95.8	87.6	86.1
BCTR-3	Yes	BS CXR	89.0	96.4	71.4	96.0	87.4	86.2
BCTR (Ours)	No	BS CXR	**90.8**	**96.6**	73.6	**96.8**	**89.6**	88.1

Implementation. At the training stage, two networks are trained as follows.

First, the classification network was trained on the TBX11K and four small TB datasets for 24 epochs, with the batch size of 64. An SGD optimizer with a momentum of 0.9 was used, and the learning rate was 0.01 for the first 12 epochs and 0.001 for the last 12 epochs.

Then, the image synthesis network was trained on our DES dataset, where the batch size and epoch were set to 1 and 250, respectively. The Adam optimizer was used with the learning rate of 0.001. During training the image synthesis network, the input CXR was also fed into the trained classification network to obtain the multi-scale diagnosis-oriented features, and then the multi-scale diagnosis-oriented features were introduced into the image synthesis network to guide the training process.

Finally, the CXRs from the TB training datasets were input into the trained the image synthesis model to obtain the corresponding bone-suppressed soft-tissue images. Then, the classification network was trained again based on these bone-suppressed images. All the networks were implemented based on Pytorch and trained on a workbench equipped with a TITANX with 12G memory.

3 Experiments

Experimental Setup. In the *first* group of experiments, we evaluate the tuberculosis diagnosis performance of the proposed BCTR and several competing

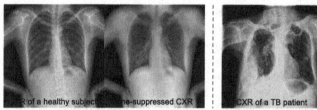

(a) CXR images before and after bone suppression of two subjects from TBX11K

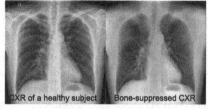

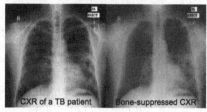

(b) CXR images before and after bone suppression of two subjects from DA+DB

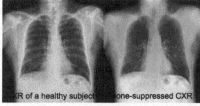

(c) CXR images before and after bone suppression of two subjects from MC+SH

Fig. 2. Visual results of the proposed image synthesis model for bone suppression on different TB datasets.

methods. Six metrics are used to evaluate the performance of TB recognition, including ACC (accuracy), AUC (of TB), SEN (sensitivity), SPE (specificity), AP (average precision), and AR (average recall) [15].

We first compared our BCTR with four state-of-the-art methods that perform simultaneous TB diagnosis and TB area detection, including (1) **SSD** [20], (2) **RetinaNet** [21] and (3) **Faster R-CNN** [22] trained from scratch, as well as their counterparts pretrained on ImageNet and (4) **FCOS** [23] that is also pretrained on ImageNet. The backbone model of SSD uses VGGNet-16, while the backbone models of RetinaNet, Faster R-CNN and FCOS use ResNet50 with feature pyramid FPN (i.e., ResNet50 w/FPN). The classification results of those benchmark methods are reported in [15]. We also compared with another popular classification model, such as DenseNet121 [24]. For a fair comparison, we used the same data split strategy and evaluation metrics as in [15] for five TB datasets. For the private DES dataset, we randomly selected 404, 50 and 50 cases as the training set, validation set and test set, respectively.

Two strategies are used in our BCTR method to improve TB recognition, including *bone suppression* and *task-specific feature transfer*. To validate their effectiveness, we also perform ablation experiments by comparing our BCTR with its three variants: (1) **BCTR-1** that uses the ResNet18 architecture but without bone suppression and task-specific feature transfer, (2) **BCTR-2** that uses a ResNet18+UNet with bone suppression but without task-specific feature transfer, and (3) **BCTR-3** that *finetunes* the classification network using the bone-suppressed images instead of *"train from scratch"* as we do in BCTR.

In the *second* group of experiments, we evaluate the bone suppression performance of our image synthesis network, and show the visual results on five TB datasets and the quantitative results on the private DES dataset. Two metrics are used to measure the quality of synthetic bone-suppressed images, including PSNR and SSIM [25].

Evaluation of Tuberculosis Recognition. Table 2 summarizes the results of different methods in tuberculosis recognition. The following observations can be drawn from Table 2. *First*, the proposed BCTR produces the best ACC, AUC, SPE and AP results among ten competing methods, validating its effectiveness in TB recognition. *Second*, compared with Faster R-CNN, our method achieves relatively lower SEN and AR results. This may be due to the imbalanced number of samples in the training set (with fewer TB subjects). On the other hand, the Faster R-CNN performs simultaneous classification and TB detection, which may also help improve sensitivity. *Besides*, our BCTR yields consistently better classification performance, compared with its two variants BCTR-1 (without bone suppression and without any guidance), BCTR-2 (with bone suppression but without any guidance). This implies that the two strategies (i.e., bone suppression and task-specific feature transfer) used in the proposed BCTR framework are beneficial for CXR-based tuberculosis recognition. *In addition*, the BCTR also achieves better results compared to BCTR-3 that finetunes the classification network in the second training process. This implies that using bone-suppressed CXRs to finetune a classification network pretrained on CXRs may not produce good results, which is different from previous findings that finetunning usually improves the classification performance [26]. This may be due to the significant heterogeneity between CXRs and bone-suppressed CXRs.

Evaluation of Synthetic Bone-Suppressed Images. In this work, we employ an image synthesis network to generate bone-suppressed CXRs for TB recognition. Therefore, the quality of those synthetic bone-suppressed CXRs is essential for the recognition performance. We now evaluate the performance of the proposed image synthesis network (trained on the private DES training data), by applying it to the six datasets to quantitatively and qualitatively evaluate its bone suppression performance.

On the DES test set, our image synthesis model achieves the quantitative results on the generated bone-suppressed CXR images, i.e., with an SSIM of 0.950 and a PSNR of 34.8 dB. Since there is no gold standard bone-suppressed image for the five tuberculosis datasets, we simply show the original CXR scans and their corresponding synthetic bone-suppressed CXRs of a healthy subject and a TB patient in Fig. 2 for the qualitative evaluation. As shown in Fig. 2, our model achieves reasonable bone suppression results on images from different categories, with good generalization performance. Figure 2 also suggests that the generated bone-suppressed images are somewhat blurred. The possible reason is that the image synthesis network only uses the basic UNet architecture, and may not have strong ability to capture high-frequency information of images.

4 Conclusion and Future Work

In this paper, we have presented a BCTR method for CXR-based tuberculosis recognition. We proposed to take advantage of the bone suppression technique to enhance the texture abnormalities in CXRs, and utilize task-specific features to guide the bone suppression network to generate bone-suppressed CXRs for TB diagnosis. Experimental results on multiple TB datasets demonstrate that our BCTR is superior to previous state-of-the-art methods for tuberculosis diagnosis, and also can generate reasonable bone-suppressed images.

In the current work, we initially verified the effectiveness of using bone suppression (with a UNet architecture) to identify TB patients. It's desired to develop more advanced image synthesis networks to improve the quality of bone-suppressed CXRs, which will be our future work. Besides, the task-specific features are simply transferred to the image synthesis network via channel-wise addition. In the future, we will design smarter knowledge transfer strategies to further improve the performance of tuberculosis recognition.

Acknowledgements. Y. Liu, and W. Yang were partially supported by the National Natural Science Foundation of China (No. 81771916) and the Guangdong Provincial Key Laboratory of Medical Image Processing (No. 2014B-030301042).

References

1. Jaeger, S., et al.: Automatic tuberculosis screening using chest radiographs. IEEE Trans. Med. Imaging **33**(2), 233–245 (2013)
2. Konstantinos, A.: Testing for tuberculosis (2010)
3. Van Cleeff, M., Kivihya-Ndugga, L., Meme, H., Odhiambo, J., Klatser, P.: The role and performance of chest x-ray for the diagnosis of tuberculosis: a cost-effectiveness analysis in Nairobi, Kenya. BMC Infect. Dis. **5**(1), 1–9 (2005)
4. Li, F., Engelmann, R., Pesce, L.L., Doi, K., Metz, C.E., MacMahon, H.: Small lung cancers: improved detection by use of bone suppression imaging–comparison with dual-energy subtraction chest radiography. Radiology **261**(3), 937–949 (2011)
5. Maduskar, P., Hogeweg, L., Philipsen, R., Schalekamp, S., Van Ginneken, B.: Improved texture analysis for automatic detection of tuberculosis (TB) on chest radiographs with bone suppression images. In: Medical Imaging 2013: Computer-Aided Diagnosis, vol. 8670, p. 86700H. International Society for Optics and Photonics (2013)
6. Schalekamp, S., van Ginneken, B., Schaefer-Prokop, C., Karssemeijer, N.: Impact of bone suppression imaging on the detection of lung nodules in chest radiographs: analysis of multiple reading sessions. In: Medical Imaging 2013: Image Perception, Observer Performance, and Technology Assessment, vol. 8673, p. 86730Y. International Society for Optics and Photonics (2013)

7. Miyoshi, T., et al.: Effectiveness of bone suppression imaging in the detection of lung nodules on chest radiographs. J. Thorac. Imaging **32**(6), 398–405 (2017)
8. Gordienko, Y., et al.: Deep learning with lung segmentation and bone shadow exclusion techniques for chest X-ray analysis of lung cancer. In: Hu, Z., Petoukhov, S., Dychka, I., He, M. (eds.) ICCSEEA 2018. AISC, vol. 754, pp. 638–647. Springer, Cham (2019). https://doi.org/10.1007/978-3-319-91008-6_63
9. Loog, M., van Ginneken, B., Schilham, A.M.: Filter learning: application to suppression of bony structures from chest radiographs. Med. Image Anal. **10**(6), 826–840 (2006)
10. Oh, D.Y., Yun, I.D.: Learning bone suppression from dual energy chest X-rays using adversarial networks. arXiv preprint arXiv:1811.02628 (2018)
11. Zarshenas, A., Liu, J., Forti, P., Suzuki, K.: Separation of bones from soft tissue in chest radiographs: anatomy-specific orientation-frequency-specific deep neural network convolution. Med. Phys. **46**(5), 2232–2242 (2019)
12. Eslami, M., Tabarestani, S., Albarqouni, S., Adeli, E., Navab, N., Adjouadi, M.: Image to images translation for multi-task organ segmentation and bone suppression in chest X-ray radiography. arXiv preprint arXiv:1906.10089 (2019)
13. Pan, Y., Liu, M., Lian, C., Xia, Y., Shen, D.: Disease-image specific generative adversarial network for brain disease diagnosis with incomplete multi-modal neuroimages. In: Shen, D., et al. (eds.) MICCAI 2019. LNCS, vol. 11766, pp. 137–145. Springer, Cham (2019). https://doi.org/10.1007/978-3-030-32248-9_16
14. Liu, Y., et al.: Joint neuroimage synthesis and representation learning for conversion prediction of subjective cognitive decline. In: Martel, A.L., et al. (eds.) MICCAI 2020. LNCS, vol. 12267, pp. 583–592. Springer, Cham (2020). https://doi.org/10.1007/978-3-030-59728-3_57
15. Liu, Y., Wu, Y.H., Ban, Y., Wang, H., Cheng, M.M.: Rethinking computer-aided tuberculosis diagnosis. In: Proceedings of the IEEE/CVF Conference on Computer Vision and Pattern Recognition, pp. 2646–2655 (2020)
16. Jaeger, S., Candemir, S., Antani, S., Wáng, Y.X.J., Lu, P.X., Thoma, G.: Two public chest x-ray datasets for computer-aided screening of pulmonary diseases. Quant. Imaging Med. Surg. **4**(6), 475 (2014)
17. Chauhan, A., Chauhan, D., Rout, C.: Role of gist and PHOG features in computer-aided diagnosis of tuberculosis without segmentation. PLoS ONE **9**(11), e112980 (2014)
18. He, K., Zhang, X., Ren, S., Sun, J.: Identity mappings in deep residual networks. In: Leibe, B., Matas, J., Sebe, N., Welling, M. (eds.) ECCV 2016. LNCS, vol. 9908, pp. 630–645. Springer, Cham (2016). https://doi.org/10.1007/978-3-319-46493-0_38
19. Ronneberger, O., Fischer, P., Brox, T.: U-Net: convolutional networks for biomedical image segmentation. In: Navab, N., Hornegger, J., Wells, W.M., Frangi, A.F. (eds.) MICCAI 2015. LNCS, vol. 9351, pp. 234–241. Springer, Cham (2015). https://doi.org/10.1007/978-3-319-24574-4_28
20. Liu, W., et al.: SSD: single shot multibox detector. In: Leibe, B., Matas, J., Sebe, N., Welling, M. (eds.) ECCV 2016. LNCS, vol. 9905, pp. 21–37. Springer, Cham (2016). https://doi.org/10.1007/978-3-319-46448-0_2
21. Lin, T.Y., Goyal, P., Girshick, R., He, K., Dollár, P.: Focal loss for dense object detection. In: Proceedings of the IEEE International Conference on Computer Vision, pp. 2980–2988 (2017)
22. Ren, S., He, K., Girshick, R., Sun, J.: Faster R-CNN: towards real-time object detection with region proposal networks. arXiv preprint arXiv:1506.01497 (2015)

23. Tian, Z., Shen, C., Chen, H., He, T.: FCOS: fully convolutional one-stage object detection. In: Proceedings of the IEEE/CVF International Conference on Computer Vision, pp. 9627–9636 (2019)
24. Huang, G., Liu, Z., Van Der Maaten, L., Weinberger, K.Q.: Densely connected convolutional networks. In: Proceedings of the IEEE Conference on Computer Vision and Pattern Recognition, pp. 4700–4708 (2017)
25. Wang, Z., Bovik, A.C., Sheikh, H.R., Simoncelli, E.P.: Image quality assessment: from error visibility to structural similarity. IEEE Trans. Image Process. **13**(4), 600–612 (2004)
26. Minaee, S., Kafieh, R., Sonka, M., Yazdani, S., Soufi, G.J.: Deep-covid: Predicting covid-19 from chest x-ray images using deep transfer learning. Med. Image Anal. **65**, 101794 (2020)

Template-Based Inter-modality Super-Resolution of Brain Connectivity

Furkan Pala[1] , Islem Mhiri[1,2] , and Islem Rekik[1(✉)]

[1] BASIRA Lab, Faculty of Computer and Informatics, Istanbul Technical University,
Istanbul, Turkey
irekik@itu.edu.tr
[2] ENISo Université de Sousse, Ecole Nationale d'Ingénieurs de Sousse,
LATIS-Laboratory of Advanced Technology and Intelligent Systems,
4023 Sousse, Tunisia
https://basira-lab.com/

Abstract. Brain graph synthesis becomes a challenging task when generating brain graphs across *different modalities*. Although promising, existing multimodal brain graph synthesis frameworks based on deep learning have several limitations. First, they mainly focus on predicting intra-modality graphs, overlooking the rich multimodal representations of brain connectivity (inter-modality). Second, while few techniques work on super-resolving low-resolution brain graphs within a single modality (i.e., intra), inter-modality graph super-resolution remains unexplored though this avoids the need for costly data collection and processing. More importantly, all these works need large amounts of training data which is not always feasible due to the scarce neuroimaging datasets especially for low resource clinical facilities and rare diseases. To fill these gaps, we propose an inter-modality super-resolution brain graph synthesis (TIS-Net) framework trained on one population driven atlas namely–connectional brain template (CBT). Our TIS-Net is grounded in three main contributions: (i) predicting a target representative template from a source one based on a novel graph generative adversarial network, (ii) generating high-resolution representative brain graph without resorting to the time consuming and expensive MRI processing steps, and (iii) training our framework using one shot learning where we estimate a representative and well centered CBT (i.e., one shot) with shared common traits across subjects. Moreover, we design a new Target Preserving loss function to guide the generator in learning the structure of target the CBT more accurately. Our comprehensive experiments on predicting a target CBT from a source CBT using our method showed the outperformance of TIS-Net in comparison with its variants. TIS-Net presents the first work for graph synthesis based on one representative shot across varying modalities and resolutions, which handles graph size, and structure variations. Our Python TIS-Net code is available on BASIRA GitHub at https://github.com/basiralab/TIS-Net.

F. Pala and I. Mhiri—Co-first authors.

I. Rekik et al. (Eds.): PRIME 2021, LNCS 12928, pp. 70–82, 2021.
https://doi.org/10.1007/978-3-030-87602-9_7

1   Introduction

Multimodal brain imaging has shown tremendous potential for neurodegenerative disorder diagnosis, where each imaging modality offers specific information for learning more comprehensive and informative data representations [1]. Although many imaging modalities can aid in precise clinical decision making, fast diagnosis is limited by high acquisition cost and long processing time [2].

To address such limitations, several deep learning studies have investigated multimodal MRI synthesis [3]. Some methods either generate a modality from another (i.e., cross-modality) or map both modalities to a commonly shared domain. Notably, generative adversarial networks (GANs) [4,5] have achieved great success in predicting medical images of different brain image modalities from a given modality. Although meaningful and powerful clinical representations were obtained from the previous studies, many challenges still exist. For instance, many techniques do not make efficient use of, or even fail to manage, non-Euclidean structured data (i.e., geometric data), such as graphs and manifolds [6]. Especially the brain which is a complex non-linear structure. Thus, a deep learning model that preserves graph-based data representation topology is a promising research path.

Therefore, researchers have investigated the geometric deep learning techniques to learn the deep graph-structure [7]. Particularly, deep graph convolutional networks (GCNs) have imbued the field of network neuroscience research through various tasks such as studying the mapping between human connectome and disease outcome. Recent studies have relied on using GCN to predict a target brain graph from a source brain graph [8].

For instance, [9] introduced MultiGraphGAN architecture, which predicts multiple brain graphs from a single brain graph while preserving the topological structure of each target predicted graph. Moreover, [10] defined a multi-GCN-based generative adversarial network (MGCN-GAN) to synthesize individual structural connectome from a functional connectome. While yielding outstanding performance, all the aforementioned methods can only transfer brain graphs from *one modality* to another while preserving the same resolution and topology, limiting their generalizability to *cross-modality* brain graph synthesis at *different resolutions and topologies* (i.e., node size, topology). Moreover, training such frameworks is computationally expensive and might not work on *scarce* neuroimaging datasets especially for rare diseases [11] and countries with limited clinical resources [12].

Recently, few-shot learning (FSL) methods have emerged as an alternative to train deep learning models in which the aim is to adjust the parameters of the model using only a few training samples [13]. In generative learning, FSL tries to create a generalization from the given few samples. Specifically, one-shot learning lets the generative model learn and generate new samples using only one training sample [14]. In fact, this technique used one sample to learn a new concept that can generalize to unseen distributions of testing samples. Several studies developed new ways of leveraging FSL in medical image-based learning tasks. For instance, [15] recommended a labelled medical image generation method that cultivated previous studies by exploiting one-shot and semi-supervised learning. Also, [16] put forward an overfitting-robust GAN framework designed for segmentation of 3D multi-modal medical images which was trained using FSL. Even training with lesser amount of data reduces the cost in terms of computational time, these works resort to data augmentation strategies or generative models to better estimate the unseen distributions of the whole population. However, due to the heterogeneity of the whole data, training on a single random sample could not reflect the whole data distribution. Therefore, we ask if we can train FSL models with a *representative* and *centered single sample* that captures the shared traits across training subjects and achieve great results without any augmentation. Moreover, this population template-based training does not require costly optimization. This will be a challenging task to tackle.

To address the above challenges and motivated by the recent development of graph neural network-based solutions (GNN), we propose a template-based inter-modality super-resolution of brain graphs (TIS-Net) method using one shot learning. First, to handle data heterogeneity within each specific modality, we learn the pairwise similarities between training brain graphs and map them into different subspaces using Single Cell Interpretation via Multikernel Learning (SIMLR) [17]. This clustering step enables to explore the underlying data distribution prior to the connectional brain template (CBT) estimation. Second, we estimate the cluster-specific CBTs which are locally centered and representative in the corresponding populations. Next, we integrate these cluster-specific CBTs to one global CBT which reflects the characteristic of the whole training population. More precisely, we used Deep Graph Normalizer Network (DGN) [18] for CBT generation which outperformed the seminal methods including [19,20]. Third, we propose a GAN architecture that takes a representative and well-centered CBT from a source domain and predicts a representative and well-centered CBT for the target domain in a one-shot fashion. In detail, the prediction task is the combination of a generator and discriminator designed using edge-based GCN layers and they are trained in an adversarial manner. Lastly, we introduce a target preserving (TP) loss function which enables the model to learn source manifold more comprehensively.

The main contributions of our work are four-fold.

1. *On a methodological level.* TIS-Net presents the first work for graph synthesis across varying modalities and resolutions using one representative shot learning, which can also be leveraged for boosting neurological disorder diagnosis.
2. *On a clinical level.* By estimating super-resolution representative connectional brain templates, one can readily spot a connectional fingerprint of a disorder (i.e., a set of altered brain connectivities) [21].
3. *On a computational level.* Our method generates new brain graphs with a reduced computational time thanks to one-shot CBT based learning. Hence, more healthcare facilities with limited equipment and tools can benefit from examining brain anomalies using *generated* high-resolution brain connectivity maps using our trained model.

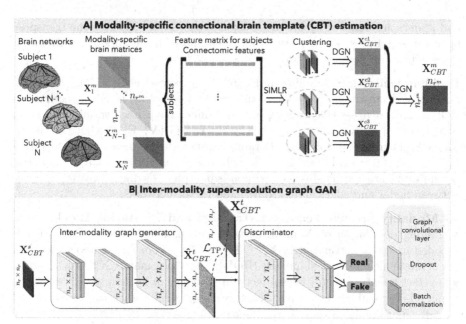

Fig. 1. *Illustration of the proposed template-based inter-modality super-resolution of brain connectivity using one shot learning.* **A|** **Modality-specific connectional brain template estimation.** For each training subject i in modality m, we vectorize the upper triangular part of its connectivity matrix $\mathbf{X}_i^m$. To handle the heterogeneous distribution of the data, we concatenate all feature vectors into a data feature matrix which we cluster similar brain networks into non-overlapping subspaces using SIMLR framework [17]. Next, for each cluster, a local centered CBT estimated using deep graph normalizer (DGN) [18], then we merge all these local cluster-specific CBTs into a global population-centered CBT. **B|** **Inter-modality super-resolution graph GAN.** Next, we propose an inter-modality super-resolution graph GAN which is trained to transform a single source template $\mathbf{X}_{CBT}^s$ (e.g., morphological) into a target template $\hat{\mathbf{X}}_{\mathbf{CBT}}^{\mathbf{t}}$ (e.g., functional) with different structural properties. Our framework is trained in an end-to-end manner and one-shot fashion by optimizing a novel Target-Preserving (TP) loss function which guides the generator in learning the target domain more effectively.

4. *On a generic level.* Our framework is a generic method as it can be applied to predict brain graphs derived from any neuroimaging modality and at any higher resolution.

2 Proposed Method

This section presents the key steps of our CBT-based one shot learning for brain graph inter-modality super-resolution framework. Matrices are denoted by boldface capital letters,e.g., $\mathbf{X}$, vectors by boldface lowercase letters, e.g., $\mathbf{x}$, and scalars are denoted by lowercase letters, e.g., x. $\mathbf{X}^T$ denotes the transpose operator. Fig. 1 presents an overview for the proposed framework in several major steps, which are detailed below.

Problem Definition. A brain graph can be represented as $\mathbf{G}(\mathbf{V}, \mathbf{E})$, where each node in $\mathbf{V}$ denotes a brain region of interest and each edge in $\mathbf{E}$ connecting two ROIs k and l denotes the strength of their connectivity. Each training subject i is represented by two brain graphs $\{\mathbf{G}_i^s(\mathbf{V}_i^s, \mathbf{E}_i^s), \mathbf{G}_i^t(\mathbf{V}_i^t, \mathbf{E}_i^t)\}$, where $\mathbf{G}^s$ is the source brain graph (morphological brain network) with n_r nodes, $\mathbf{G}^t$ is the target brain graph (functional brain network) with $n_{r'}$ nodes where $n_r \neq n_{r'}$. These two brain graphs are specifically deemed non-isomorphic, having no correspondence between nodes and edges across source and target networks.

A- Modality-Specific Feature Extraction and Clustering Block. Given a training population of N brain networks in modality m, each network i is encoded in a symmetric matrix $\mathbf{X}_i^m \in \mathbb{R}^{n_{r^m} \times n_{r^m}}$, where n_{r^m} denotes the number of anatomical regions of interest (ROIs). Since each matrix $\mathbf{X}_i^m$ is symmetric, we extract a feature vector for subject i in modality m by simply vectorizing its upper off-diagonal triangular part. Next, we horizontally stack feature vectors of all subjects to define a data feature matrix of size $N \times \frac{n_{r^m} \times (n_{r^m} - 1)}{2}$ (Fig. 1A). Next, we disentangle the heterogeneous distribution of the brain graphs by clustering similar brain connectomes into non-overlapping subspaces using Single Cell Interpretation via Multikernel Learning (SIMLR) framework [17]. This clustering step enables to explore the underlying data distribution prior to the CBT estimation.

B- Population-Based Connectional Brain Template (CBT) Learning Block. When we look more closely at how such techniques are implemented in practice, we run into two challenges. The first one is the time-consuming training step due to the huge size of the dataset. The second one is the performance of the deep models which can be extremely sensitive to changes in training and test set distributions caused by local or global perturbations. To overcome these issues, we aim to estimate a centered and representative brain connectional template for both source and target modalities (Fig. 1A), which has enough prior knowledge about the learning domain. The main idea behind our CBT estimation is to first estimate cluster-specific CBTs to disentangle brain graph data heterogeneity, then map all of the cluster-specific CBTs to generate a global representative center of all individuals in the population. To deal with this process, we use the deep graph normalizer network (DGN) [18]. DGN combines two or more weighted graphs into a population center graph (i.e., CBT). This learning challenge is primarily based on mapping connection patterns for each node in each graph in the provided modality onto a high-dimensional vector representation, specifically a node feature vector. In order to preserve the unique domain-specific topological properties of the brain graphs during the mapping process, a topology-constrained normalization loss function is introduced to penalize the deviation from the target population topology. Next, we extract the CBT edges from the pairwise connection of node embeddings. Finally, to learn the cluster-specific CBT encoded in a matrix $\mathbf{X}_{CBT}^{m_c}$, we minimize the Frobenius distance to a random set C of subjects belonging to the same cluster to avoid overfitting as follows: $\arg\min_{\mathbf{X}_{CBT}^{m_c}} \sum_{c \in C} \left\| \mathbf{X}_{CBT}^{m_c} - \mathbf{X}_c^{m_c} \right\|_F$

Ultimately, the final training CBT is computed by mapping all cluster-specific CBTs using DGN. Using the proposed strategy, we estimate a training network atlas $\mathbf{X}_{CBT}^s$ for source brain graphs and $\mathbf{X}_{CBT}^t$ for target brain graphs.

C- Adversarial Inter-Modality Graph Generator Block. Following the estimation of the one representative shot step, we design an inter-modality graph GAN that handles shifts in graph resolution (i.e., node size variation).

i- Inter-modality Super-Resolution Brain Graph Generator. Inspired by the dynamic edge convolution proposed in [22] and the U-net architecture [23] with skip connections, we design an inter-modality graph generator G_{CBT} which is composed of three edge-based GCN layers regularized by batch normalization and dropout for each layer's output (Fig. 1B). Specifically, G_{CBT} takes as input the source $\mathbf{X}_{CBT}^s$ of size $n_r \times n_r$ and outputs the predicted $\hat{\mathbf{X}}_{CBT}^t$ of size $n_{r'} \times n_{r'}$, where $n_r \neq n_{r'}$. Particularly, owing to dynamic graph-based edge convolution operation [22], each edge-based GCN layer comprises a unique dynamic filter that outputs edge-specific weight matrix dictating the information flow between nodes k and l to learn a comprehensive node feature vector. Then, we define a mapping function $T_r : \mathbb{R}^{n_r \times n_{r'}} \mapsto \mathbb{R}^{n_{r'} \times n_{r'}}$ which takes as input the embedded matrix of the whole graph in the latest GCN layer of size $n_r \times n_{r'}$ and outputs the generated target CBT of size $n_{r'} \times n_{r'}$ (see subsection: *Graph super-resolution based on dynamic edge convolution*).

ii- Inter-modality super-Resolution Graph Discriminator Based on Adversarial Training. Our inter-modality generator G_{CBT} is trained adversarially against a discriminator network D_{CBT} (Fig. 1B). To discriminate between the predicted and ground truth target CBT graph, we design a two-layer graph neural network [22]. Our discriminator D_{CBT} takes as input the target $\mathbf{X}_{CBT}^t$ and the generator output $\hat{\mathbf{X}}_{CBT}^t$, and outputs a value between 0 and 1, measuring the realness of the generator output. To improve the quality of our discriminator's performances, we add the adversarial loss function to maximize the discriminator's output value for the $\mathbf{X}_{CBT}^t$ and minimize it for $\hat{\mathbf{X}}_{CBT}^t$.

Graph Super-Resolution Based on Dynamic Edge Convolution. In both DGN and GAN models, we use a dynamic graph-based edge convolution process in each proposed GCN layer [22]. Notably, we define h the layer index in the neural network and d_h the output dimension of the corresponding layer. Each layer h includes a filter generating network $F^h : \mathbb{R} \mapsto \mathbb{R}^{d_h \times d_{h-1}}$ which dynamically creates a weight matrix for filtering message passing between ROIs k and l given the edge weight e_{kl}. At this point e_{kl} is edge feature (i.e., connectivity weight) that measures the relationship between ROIs k and l. In our framework, the goal of each layer is to produce the graph convolution result which can be observed as a filtered signal $\mathbf{z}^h(k) \in \mathbb{R}^{d_h \times 1}$ at node k. The overall edge-conditioned convolution operation is defined as follows:

$\mathbf{z}_k^h = \Theta^h . \mathbf{z}_k^{h-1} + \frac{1}{|N(k)|} \sum_{l \in N(k)} F^h(\mathbf{e}_{kl}; \mathbf{W}^h) \mathbf{z}_l^{h-1} + \mathbf{b}^h$ where $\mathbf{z}_k^h \in \mathbb{R}^{d_h \times 1}$ is the embedding of node k in layer h, $\Theta_{lk}^h = F^h(\mathbf{e}_{kl}; \mathbf{W}^h)$ represents the dynamically generated edge-specific weights by F^h. $\mathbf{b}^h \in \mathbb{R}^{d_h}$ denotes a network bias and $N(k)$ denotes the neighbors of node k.

Given the learned embedding $\mathbf{z}_k^h \in \mathbb{R}^{d_h}$ for node k in layer h, we determine the embedding of the whole graph in layer h as $\mathbf{Z}^h \in \mathbb{R}^{n_r \times d_h}$ where n_r is the number of nodes. We would like to point out to the reader that any change in resolution may be readily represented as a transformation $\mathcal{T}_r : \mathbb{R}^{n_r \times d_h} \mapsto \mathbb{R}^{d_h \times d_h}$ where $\mathcal{T}_r$ is formulated as follows: $\mathcal{T}_r = (\mathbf{Z}^h)^T \mathbf{Z}^h$. Therefore, super-resolution is only defined by setting the desired target graph resolution d_h. In our case, we fix d_h of the latest layer in the generator to $n_{r'}$ to output the predicted template $\hat{\mathbf{X}}_{\mathbf{CBT}}^t$ of size $n_{r'} \times n_{r'}$ (Fig. 1B).

D- Target Preserving Loss Function. Conventionally, GAN generators are optimized based on the response of their corresponding discriminators. However, during a few training epochs, the discriminator can readily differentiate between real and predicted graphs, thus the adversarial loss would be close to 0. In such a case, the generator cannot provide satisfactory results. To overcome this dilemma, we need to enforce the generator-discriminator synchronous learning during the training process. Thus, we propose a new target preserving (TP) loss which consists of three sub-losses: adversarial loss [24], L1 loss [25] and Pearson correlation coefficient (PCC) loss [10]. We define our TP loss function as follows:

$$\mathcal{L}_{\mathrm{TP}} = \lambda_1 \mathcal{L}_{adv} + \lambda_2 \mathcal{L}_{L1} + \lambda_3 \mathcal{L}_{PCC} \tag{1}$$

where $\mathcal{L}_{adv}$ denotes the adversarial loss which measures the difference between the generated and target CBTs as both generator and discriminator are iteratively optimized through the adversarial loss:

$\mathrm{argmin}_{G_{CBT}} \max_{D_{CBT}} \mathcal{L}_{adv} = \mathbb{E}_{G_{CBT}(\mathbf{x}_{CBT}^t)} [\log(D_{CBT}(G_{CBT}(\mathbf{X}_{CBT}^t))) +$

$\mathbb{E}_{G_{CBT}(\hat{\mathbf{x}}_{CBT}^t)} \left[\log \left(1 - D_{CBT} \left(G_{CBT} \left(\hat{\mathbf{X}}_{CBT}^t \right) \right) \right) \right]$

To improve the predicted brain graph quality, we propose to add an $L1$ loss term minimizing the distance between each predicted brain template $\hat{\mathbf{X}}_{CBT}^t$ and its related target $\mathbf{X}_{CBT}^t$. The $l1$ loss function is expressed as follows: $\mathcal{L}_{l1} = \left\| \mathbf{X}_{CBT}^t - \hat{\mathbf{X}}_{CBT}^t \right\|_1$. Although resistant to outliers, the $L1$ loss only considers the element-wise similarity in edge weights between the predicted and target CBTs and ignores the overall correlation between both graphs. Therefore, we involve the Pearson correlation coefficient (PCC) in our loss which quantifies the overall correlation between the predicted and target brain templates. Moreover, since greater PCC implies a higher correlation between the target and the predicted graphs, we suggest to minimize the PCC loss function as follows: $\mathcal{L}_{PCC} = 1 - PCC$.

3   Results and Discussion

Evaluation Dataset. We used three-fold cross-validation strategy to evaluate our TIS-Net framework on 100 subjects from the Southwest University Longitudinal Imaging Multimodal (SLIM) public dataset[1] where each subject has T1-w, T2-w MRI and resting-state fMRI (rsfMRI) scans. Our model is implemented using the PyTorch-Geometric library [26].

Morphological Brain Graphs (Source). For each subject, we used FreeSurfer [27] to generate the cortical morphological network from structural T1-w MRI as introduced in [28,29]. Particularly, we computed the pairwise absolute difference in cortical thickness between pairs of regions of interest to generate a 35×35 morphological connectivity matrix for each subject denoted as $\mathbf{X}^s$.

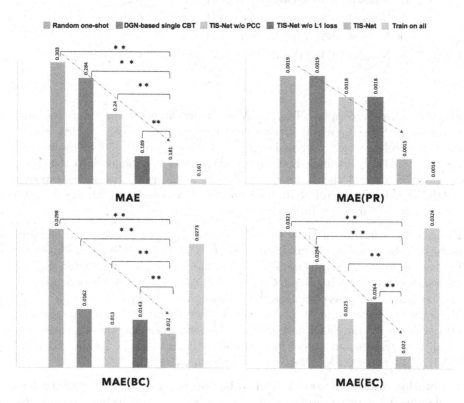

Fig. 2. *Functional brain connectivity prediction results from source morphological connectivity using different evaluation metrics.* Evaluation of inter-modality brain graph synthesis by our TIS-Net against five comparison methods. We used the mean absolute error (MAE) between the real and predicted CBTs, and the mean absolute difference between their three topological measures as evaluation metrics. (PR: PageRank centrality, BC: betweenness centrality and EC: eigenvector centrality. w/o: without.) TIS-Net significantly outperformed all benchmark methods using two-tailed paired t-test ($p < 0.001$) –excluding the "Train on all".

---

[1] http://fcon_1000.projects.nitrc.org/.

Table 1. *Evaluation of the estimated CBT by DGN* [18] *in both target and source domains using different clustering strategies.* We display the Mean Absolute Error (MAE) between the estimated CBT from the training data and all individual networks in the training population (both source and target domains) using DGN-based CBT (no clustering), DGN-based CBT (K-means = 2), DGN-based CBT (SIMLR = 2), DGN-based CBT (K-means = 3) and DGN-based CBT (SIMLR = 3). Clearly, DGN-based CBT (SIMLR = 3) achieves the minimum MAE.

Methods	MAE	
	Source domain	Target domain
DGN-based CBT w/o clustering	0.3	0.29
DGN-based CBT (K-means = 2)	0.24	0.26
DGN-based CBT (SIMLR = 2)	0.24	0.25
DGN-based CBT (K-means = 3)	0.23	0.25
DGN-based CBT (SIMLR = 3)	**0.22**	**0.23**

Functional Brain Graphs (Target). After many preprocessing steps of each resting-state fMRI applying preprocessed connectomes project quality assessment protocol, we used a whole-brain parcellation approach [30] to reconstruct the functional connectomes. Each brain rsfMRI was partitioned into 160 ROIs. Next, we computed the Pearson correlation coefficient between two average rsfMRI signals of pairs of ROIs to define the functional connectivity weights. These indicate our target brain graphs $\mathbf{X}^t$.

Parameter Setting. In order to find the best CBT, we tested our framework using different number of clusters $n_c = \{1, 2, 3, ..., 6\}$ clusters and we find that the best result is $n_c = 3$ using SIMLR. For the DGN hyperparameters, we set the random training sample set C size to 10. Also, for the prediction task, we set the generator hyperparameters as follows: $\lambda_1 = 1$, $\lambda_2 = 1$, and $\lambda_3 = 0.5$. Moreover, we chose AdamW [31] as our default optimizer and set the learning rate at 0.005 for the DGN and 0.025 for the generator and the discriminator. Finally, we trained our model for 400 epochs using a single Tesla $V100$ GPU (NVIDIA GeForce GTX TITAN with 32 GB memory).

Evaluation and Comparison Methods. We evaluate the benefit of our model against five baseline methods: (1) Random one shot where we used an *inverted* LOO-CV strategy which means that only one sample is used for training and all remaining samples are used for testing, (2) DGN-based single CBT where we used the DGN for the whole training population to generate the final CBT without clustering, (3) TIS-Net w/o PCC where we used our proposed framework without the Pearson correlation coefficient based loss, (4) TIS-Net w/o L1 loss where we used our proposed framework without the L1 loss, and (5) "Train on all" samples where we trained all subjects instead of one representative shot. As illustrated in Fig. 2, we computed the mean absolute error between the predicted and the real brain graphs. Clearly, our TIS-Net significantly (*pvalue* < 0.001

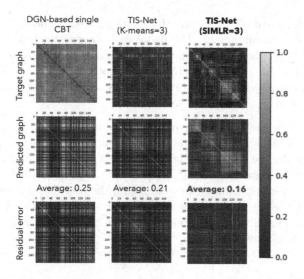

Fig. 3. *Visual comparison between the target and the predicted brain graphs.* Comparing the target to the predicted brain graphs by TIS-Net and two comparison methods using a representative testing subject.

using two-tailed paired t-test) outperformed comparison methods by achieving the lowest error between the predicted and real brain graphs across all evaluation metrics including topological properties using betweenness centrality, PageRank centrality and eigenvector centrality except train-on-all for the MAE and PageRank.

Insights into Connectional Brain Template Representativeness. To evaluate the centeredness and representativeness of the brain template produced by DGN [18] (Fig. 1A), we computed the mean absolute error between estimated network atlas and all individual networks in the training population for each domain using SIMLR clustering algorithm and K-means algorithm with different numbers of clusters (Table 1). Clearly, SIMLR consistently achieved the minimum distance across different numbers of clusters n_c. In particular, $n_c = 3$ achieved the best results. Indeed, changes in evaluation measures were negligible when varying the number of cluster $n_c > 3$. Besides, SIMLR reduced weak similarities and clustered subpopulations more accurately in comparison with K-means algorithm.

Insights Into Topological Measures. To prove the fidelity of the predicted brain graph to the real one in topology and structure, we evaluated TIS-Net using various topological measures (eigenvector, closeness, and PageRank). Fig. 2 shows that our method consistently achieves the best topology-preserving predictions compared with other baseline methods.

Insights into the Proposed TP Loss Function. To investigate the efficiency of the proposed TP loss, we trained TIS-Net with different loss functions. As illustrated in Fig. 2, our proposed TP loss outperforms its ablated versions. In fact, the $L1$ loss focuses only on minimizing the distance between two CBTs at the local level. Besides, PCC aims to maximize global connectivity patterns between the predicted and real brain templates in the one-shot training. The combination of these complementary losses scored the best results while relaxing inter-modality graph hypothesis.

Reproducibility. Beside generating functional templates, our framework could also capture the delicate variations in connectivity patterns across predicted graphs. Specifically, we display in Fig. 3, the target, predicted, and residual brain graphs for a representative testing brain graph using four different methods. The residual graph is calculated by computing the absolute difference between the target and predicted connectomes of a representative subject. An average difference value of the residual is represented on top of each graph achieving a noticeable reduction by TIS-Net.

4   Conclusion

In this paper, we proposed the first work for template-based inter-modality super-resolution of brain graphs, namely TIS-Net, which nicely handles variations in graph size, and structure using one shot learning. Our method can also be leveraged for developing precision medicine and may become a helpful component for reducing the cost of batch-based training of GNNs models. The proposed TIS-Net outperforms the baseline methods in terms of brain graph super-resolution results. TIS-Net not only predicts reliable HR functional brain networks from morphological ones but also preserves the topology of the target domain. In future work, we will extend our architecture to predict time-dependent brain graphs using one shot learning.

Acknowledgments. This work was funded by generous grants from the European H2020 Marie Sklodowska-Curie action (grant no. 101003403, http://basira-lab.com/normnets/) to I.R. and the Scientific and Technological Research Council of Turkey to I.R. under the TUBITAK 2232 Fellowship for Outstanding Researchers (no. 118C288, http://basira-lab.com/reprime/). However, all scientific contributions made in this project are owned and approved solely by the authors.

References

1. Shen, X., Tokoglu, F., Papademetris, X., Constable, R.T.: Groupwise whole-brain parcellation from resting-state FMRI data for network node identification. Neuroimage **82**, 403–415 (2013)
2. Cao, B., Zhang, H., Wang, N., Gao, X., Shen, D.: Auto-GAN: self-supervised collaborative learning for medical image synthesis. In: Proceedings of the AAAI Conference on Artificial Intelligence, pp. 10486–10493 (2020)
3. Wang, T., et al.: A review on medical imaging synthesis using deep learning and its clinical applications. J. Appl. Clin. Med. Phys. **22**, 11–36 (2021)

4. Singh, N.K., Raza, K.: Medical image generation using generative adversarial networks: a review. In: Health Informatics: A Computational Perspective in Healthcare, pp. 77–96 (2021)
5. Wang, C., et al.: DICyc: GAN-based deformation invariant cross-domain information fusion for medical image synthesis. Inf. Fus. **67**, 147–160 (2021)
6. Bronstein, M.M., Bruna, J., LeCun, Y., Szlam, A., Vandergheynst, P.: Geometric deep learning: going beyond Euclidean data. IEEE Signal Process. Mag. **34**, 18–42 (2017)
7. Bessadok, A., Mahjoub, M.A., Rekik, I.: Graph neural networks in network neuroscience. arXiv preprint arXiv:2106.03535 (2021)
8. Mhiri, I., Nebli, A., Mahjoub, M.A., Rekik, I.: Non-isomorphic Inter-modality graph alignment and synthesis for holistic brain mapping. In: Feragen, A., Sommer, S., Schnabel, J., Nielsen, M. (eds.) IPMI 2021. LNCS, vol. 12729, pp. 203–215. Springer, Cham (2021). https://doi.org/10.1007/978-3-030-78191-0_16
9. Bessadok, A., Mahjoub, M.A., Rekik, I.: Topology-aware generative adversarial network for joint prediction of multiple brain graphs from a single brain graph. In: Martel, A.L., et al. (eds.) MICCAI 2020. LNCS, vol. 12267, pp. 551–561. Springer, Cham (2020). https://doi.org/10.1007/978-3-030-59728-3_54
10. Zhang, L., Wang, L., Zhu, D.: Recovering brain structural connectivity from functional connectivity via multi-GCN based generative adversarial network. In: Martel, A.L., et al. (eds.) MICCAI 2020. LNCS, vol. 12267, pp. 53–61. Springer, Cham (2020). https://doi.org/10.1007/978-3-030-59728-3_6
11. Schaefer, J., Lehne, M., Schepers, J., Prasser, F., Thun, S.: The use of machine learning in rare diseases: a scoping review. Orphanet J. Rare Dis. **15**, 1–10 (2020). https://doi.org/10.1186/s13023-020-01424-6
12. Piette, J.D., et al.: Impacts of e-health on the outcomes of care in low-and middle-income countries: where do we go from here? Bull. World Health Organ. **90**, 365–372 (2012)
13. Kadam, S., Vaidya, V.: Review and analysis of zero, one and few shot learning approaches. In: Abraham, A., Cherukuri, A.K., Melin, P., Gandhi, N. (eds.) ISDA 2018 2018. AISC, vol. 940, pp. 100–112. Springer, Cham (2020). https://doi.org/10.1007/978-3-030-16657-1_10
14. Rezende, D., Danihelka, I., Gregor, K., Wierstra, D., et al.: One-shot generalization in deep generative models, pp. 1521–1529 (2016)
15. Zhao, A., Balakrishnan, G., Durand, F., Guttag, J.V., Dalca, A.V.: Data augmentation using learned transformations for one-shot medical image segmentation, pp. 8543–8553 (2019)
16. Mondal, A.K., Dolz, J., Desrosiers, C.: Few-shot 3D multi-modal medical image segmentation using generative adversarial learning. arXiv preprint arXiv:1810.12241 (2018)
17. Wang, B., Ramazzotti, D., De Sano, L., Zhu, J., Pierson, E., Batzoglou, S.: SIMLR: a tool for large-scale genomic analyses by multi-kernel learning. Proteomics **18**, 1700232 (2018)
18. Gurbuz, M.B., Rekik, I.: Deep graph normalizer: a geometric deep learning approach for estimating connectional brain templates. In: Martel, A.L., et al. (eds.) MICCAI 2020. LNCS, vol. 12267, pp. 155–165. Springer, Cham (2020). https://doi.org/10.1007/978-3-030-59728-3_16
19. Rekik, I., Li, G., Lin, W., Shen, D.: Estimation of brain network atlases using diffusive-shrinking graphs: application to developing brains. In: Niethammer, M., et al. (eds.) IPMI 2017. LNCS, vol. 10265, pp. 385–397. Springer, Cham (2017). https://doi.org/10.1007/978-3-319-59050-9_31

20. Dhifallah, S., Rekik, I., Initiative, A.D.N., et al.: Estimation of connectional brain templates using selective multi-view network normalization. Med. Image Anal. **59**, 101567 (2020)

21. Mhiri, I., Mahjoub, M.A., Rekik, I.: Supervised multi-topology network cross-diffusion for population-driven brain network atlas estimation. In: Martel, A.L., et al. (eds.) MICCAI 2020. LNCS, vol. 12267, pp. 166–176. Springer, Cham (2020). https://doi.org/10.1007/978-3-030-59728-3_17

22. Simonovsky, M., Komodakis, N.: Dynamic edge-conditioned filters in convolutional neural networks on graphs. In: Proceedings of the IEEE Conference on Computer Vision and Pattern Recognition, pp. 3693–3702 (2017)

23. Ronneberger, O., Fischer, P., Brox, T.: U-Net: convolutional networks for biomedical image segmentation. In: Navab, N., Hornegger, J., Wells, W.M., Frangi, A.F. (eds.) MICCAI 2015. LNCS, vol. 9351, pp. 234–241. Springer, Cham (2015). https://doi.org/10.1007/978-3-319-24574-4_28

24. Goodfellow, I.J., et al.: Generative adversarial networks. arXiv preprint arXiv:1406.2661 (2014)

25. Gürler, Z., Nebli, A., Rekik, I.: Foreseeing brain graph evolution over time using deep adversarial network normalizer. In: Rekik, I., Adeli, E., Park, S.H., Valdés Hernández, M.C. (eds.) PRIME 2020. LNCS, vol. 12329, pp. 111–122. Springer, Cham (2020). https://doi.org/10.1007/978-3-030-59354-4_11

26. Fey, M., Lenssen, J.E.: Fast graph representation learning with Pytorch geometric. arXiv preprint arXiv:1903.02428 (2019)

27. Fischl, B.: Freesurfer. Neuroimage **62**, 774–781 (2012)

28. Mahjoub, I., Mahjoub, M.A., Rekik, I.: Brain multiplexes reveal morphological connectional biomarkers fingerprinting late brain dementia states. Sci. Rep. **8**, 1–14 (2018)

29. Raeper, R., Lisowska, A., Rekik, I.: Cooperative correlational and discriminative ensemble classifier learning for early dementia diagnosis using morphological brain multiplexes. IEEE Access **6**, 43830–43839 (2018)

30. Dosenbach, N.U., et al.: Prediction of individual brain maturity using fMRI. Science **329**, 1358–1361 (2010)

31. Loshchilov, I., Hutter, F.: Fixing weight decay regularization in Adam (2018)

Adversarial Bayesian Optimization for Quantifying Motion Artifact Within MRI

Anastasia Butskova[1], Rain Juhl[1], Dženan Zukić[2], Aashish Chaudhary[2], Kilian M. Pohl[1,3], and Qingyu Zhao[1(✉)]

[1] Stanford University, Stanford, CA, USA
qingyuz@stanford.edu
[2] Kitware Inc., Carrboro, NC, USA
[3] SRI International, Menlo Park, CA, USA

Abstract. Subject motion during an MRI sequence can cause ghosting effects or diffuse image noise in the phase-encoding direction and hence is likely to bias findings in neuroimaging studies. Detecting motion artifacts often relies on experts visually inspecting MRIs, which is subjective and expensive. To improve this detection, we develop a framework to automatically quantify the severity of motion artifact within a brain MRI. We formulate this task as a regression problem and train the regressor from a data set of MRIs with various amounts of motion artifacts. To resolve the issue of missing fine-grained ground-truth labels (level of artifacts), we propose Adversarial Bayesian Optimization (ABO) to infer the distribution of motion parameters (i.e., rotation and translation) underlying the acquired MRI data and then inject synthetic motion artifacts sampled from that estimated distribution into motion-free MRIs. After training the regressor on the synthetic data, we applied the model to quantify the motion level in 990 MRIs collected by the National Consortium on Alcohol and Neurodevelopment in Adolescence. Results show that the motion level derived by our approach is more reliable than the traditional metric based on Entropy Focus Criterion and manually defined binary labels.

1 Introduction

Due to the relatively long scanning time required for MRI sequences, subject motion creates one of the most common artifacts in MR imaging that results in blurring and ghosting in the image [1]. Fast and accurate quantification of motion artifacts within MRIs, therefore, becomes critical in most clinical settings as it generates real-time feedback on whether a re-scan is needed while the subject is still in the scanner, thereby reducing the need for scheduling additional scanning sessions [2]. Moreover, in large-scale neuroimaging studies, such quantification can be used to minimize the risk of motion-related confounders biasing findings by either confining analysis to MRIs with minimum motion artifacts or controlling for the level of motion artifact in statistical models [3,4].

© Springer Nature Switzerland AG 2021
I. Rekik et al. (Eds.): PRIME 2021, LNCS 12928, pp. 83–92, 2021.
https://doi.org/10.1007/978-3-030-87602-9_8

In most existing neuroimaging studies, detection of motion artifacts in T1-weighted MRI relies on experts visually reviewing scans and labeling an MRI as either motion-free or with motion artifacts [5]. Such binary (or categorical) labeling is often subjective and inaccurate, especially when the scans are from multiple sites with various imaging protocols. An alternative to manual labeling is to use Entropy Focus Criterion (EFC) [6] to quantify the severity of motion artifact, but this metric is only suitable for detecting large-scale motions. To improve the quantification, prior works have tried to automate motion detection based on machine learning models. Lorch et al. [2] proposed a framework based on Random Forests to detect head and respiratory motion in MRIs, but their framework was only evaluated on synthetic data. Other works [7,8] designed a convolutional neural network to classify image patches and then used a special voting mechanism to determine the presence of motion artifact in the entire MRI. However, these classification models always assume availability of accurate ground-truth labels, which, as mentioned above, are difficult and expensive to produce for large multi-site studies. Moreover, these models only provide binary predictions, which cannot quantify detailed severity of the artifact. To resolve these issues, we aim to develop a regression framework to quantify the fine-grained level of motion artifact in an MRI via a convolutional neural network.

We are specifically interested in designing a training scheme to learn the regression network based on binary (presence or absence of artifact) and potentially noisy ground-truth labels derived by manual labeling. To generate training data with accurate and continuous labels (level of artifact), we injected synthetic artifacts induced by rigid motion into motion-free MRIs. To do so, we first developed a novel method called Adversarial Bayesian Optimization (ABO) to infer the distribution of motion parameters, i.e., the range of rotation and translation in 3D, underlying the motion observed in the real data set. We then generated a simulated data set with synthetic artifacts controlled by the estimated motion range, so that the regressor network could be trained in a traditional supervised fashion. We tested the proposed framework to quantify the severity of motion artifact in 990 T1-weighted MRIs from the National Consortium on Alcohol and Neurodevelopment in Adolescence (NCANDA) [9]. Results indicate that the proposed ABO could accurately estimate the range of motion parameters underlying a given data set. The regressor trained based on this estimated range more accurately quantified the level of motion in MRIs compared to traditional metrics based on EFC.

2   Methods

Let $\mathcal{X}_c$ be a set of motion-free MRIs and $\mathcal{X}_m$ be a set of MRIs with various level of motion, our goal is to derive a function q, such that for an MRI $x \in \mathcal{X}_m$, $y = q(x)$ quantifies the level of motion. To train the regressor q without ground-truth y, a typical approach is to rely on data simulation. Here, we leverage the simulator function $\hat{x} = s(x; \tau)$ proposed in [10]. The simulator can inject "k-space" motion artifacts controlled by a set of parameters τ (encoding rotation

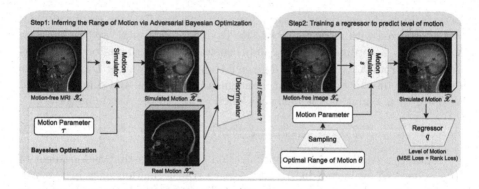

Fig. 1. Model Architecture: Step 1. Adversarial Bayesian Optimization (ABO) is used to infer the distribution of motion parameters underlying the real data set with motion artifacts $\mathcal{X}_m$; Step 2. A regressor is trained on a simulated data set, whose motion artifacts are sampled from the distribution estimated in Step 1.

and translation of rigid motion) into a motion-free image $x \in \mathcal{X}_c$ to generate a synthetic motion-corrupted MRI $\hat{x} \in \hat{\mathcal{X}}_m$. As the ground-truth motion of $\hat{x}$ is known (related to τ), one can learn q using traditional supervised training.

However, due to the variability in imaging protocols and characteristics of the study cohorts, one typically does not know the range of motion in $\mathcal{X}_m$, so the model trained on the simulated set $\hat{\mathcal{X}}_m$ might not generalize well. To resolve this issue, we propose the following 2-step training strategy (Fig. 1): (1) infer the underlying distribution of motion parameters $p(\tau)$ associated with the real motion artifacts in $\mathcal{X}_m$; and (2) train the regressor q on a simulated data set $\hat{\mathcal{X}}_m$, where τ (and also y) for each simulated MRI is sampled from $p(\tau)$.

2.1 Inferring Distribution of Motion Parameters via Adversarial Bayesian Optimization

Let $\tau = \{\tau_1, ..., \tau_6\}$ be the amount of rotation and translation of the simulated motion in the 3 spatial directions, and let $\theta = \{\theta_1, ..., \theta_6\}$ be the maximum transformation in those 6°. We then assume the transformation parameters follow independent uniform distributions $p(\tau; \theta) = \prod_k p(\tau_k; \theta_k) = \prod_k \mathcal{U}(-\theta_k, \theta_k)$. As such, $p(\tau; \theta)$ can be used to characterize a data set with various level of motion artifacts.

We now aim to infer the maximum transformation θ associated with $\mathcal{X}_m$. We do so by adversarially training a discriminator network D with parameters ϕ, such that it can not distinguish the real data $\mathcal{X}_m$ from simulated data $\hat{\mathcal{X}}_m$, which are generated by injecting synthetic motion (sampled from $p(\tau; \theta)$) into motion-free MRIs $\mathcal{X}_c$. This objective function can be written as

$$\max_{\theta} \min_{\phi} \mathbb{E}_{x \sim X_m} [\log D_\phi(x)] + \mathbb{E}_{\tau \sim p(\tau; \theta), x \sim \mathcal{X}_c} [\log(1 - D_\phi(s(x; \tau)))]. \quad (1)$$

As typically performed in adversarial learning, this loss function can be alternatively optimized with respect to $\boldsymbol{\theta}$ and $\boldsymbol{\phi}$. To update $\boldsymbol{\phi}$ given $\boldsymbol{\theta}$, we first simulate $\hat{\boldsymbol{X}}_m$ from $\boldsymbol{X}_c$ according to $p(\boldsymbol{\tau};\boldsymbol{\theta})$ and then train the discriminator D to differentiate $\hat{\boldsymbol{X}}_m$ from $\boldsymbol{X}_m$. To update $\boldsymbol{\theta}$ given a trained discriminator D, we optimize $\boldsymbol{\theta}$ so that the discriminator cannot distinguish between real and simulated data. We do so by maximizing the following function to fool the discriminator.

$$f(\boldsymbol{\theta}) = \mathbb{E}_{\boldsymbol{\tau} \sim p(\boldsymbol{\tau};\boldsymbol{\theta}), x \sim \boldsymbol{\mathcal{X}}_c} \left[\log(1 - D_{\phi}(s(\boldsymbol{x};\boldsymbol{\tau})))\right]. \tag{2}$$

However, a challenge in the optimization of Eq. (2) is that the gradient of $f(\boldsymbol{\theta})$ is hard to evaluate (due to the k-space transformation of x in $s(\boldsymbol{x};\tau)$), so traditional back-propagation can not be directly applied. Here, we resort to Bayesian optimization to optimize Eq. (2).

Bayesian Optimization. [11] is a machine-learning-based approach for finding the maximum of f without estimating its gradient. To do so, Bayesian optimization formulates a probability distribution over the functional $f(\cdot)$ as a Gaussian Process and repeatedly evaluates function values at various points. Let $\boldsymbol{\Theta} = [\boldsymbol{\theta}^1, ..., \boldsymbol{\theta}^t]$ be the first t evaluated points, and $\boldsymbol{f} = [f(\boldsymbol{\theta}^1), ..., f(\boldsymbol{\theta}^t)]$ be their functional values. Let $k(\cdot,\cdot)$ denote the covariance function [12] associated with the Gaussian Process. Then the functional value at an unobserved point $\boldsymbol{\theta}^*$ follows a normal distribution $f(\boldsymbol{\theta}^*) \sim \mathcal{N}(\mu[\boldsymbol{\theta}^*], \sigma^2[\boldsymbol{\theta}^*])$, where

$$\mu[\boldsymbol{\theta}^*] = K(\boldsymbol{\theta}^*, \boldsymbol{\Theta})K(\boldsymbol{\Theta}, \boldsymbol{\Theta})^{-1}\boldsymbol{f} \tag{3}$$

$$\sigma^2[\boldsymbol{\theta}^*] = K(\boldsymbol{\theta}^*, \boldsymbol{\theta}^*) - K(\boldsymbol{\theta}^*, \boldsymbol{\Theta})K(\boldsymbol{\Theta}, \boldsymbol{\Theta})^{-1}K(\boldsymbol{\Theta}, \boldsymbol{\theta}^*), \tag{4}$$

Here, $K(\boldsymbol{\Theta}, \boldsymbol{\Theta})$ is a $t \times t$ matrix with each element (i, j) recording $k(\boldsymbol{\theta}^i, \boldsymbol{\theta}^j)$, and likewise $K(\boldsymbol{\theta}^*, \boldsymbol{\Theta})$ is a $t \times 1$ vector. Given such a probability distribution, an acquisition function is used to select the next evaluation point with the highest chance to get closer to the global maximum. Here we use the Upper Confidence Bound (UCB) [12] as the acquisition function

$$\arg \max_{\boldsymbol{\theta}^*} \text{UCB}(\boldsymbol{\theta}^*) = \mu[\boldsymbol{\theta}^*] + \beta\sigma[\boldsymbol{\theta}^*]; \tag{5}$$

That is, the UCB favors regions with either high expected values $\mu[\boldsymbol{\theta}^*]$ or high uncertainty (for exploring scarcely evaluated regions). As such, we repeatedly use Eqs. (3), (4), and (5) to evaluate new points until reaching a certain number of iterations. Lastly, we embed the Bayesian optimization strategy into the adversarial training scheme, resulting in the Adversarial Bayesian Optimization (see Algorithm 1).

2.2   Training a Regressor for Quantifying Level of Motion

We now simulate the data set $\hat{\boldsymbol{\mathcal{X}}}_m$ to train the regressor q. To generate a simulated $\hat{x}$ and its corresponding ground truth y, we first randomly select $x \in \boldsymbol{\mathcal{X}}_c$ and sample y from $\mathcal{U}(0, 1)$. Then we construct $\hat{x} = s(\boldsymbol{x}; y\boldsymbol{\theta})$, i.e. an image with

Algorithm 1. Adversarial Bayesian Optimization (ABO)

Initialize $\boldsymbol{\theta}$

for $n = 1$ to N **do**

    Simulate $\hat{\boldsymbol{X}}_m$ by sampling MRI from $\boldsymbol{X}_c$ and sampling motion from $p(\boldsymbol{\tau}; \boldsymbol{\theta})$

    Train the discriminator D on $\boldsymbol{X}_m$ and $\hat{\boldsymbol{X}}_m$

    Bayesian Optimization to Update $\boldsymbol{\theta}$:

        for $t = 1$ to T **do**

            Evaluate $f(\boldsymbol{\theta}^t)$ via Eq. (2)

            Evaluate $\mu[\boldsymbol{\theta}^*]$ and $\sigma^2[\boldsymbol{\theta}^*]$ via Eqs. (3) and (4)

            Derive $\boldsymbol{\theta}^{t+1}$ by optimizing Eq. (5)

        end for

        $\boldsymbol{\theta} \leftarrow \arg\max(f(\boldsymbol{\theta}^1)...f(\boldsymbol{\theta}^T))$

end for=0

---

motion artifact at level y, and the maximum possible motion is guaranteed to be $\boldsymbol{\theta}$. Next, we train q on the simulated image and label pairs. As we intend to use q to rank the motion level across images, we design the training loss function as a combination of MSE loss and a rank loss [13]. Given two training images $\hat{\boldsymbol{x}}_1, \hat{\boldsymbol{x}}_2$ with $y_1 < y_2$, the objective function is

$$\arg\min_q \ (y_1 - q(\hat{\boldsymbol{x}}_1))^2 + (y_2 - q(\hat{\boldsymbol{x}}_2))^2 + \text{ReLU}(q(\hat{\boldsymbol{x}}_1) - q(\hat{\boldsymbol{x}}_2)), \qquad (6)$$

which encourages the predictions to follow the correct rank order.

3   Experimental Setup

Data Set. We used the T1-weighted MRI from the National Consortium on Alcohol and Neurodevelopment in Adolescence (NCANDA) [9,14], which acquired MRIs at 5 different sites across the US (denoted in the following as Site A through Site E). Up to year 6 of the study, all MRIs went through manual quality control, and 159 MRIs were labeled as 'unusable' due to motion artifacts, which were regarded as the real motion data set $\boldsymbol{X}_m$. To construct $\boldsymbol{X}_c$, we first collected the 831 MRIs from all subjects that were labeled as 'usable' by manual quality control[1]. Considering imperfect human labeling, we selected 200 MRIs with lowest Entropy Focus Criterion (EFC) as $\boldsymbol{X}_c$. All MRIs analyzed in this study had a resolution of $256 \times 256 \times 146$ or $256 \times 256 \times 160$ but were downsized to the size of 128×128 in the axial plane and cropped in the z-direction to reserve 128 axial slices containing the brain region.

Validation. To test if the proposed ABO can accurately imply the range of motion parameters associated with a data set, we applied it to simulated data sets with known ground-truth $\boldsymbol{\theta}$. We did so by setting all the 6 parameters $\{\theta_1...\theta_6\}$ to an equal number and varied this number in the range of $\{0.1, 0.2, 0.5, 1, 2, 4\}$. We then ran the optimization using a search range of $[0,$

---

[1] Only baseline scan was used if longitudinal scans were available.

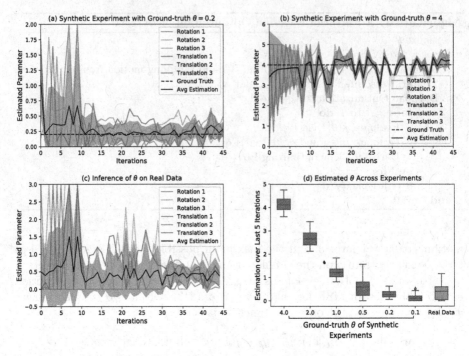

Fig. 2. (a–c). Estimated range of motion $\boldsymbol{\theta}$ along iterations of Adversarial Bayesian Optimization (ABO) for synthetic and real data sets. (d) Distribution of the 6 estimated motion parameters over the last 5 iterations of ABO for synthetic (blue) and real (orange) data sets. (Color figure online)

5] for each motion parameter, initialized $\boldsymbol{\theta}$ with 5 searches in that range, and performed additional search 45 times with $\beta = 2.58$ in the UCB, which decayed by a factor of 0.95 after the 5th iteration. After this validation, we applied ABO to infer $\boldsymbol{\theta}$ associated with the real motion set $\boldsymbol{\mathcal{X}}_m$.

Next, given the estimated $\boldsymbol{\theta}$, we simulated $\hat{\boldsymbol{\mathcal{X}}}_m$ containing 200 synthetic MRIs with ground-truth y. We additionally simulated 200 MRIs as a separate validation data set. To train the regressor q, we designed a simple feed-forward convolutional neural work containing 4 double convolutional layers of feature-channel dimension increasing from 32 to 256 with ReLU activations and max-pooling layers in between. Each convolutional layer used a kernel size of 3×3, stride of 1, and zero padding. Adam optimizer was used with a learning rate of 1e-4. After training, we applied q to the validation set to measure the prediction accuracy of y based on R^2 coefficients. We also compared the proposed loss function in Eq. (6) with the vanilla regression simply based on the MSE loss. Finally, we applied q to quantify the level of motion for the real data in $\boldsymbol{\mathcal{X}}_m$ and all usable baseline scans of NCANDA. The code for ABO and regression is available at https://github.com/anastasb/MRI-Motion-Artifact-Quantification.

4   Results

Results on Adversarial Bayesian Optimization. Figure 2 shows the inference of θ for synthetic data set $\hat{\mathcal{X}}_m$ with known ground-truth θ and for real data $\mathcal{X}_m$. When $\hat{\mathcal{X}}_m$ was constructed by ground-truth $\theta = 0.2$, all 6 estimated parameters converged around the ground-truth value with diminishing variance (Fig. 2(a)). The results replicated when we tested the optimization on $\hat{\mathcal{X}}_m$ with a larger ground-truth $\theta = 4$ (Fig. 2(b)). The first 6 columns in Fig. 2(d) indicate that the proposed ABO approach consistently recovered the correct range of motion θ for data sets with various ground-truth θ.

Next, we applied ABO to the real motion data set $\mathcal{X}_m$ to infer its underlying motion parameters θ (Fig. 2(c)), and the estimated parameters converged around 0.5. We also observe from Fig. 2(d) that the distribution of the 6 estimated parameters (over the last 5 iterations) for the real set $\mathcal{X}_m$ was comparable to the distribution for the simulated set $\hat{\mathcal{X}}_m$ with ground-truth $\theta = 0.5$. We therefore set $\theta = 0.5$ for all 6 motion parameters for the real motion set $\mathcal{X}_m$.

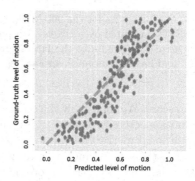

Fig. 3. Inference of motion level on the validation set.

Results on Motion Level Regression. After training q on a simulated set $\hat{\mathcal{X}}_m$ with $\theta = 0.5$, we tested it on the hold-out set of 200 simulated test images. Figure 3 shows that the predicted motion level generally aligned with the ground truth ($R^2 = 0.52$), although the prediction seemed to be overly sensitive for small motion artifacts (e.g., < 0.2). However, this result was still more accurate than the vanilla regression model that only incorporated the MSE loss but without using the ranking loss ($R^2 = 0.43$).

Finally, we applied the trained regressor to estimate the level of motion for all 990 MRIs. Figure 4 displays MRIs from each site with the highest and lowest

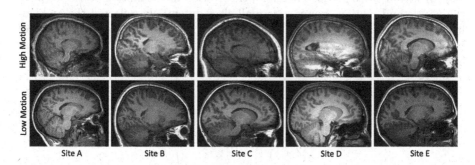

Fig. 4. MRIs from each site with the highest and lowest motion score derived by our regression model.

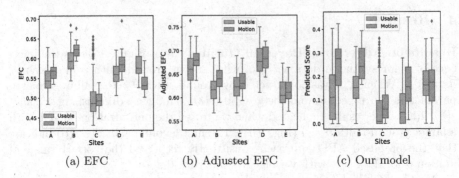

(a) EFC   (b) Adjusted EFC   (c) Our model

Fig. 5. Motion scores for the 990 MRIs of NCANDA defined by (a) EFC, (b) adjusted EFC (EFC computed over voxels with positive intensities), and (c) our model prediction. Our model results in the largest group separation in the scores for Sites A, B, and D, and has less variance across sites.

motion score. We then show the estimated scores for both usable MRIs and MRIs with motion across 5 sites in Fig. 5(c), and compared them with the scores defined by traditional metrics for motion quantification based on EFC (Fig. 5(a)+(b)). Note that the scale of motion level was incommensurate across approaches. We observe that the EFC-based scores exhibited a large variance across sites, which was significantly larger than the group difference between the usable and motion groups. This finding rendered the EFC metrics unsuitable for multi-site studies. On the other hand, the cross-site variance was largely reduced by our approach.

Compared to the EFC-based scores, the scores derived by the regressor network had a significantly larger separation between the usable and motion groups for sites A, B, and D ($p < 0.01$ two-way ANOVA). This indicates our approach could better quantify motion artifacts within MRIs. Note, although we expected a significant separation between the two groups, we did not regard the binary group assignment as ground-truth due to errors in manual quality control. This was supported by Fig. 6 showing two example MRIs defined as 'usable' in quality control but received high motion scores by our model. A follow-up visual inspection concluded that there were substantial motion artifacts (e.g., ringing effects) within these MRIs. Therefore, the overlap in the motion scores between the usable and motion groups across all sites potentially reflected the ambiguity and imperfection of manual labeling. Such noisy manual labels might prohibit effective training of supervised binary classification and supported that our regression-based framework might be more effective in quantifying motion artifacts.

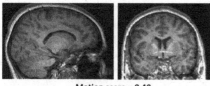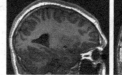

Motion score = 0.40   Motion score = 0.36

Fig. 6. Examples of MRIs which manual quality control labeled as usable but our estimator gave a high motion score.

5 Conclusion

We presented a framework for quantifying the severity of artifact induced by subject motion in T1-weighted MRIs. Being a part of the framework, we proposed Adversarial Bayesian Optimization, a novel approach for performing adversarial training when the generator component (i.e., the motion simulator in our context) is non-differentiable. The proposed motion-quantification framework can be used either as a first or second rater in quality control workflows. Here we used it as a second rater and found some instances of motion artifact which escaped manual quality control (Fig. 6). Our future research direction is to develop a method to improve the quality of MRIs with motion artifacts. In other words, given a motion-corrupted MRI, we aim to reconstruct the corresponding motion-free MRI. For that, the method presented herein was a prerequisite.

Acknowledgment. This research was supported in part by NIH U24 AA021697, MH119022, and Stanford HAI AWS Cloud Credit. The data were part of the public NCANDA data release NCANDA_PUBLIC_BASE_STRUCTURAL_V01 [15], whose collection and distribution were supported by NIH funding AA021697, AA021695, AA021692, AA021696, AA021681, AA021690, and AA02169.

References

1. Zaitsev, M., Maclaren, J., Herbst, M.: Motion artefacts in MRI: a complex problem with many partial solutions. J. Magn. Reson. Imaging: JMRI **42**, 887–901 (2015)
2. Lorch, B., Vaillant, G., Baumgartner, C., Bai, W., Rueckert, D., Maier, A.: Automated detection of motion artefacts in MR imaging using decision forests. J. Med. Eng. 1–9 (2017)
3. Reuter, M., Tisdall, M., Qureshi, A., Buckner, R., Kouwe, A., Fischl, B.: Head motion during MRI acquisition reduces gray matter volume and thickness estimates. NeuroImage **107**, 107–115 (2015)
4. Johnstone, T.: Motion correction and the use of motion covariates in multiple-subject fMRI analysis. Human Brain Mapp. **27**, 779–88 (2006)
5. Backhausen, L., Herting, M., Buse, J., Roessner, V., Smolka, M., Vetter, N.: Quality control of structural MRI images applied using freesurfer–a hands-on workflow to rate motion artifacts. Front. Neurosci. **10**, 1–10 (2016)

6. Atkinson, D., Hill, D., Stoyle, P., Summers, P., Keevil, S.: Automatic correction of motion artifacts in magnetic resonance images using an entropy focus criterion. IEEE Trans. Med. Imaging **16**, 903–910 (1997)
7. Küstner, T., et al.: Automated reference-free detection of motion artifacts in magnetic resonance images. Magn. Reson. Mater. Phys. Biol. Med. **31**, 243–256 (2017). https://doi.org/10.1007/s10334-017-0650-z
8. Fantini, I., Rittner, L., Yasuda, C., Lotufo, R.: Automatic detection of motion artifacts on MRI using deep CNN. In: 2018 International Workshop on Pattern Recognition in Neuroimaging (PRNI), pp. 1–4 (2018)
9. Brown, S.A., et al.: The national consortium on alcohol and neurodevelopment in adolescence (NCANDA): a multisite study of adolescent development and substance use. J. Stud. Alcohol Drugs **76**(6), 895–908 (2015)
10. Pérez-García, F., Sparks, R., Ourselin, S.: Torchio: a python library for efficient loading, preprocessing, augmentation and patch-based sampling of medical images in deep learning. Comput. Methods Programs Biomed. **208**, 106236 (2021)
11. Frazier, P.I.: A tutorial on Bayesian optimization arxiv:1807.02811 (2018)
12. Snoek, J., Larochelle, H., Adams, R.P.: Practical Bayesian optimization of machine learning algorithms (2012) arXiv:1206.2944
13. Chen, W., Liu, T.Y., Lan, Y., Ma, Z.M., Li, H.: Ranking measures and loss functions in learning to rank. In: Bengio, Y., Schuurmans, D., Lafferty, J., Williams, C., Culotta, A. (eds.) Advances in Neural Information Processing Systems, vol. 22, Curran Associates, Inc. (2009)
14. Pfefferbaum, A., et al.: Adolescent development of cortical and white matter structure in the NCANDA sample: role of sex, ethnicity, puberty, and alcohol drinking. Cereb. Cortex **26**, 4101–4121 (2015)
15. Pohl, K.M., Sullivan, E.V., Pfefferbaum, A.: The NCANDA_PUBLIC_BASE_STRUCTURAL_V01 data release of the national consortium on alcohol and neurodevelopment in adolescence (NCANDA). In: Sage Bionetworks Synapse (2017)

False Positive Suppression in Cervical Cell Screening via Attention-Guided Semi-supervised Learning

Xiaping Du, Jiayu Huo, Yuanfang Qiao, Qian Wang, and Lichi Zhang[✉]

School of Biomedical Engineering, Shanghai Jiao Tong University, Shanghai, China
lichizhang@sjtu.edu.cn

Abstract. Cervical cancer is one of the primary factors that endanger women's health, and Thinprep cytologic test (TCT) is the common testing tool for the early diagnosis of cervical cancer. However, it is tedious and time-consuming for pathologists to assess and find abnormal cells in many TCT samples. Thus, automatic detection of abnormal cervical cells is highly demanded. Nevertheless, false positive cells are inevitable after automatic detection. It is still a burden for the pathologist if the false positive rate is high. To this end, here we propose a semi-supervised cervical cell diagnosis method that can significantly reduce the false positive rate. First, we incorporate a detection network to localize the suspicious abnormal cervical cells. Then, we design a semi-supervised classification network to identify whether the cervical cells are truly abnormal or not. To boost the performance of the semi-supervised classification network, and make full use of the localizing information derived from the detection network, we use the predicted bounding boxes of the detection network as an additional constraint for the attention masks from the classification network. Besides, we also develop a novel consistency constraint between the teacher and student models to guarantee the robustness of the network. Our experimental results show that our network can achieve satisfactory classification accuracy using only a limited number of labeled cells, and also greatly reduce the false positive rate in cervical cell detection.

Keywords: Cervical cell detection · False positive reduction · Semi-supervised classification

1 Introduction

Cervical cancer is one of the most common cancers among women [2]. The disease can be prevented if diagnosed and intervened in the early stage through screening [14]. Thinprep cytologic test (TCT), which is a widely recognized screening technique for cervical cancer since the 1990s, can help the pathologist to find abnormal squamous cervical cells that may turn into cancer [1,9]. The abnormal squamous cervical cells include atypical squamous cells of undetermined

© Springer Nature Switzerland AG 2021
I. Rekik et al. (Eds.): PRIME 2021, LNCS 12928, pp. 93–103, 2021.
https://doi.org/10.1007/978-3-030-87602-9_9

significance (ASC-US), atypical squamous cells, cannot exclude HSIL (ASC-H), low-grade squamous intraepithelial lesion (LSIL), high-grade squamous intraepithelial lesion (HSIL), and invasive squamous cancer [13]. The normal or negative cervical cells, on the other hand, have no cytological abnormalities. Usually, there are thousands of squamous cervical cells in a single whole-slide TCT sample in screening. Finding abnormal cells in such a huge number of cells is essential but time-consuming, and is greatly influenced by the subjective judgment of the pathologist. Therefore, developing an automatic TCT screening system for finding abnormal cervical cells is highly demanded.

With the development of deep learning in image detection and classification fields [22,23], many attempts have been made in automatic abnormal cervical cell detection [11,15,17,18,20,21]. For example, Yi *et al.* [20] introduced an automatic cervical cell detection method based on DenseNet [6] with abundant data augmentation and data balancing strategies. Taha *et al.* [18] extracted features by a pre-trained Convolutional Neural Network (CNN) and sent the output features to an SVM classifier for the final classification. Rehman *et al.* [15] designed an automatic cervical cell classification system based on CNN, which can extract deep features of the cell image patch. But its input is limited to small single-cell patches, which is not practical for the scanned whole-slide images (WSI) containing thousands of cells. Sompawong *et al.* [17] utilized Mask R-CNN [3] to classify normal and abnormal cervical cells. However, the inputs require segmentation of the cell nucleus, which is also heavy workload in labeling. Li *et al.* [11] refined the detection results with a deep verification network. The classification model is a fully supervised ResNet [4] that needs massive labeled training data.

It is also worth mentioning that all aforementioned detection works can inevitably produce false positive results. Some of the normal cells are mistakenly classified as abnormal, and therefore require the pathologist for manual check to ensure the accuracy of the final diagnostic result. Our paper aims to suppress the false positives of the detected abnormal cells by developing a classification model for cells detected, so that the workload of such manual screening can be reduced. However, sufficient annotations are usually required when training a classification model with good generalization ability, but the cost of collecting such data is also high, sometimes even impractical.

To resolve this issue, our false positive suppressing method is developed based on the semi-supervised learning technology, which can train the model using only limited labeled data, and the rest as unlabeled. Some existing studies, such as Mean Teacher (MT) which consists of the teacher and student models [7,19], have already shown their significance. The parameters of the teacher model are updated from the student model through the exponential moving average (EMA) strategy. By the enforced consistency between two models, the MT method can therefore construct the model using both labeled and unlabeled data in our framework. Besides, we incorporate the predicted boxes of the detection network as the attention guidance information, which can guide the MT network to focus more on the target cells in the image. We also design a constrained loss to ensure the consistency of the attention maps from the two models after inputs with different perturbations fed to the model.

2   Method

The proposed framework is shown in Fig. 1. This framework includes two stages: one for detecting suspicious cervical cells and the other for the proposed false positive suppression network, which is based on the pre-stage doubtful cervical cells detection results.

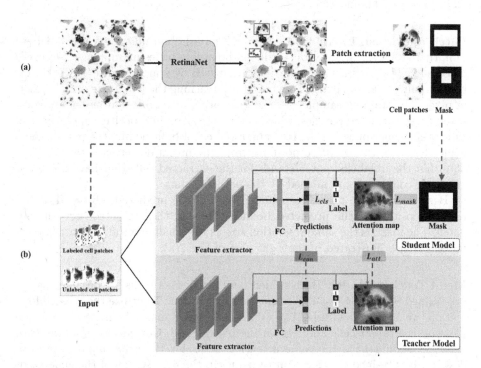

Fig. 1. Overview of the proposed framework. (a) The detection pipeline based on RetinaNet. (b) The detailed architecture of the proposed classification network. (Color figure online)

2.1   Suspicious Cervical Cells Detection

In the detection stage, we adopt RetinaNet [12] as our detection model, which has demonstrated its effectiveness in this field. Once the detection model is constructed, it can automatically locate the suspicious cervical cells by providing their bounding boxes with the corresponding confidence scores. The suspicious cervical cells consist of the true abnormal cases, and those false positive ones that are further eliminated in the subsequent stage. Note that the predicted bounding box indicates the precise location of each suspicious abnormal cervical cell in the form of the coordinate of the upper left corner, the height and the width of bounding box. In Fig. 1(a), examples of the predicted bounding boxes are shown in red.

2.2 False Positive Suppression Network

The pipeline of the novel semi-supervised network is shown in Fig. 1(b). The cervical cell patches and masks generated by the previous detection model are the inputs of the network, while the outputs are the probabilities of the normal/abnormal classification.

Patch and Mask Extraction. After the initial abnormal cervical cell detection, there are typically several predicted bounding boxes in an image. For each detected cell, we extract its corresponding patch by starting with the center of the bounding box, and expand outward until reaching the patch size of 224×224 in pixel. The training samples for the false positive suppression model are generally those extracted patches, which can be classified into positive and negative cases: the former ones are the true abnormal cells confirmed by the pathologist, while the latter are the false positives from the initial detection network. Also note that the unlabeled patch stands for the extracted cells that have not been manually confirmed and labeled by the pathologists.

Besides, we make the corresponding mask of the predicted cell based on its relative position of the predicted bounding box in the extracted cervical cell patch. The mask is the same resolution size as the small cervical cell patch. This process is also illustrated in the right part of Fig. 1(a).

Mean Teacher Model. Both student and teacher networks are designed using pre-trained SE-ResNeXt-50 [5] for feature extraction. To utilize the mask information, we adopt the Grad-CAM [16] mechanism to make the network to focus more on the target cells and less on the background. Considering that the two networks should have the same attention maps for the same inputs, we also design a constrained loss function to maintain the consistency of the generated attention maps from the two networks.

Let $D_L = \{(x_i, y_i)\}, i = 1, ..., M$ and $D_U = \{(x_i)\}, i = M + 1, ..., N$ denote the labeled dataset and the unlabeled dataset, respectively. Here x_i represents the cervical cell patch, and y_i represents the category label of the cervical cell patch. The MT model consists of the student and the teacher models. The parameters of the teacher model θ'_t at training step t are optimized by the Exponential Moving Average (EMA) [19] strategy:

$$\theta'_t = \alpha\theta'_{t-1} + (1 - \alpha)\theta_t, \tag{1}$$

where θ_t denotes the parameters of the student mode in the training step t, and θ_t is updated by the way of backward propagation. θ'_{t-1} denotes the parameter of the teacher model in the training step $t-1$, and α decays as the epoch increases.

The purpose of the MT framework is to enforce the final outputs of two models to be consistent with the same inputs under different perturbations. There are two basic losses to supervise the MT model: (1) the cross-entropy loss to minimize the predicted errors of labeled data in the student network, and (2) the consistency loss to enforce consistent predictions between the student and teacher model. The cross-entropy is written as:

$$L_{cls} = \frac{1}{M} \sum_{i=1}^{M} CE(y_i, f(x_i, \theta)), \qquad (2)$$

where $f(x_i, \theta)$ denotes the predictions of the student model and CE means cross entropy loss, and this supervised loss L_{cls} is only for labeled data. On the other hand, The consistency loss is written as:

$$L_{con} = \frac{1}{N} \sum_{i=1}^{N} MSE(f(x_i, \theta), f(x_i', \theta')), \qquad (3)$$

where $f(x_i', \theta')$ denotes the predictions of the teacher model, and MSE for the means mean square error.

Mask Guided Attention Map. As previously mentioned, there is one mask generated for every cervical cell patch, which is used to further improve the MT model by following an attention-guidance mechanism. Specifically, since we have the masks to locate the cells under study in the patches, we can make the attention map from the model to be constrained within the masks, and therefore make the model to focus on the target cells.

Note that since the mask is generated from the detected bounding box, there is only a rectangle to indicate the approximate location of the cell. Hence the mask is not the real shape of the cell, and we cannot use it directly to restrict the attention map. In this way, we design a relaxed constraint instead, which is $S(x) = 1/[1 + e^{-100(x-0.9)}]$ as the transformed sigmoid function. It can turn the value to be either close to 1 or 0 based on the threshold of 0.9. The loss function based on the mask guidance is written as follows:

$$L_{mask} = \frac{1}{M} \sum_{i=1}^{M} MSE(S(AM_i), Mask_i), \qquad (4)$$

where AM represents the attention map of the student model, and $S(AM)$ is the soft mask of one attention map after the transform of $S(x)$.

Attention Constraint. Because the inputs of the student and teacher networks are the same images, the principle of MT is to enforce the outputs of two models to be consistent. In this way, the attention maps both obtained from the student and teacher model should also be consistent. The final loss function between the two attention maps is:

$$L_{att} = \frac{1}{N} \sum_{i=1}^{N} \frac{1}{pq} \sum_{j} \sum_{k} MSE(AM(j,k), AM^{'}(j,k)), \tag{5}$$

where $AM(p,q)$ denotes the value of the attention map from the student model located at (j, k), $AM(j,k)^{'}$ denotes the value of the attention map from the teacher model located at (j,k), p and q represent the length and width of the attention map respectively, and N is the number of labeled and unlabeled cell patches.

Total Loss. The total loss for our framework is written as follows:

$$L_{total} = L_{cls} + \lambda(L_{con} + L_{mask} + L_{att}), \tag{6}$$

where λ is a loss weight calculated by the ramp-up function [10].

3   Experimental Results

Dataset. For the first stage of suspicious cervical cell detection, our training dataset includes 9000 images with the size of 1024 × 1024 pixels from WSIs. All abnormal cervical cells are manually annotated in the form of bounding boxes. The testing dataset has 2050 images, including 1000 positive images with abnormal cervical cells, and 1050 negative images that all cells are normal. Then, for the false positive suppression model, we have extracted the cell patches from both positive and negative images after the suspicious cell detection. We have prepared 5000 labeled positive cell patches and 5000 negative cell patches. The ratio of cell patches in the training, validation and testing dataset is 7:1:2.

Implementation Details. The backbone of the suspicious cell detection network is RetinaNet [12] with ResNet-50 [4]. The backbone model of the proposed semi-supervised classification network is SE-ResNeXt-50 [5]. Adam [8] is the optimizer of both detection and semi-supervised models. The started learning rate is $1e-4$. The consistency loss, mask loss and attention map loss are added into training at the 20th epoch. Besides, to increase the model robustness, some data augmentations such as horizontal and vertical flipping, brightness and contrast change are added for the semi-supervised learning.

Evaluation of the Proposed Semi-supervised Network. We compare the classification performance of the proposed semi-supervised network and the fully supervised (FS) method with different amount of labeled and unlabeled data. For fair comparison, the backbones of all methods are the same SE-ResNeXt-50.

We randomly divide the labeled data into 10 groups with equal image number, each group includes 700 true positive abnormal cell patches and 700 false positive abnormal cell patches. D_L is the number of groups that the labeled data are used in the experiment, and D_U stands for the unlabeled data. Since the fully supervised method can only use the labeled data, we increase the number of labeled data gradually to observe the improving classification performance. For the semi-supervised method, we remove the label information from 8 groups of labeled data to make them the same with the unlabeled data. Hence we use 2 groups of labeled data and 8 groups of unlabeled data to evaluate the advantages of the proposed method based on MT.

Table 1. Comparison of fully supervised classification and the proposed method based on MT.

Method $(D_L:D_U)$	ACC	AUC	Precison	F1
FS. (1:0)	0.749	0.828	0.750	0.749
FS. (2:0)	0.787	0.860	0.793	0.786
FS. (3:0)	0.805	0.877	0.806	0.805
FS. (4:0)	0.826	0.901	0.830	0.825
FS. (5:0)	0.841	0.912	0.843	0.841
FS. (6:0)	0.873	0.936	0.874	0.873
FS. (7:0)	0.885	0.943	0.886	0.885
FS. (8:0)	0.887	0.946	0.888	0.886
FS. (9:0)	0.890	0.945	0.890	0.889
FS. (10:0)	**0.893**	**0.949**	**0.893**	**0.892**
MT (2:8)	0.876	0.940	0.876	0.876
MT+L_{att} (2:8)	0.883	0.943	0.883	0.882
MT+L_{mask} (2:8)	0.880	0.942	0.880	0.880
MT+L_{att}+L_{mask} (2:8) (Proposed Method)	**0.886**	**0.946**	**0.886**	**0.886**

Table 1 shows the experimental results of the proposed method, the fully supervised method and the MT method. It is observed that the performance of fully supervised experiments improves with the increasing number of labeled data. The performance of the basic MT method using 2 groups of labeled data and 8 groups of unlabeled data is comparable with the fully supervised classification network using 6 groups of labeled data. Meanwhile, the proposed method with the same training data of MT method can be comparable with the fully supervised classification network using 8 groups of labeled data. This clearly shows that our method outperforms the MT method, and it also greatly reduces the cost of label data preparation compared with the fully supervised method.

We also visualize the generated attention maps of different networks in Fig. 2 to further demonstrate their effectiveness. It can be observed that both MT in Fig. 2(d) and the proposed method in Fig. 2(e) focus on the cells better than the fully supervised method with the same amount of the labeled data as in Fig. 2(b). Compared with MT, the second and last row of attention maps in (d) and (e) show that our method has better focus on the cells.

False Positives After Suppression. Table 2 shows the results with false positives after suppression or not. Our work aims to suppress the false positives in cervical abnormal cell detection with the improved semi-supervised classification network. After suspicious abnormal cell detection, we have only made classifications for the detected cells whose confidence score is higher than 0.1 because a large number of cells whose confidence scores under 0.1 have little value for clinical diagnosis.

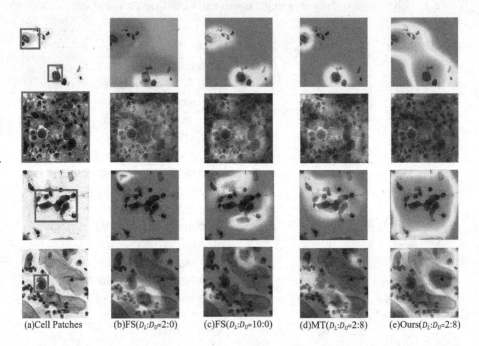

(a)Cell Patches   (b)FS(D_L:D_U=2:0)   (c)FS(D_L:D_U=10:0)   (d)MT(D_L:D_U=2:8)   (e)Ours(D_L:D_U=2:8)

Fig. 2. Attention maps of the proposed method and MT. (a) shows cell patches to be classified by the network and the red bounding box in (a) is the object cell that the network need to focus on. (b) shows attention maps from the fully supervised method (D_L: D_U = 2:0). (c) shows attention maps from the fully supervised method (D_L: D_U = 10:0). (d) shows attention maps from the MT (D_L: D_U = 2:8). (e) shows attention maps of the proposed method (D_L: D_U = 2:8).

In our experiment, we use the number of false positive cells per image to represent the FP suppression. FP_1 means the average number of false-positive cells per positive image, and FP_0 is the average number of false-positive cells per negative image. FP_{avg} means the average number of false-positive cells per image.

Table 2. Experimental results of cervical images after false positive suppressing.

Method	FP_1	FP_0	FP_{avg}	Precison	Recall	Map@IoU $= 0.5$
Retinanet	13.95	7.98	10.89	0.16	**0.85**	57.2%
Retinanet+FS.(10:0)	8.85	0.50	4.57	0.27	0.79	**61.2%**
Retinanet+FS.(2:0)	8.45	0.90	4.58	0.26	0.74	56.3%
Retinanet+MT(2:8)	**8.38**	0.75	4.47	0.26	0.75	57.9%
Retinanet+Proposed	8.68	**0.39**	**4.44**	**0.28**	0.78	60.4%

We define a predicted suspicious cell as a false positive cell if the IoU of its predicted bounding box and the annotated abnormal cell is less than 0.5. For positive images, the number of false-positive cells per image is 13.9, and the number decreases to 8.6 after our FP suppression. For negative images, the numbers are 8.0 and 0.4, which demonstrates the effectiveness of FP suppression specifically for the negative images. Though the recall of the test data is decreased from 0.85 to 0.78, it won't affect the final image classification performance since there are usually sufficient cells detected if the case is diagnosed as abnormal. As the differences of FP are greatly enlarged after our FP suppression, and negative cervical cell images account for about 90% in the actual clinical test situation, our method can greatly reduce the influences of detection errors, and aid the final diagnosis.

4   Conclusion

In our work, we propose a semi-supervised method based on MT. Our method can effectively suppress the false positive cells after detection and can achieve good performance with only limited labeled data. Besides, we introduce the detected bounding box as mask guidance and ensure the consistency between attention maps of teacher and student models. Our work has great value in potential clinical applications, and can also be further applied to other cell detection tasks in the computer-aided diagnosis of pathology images.

Acknowlegements. This work was supported by the National Natural Science Foundation of China (NSFC) grants (62001292), Shanghai Pujiang Program (19PJ1406800), and Interdisciplinary Program of Shanghai Jiao Tong University.

References

1. Abulafia, O., Pezzullo, J.C., Sherer, D.M.: Performance of ThinPrep liquid-based cervical cytology in comparison with conventionally prepared Papanicolaou smears: a quantitative survey. Gynecol. Oncol. **90**(1), 137–144 (2003)
2. Gultekin, M., Ramirez, P.T., Broutet, N., Hutubessy, R.: World health organization call for action to eliminate cervical cancer globally (2020)

3. He, K., Gkioxari, G., Dollár, P., Girshick, R.: Mask R-CNN. In: Proceedings of the IEEE International Conference on Computer Vision, pp. 2961–2969 (2017)

4. He, K., Zhang, X., Ren, S., Sun, J.: Deep residual learning for image recognition. In: Proceedings of the IEEE Conference on Computer Vision and Pattern Recognition, pp. 770–778 (2016)

5. Hu, J., Shen, L., Sun, G.: Squeeze-and-excitation networks. In: Proceedings of the IEEE Conference on Computer Vision and Pattern Recognition, pp. 7132–7141 (2018)

6. Huang, G., Liu, Z., Van Der Maaten, L., Weinberger, K.Q.: Densely connected convolutional networks. In: Proceedings of the IEEE Conference on Computer Vision and Pattern Recognition, pp. 4700–4708 (2017)

7. Huo, J., et al.: A self-ensembling framework for semi-supervised knee cartilage defects assessment with dual-consistency. In: Rekik, I., Adeli, E., Park, S.H., Valdés Hernández, M.C. (eds.) PRIME 2020. LNCS, vol. 12329, pp. 200–209. Springer, Cham (2020). https://doi.org/10.1007/978-3-030-59354-4_19

8. Kingma, D.P., Ba, J.: Adam: a method for stochastic optimization. arXiv preprint arXiv:1412.6980 (2014)

9. Koss, L.G.: Cervical (Pap) smear: new directions. Cancer 71(S4), 1406–1412 (1993)

10. Laine, S., Aila, T.: Temporal ensembling for semi-supervised learning. arXiv preprint arXiv:1610.02242 (2016)

11. Li, C., Wang, X., Liu, W., Latecki, L.J.: Deepmitosis: mitosis detection via deep detection, verification and segmentation networks. Med. Image Anal. 45, 121–133 (2018). https://doi.org/10.1016/j.media.2017.12.002. https://www.sciencedirect.com/science/article/pii/S1361841517301834

12. Lin, T.Y., Goyal, P., Girshick, R., He, K., Dollár, P.: Focal loss for dense object detection. In: Proceedings of the IEEE International Conference on Computer Vision, pp. 2980–2988 (2017)

13. Ndifon CO, A.E.G.: Atypical squamous cells of undetermined significance. StatPearls Publishing, Treasure Island (2020)

14. Patel, M.M., Pandya, A.N., Modi, J.: Cervical pap smear study and its utility in cancer screening, to specify the strategy for cervical cancer control. Natl. J. 2(1), 49 (2011)

15. Rehman, A.u., Ali, N., Taj, I., Sajid, M., Karimov, K.S., et al.: An automatic mass screening system for cervical cancer detection based on convolutional neural network. In: Mathematical Problems in Engineering 2020 (2020)

16. Selvaraju, R.R., Cogswell, M., Das, A., Vedantam, R., Parikh, D., Batra, D.: Grad-CAM: visual explanations from deep networks via gradient-based localization. In: Proceedings of the IEEE International Conference on Computer Vision, pp. 618–626 (2017)

17. Sompawong, N., et al.: Automated pap smear cervical cancer screening using deep learning. In: 2019 41st Annual International Conference of the IEEE Engineering in Medicine and Biology Society (EMBC), pp. 7044–7048 (2019)

18. Taha, B., Dias, J., Werghi, N.: Classification of cervical-cancer using pap-smear images: a convolutional neural network approach. In: Valdés Hernández, M., González-Castro, V. (eds.) MIUA 2017. CCIS, vol. 723, pp. 261–272. Springer, Cham (2017). https://doi.org/10.1007/978-3-319-60964-5_23

19. Tarvainen, A., Valpola, H.: Mean teachers are better role models: weight-averaged consistency targets improve semi-supervised deep learning results. arXiv preprint arXiv:1703.01780 (2017)

20. Yi, L., Lei, Y., Fan, Z., Zhou, Y., Chen, D., Liu, R.: Automatic detection of cervical cells using dense-cascade R-CNN. In: Peng, Y., et al. (eds.) PRCV 2020. LNCS, vol. 12306, pp. 602–613. Springer, Cham (2020). https://doi.org/10.1007/978-3-030-60639-8_50

21. Zhou, M., Zhang, L., Du, X., Ouyang, X., Zhang, X., Shen, Q., Luo, D., Fan, X., Wang, Q.: Hierarchical pathology screening for cervical abnormality. Comput. Med. Imaging Graph. **89**, 101892 (2021)

22. Zhou, M., et al.: Hierarchical and robust pathology image reading for high-throughput cervical abnormality screening. In: Liu, M., Yan, P., Lian, C., Cao, X. (eds.) MLMI 2020. LNCS, vol. 12436, pp. 414–422. Springer, Cham (2020). https://doi.org/10.1007/978-3-030-59861-7_42

23. Zou, Z., Shi, Z., Guo, Y., Ye, J.: Object detection in 20 years: a survey. arXiv preprint arXiv:1905.05055 (2019)

Investigating and Quantifying the Reproducibility of Graph Neural Networks in Predictive Medicine

Mohammed Amine Gharsallaoui[1,2] , Furkan Tornaci[1], and Islem Rekik[1]($\boxtimes$)

[1] BASIRA Lab, Faculty of Computer and Informatics,
Istanbul Technical University, Istanbul, Turkey
irekik@itu.edu.tr
[2] Ecole Polytechnique de Tunisie (EPT), Tunis, Tunisia
http://basira-lab.com

Abstract. Graph neural networks (GNNs) have gained an unprecedented attention in many domains including dysconnectivity disorder diagnosis thanks to their high performance in tackling graph classification tasks. Despite the large stream of GNNs developed recently, prior efforts invariably focus on boosting the classification accuracy while ignoring the model reproducibility and interpretability, which are vital in pinning down disorder-specific biomarkers. Although less investigated, the discriminativeness of the original input features -biomarkers, which is reflected by their learnt weights using a GNN gives informative insights about their reliability. Intuitively, the reliability of a given biomarker is emphasized if it belongs to the sets of top discriminative regions of interest (ROIs) using different models. Therefore, we define the first axis in our work as *reproducibility across models*, which evaluates the commonalities between sets of top discriminative biomarkers for a pool of GNNs. This task mainly answers this question: *How likely can two models be congruent in terms of their respective sets of top discriminative biomarkers?* The second axis of research in our work is to investigate *reproducibility in generated connectomic datasets*. This is addressed by answering this question: *how likely would the set of top discriminative biomarkers by a trained model for a ground-truth dataset be consistent with a predicted dataset by generative learning?* In this paper, we propose a reproducibility assessment framework, a method for quantifying the commonalities in the GNN-specific learnt feature maps across models, which can complement explanatory approaches of GNNs and provide new ways to assess predictive medicine via biomarkers reliability. We evaluated our framework using four multiview connectomic datasets of healthy neurologically disordered subjects with five GNN architectures and two different learning mindsets: (a) conventional training on all samples (resourceful) and (b) a few-shot training on random samples (frugal). Our code is available at https://github.com/basiralab/Reproducible-Generative-Learning.

Electronic supplementary material The online version of this chapter (https://doi.org/10.1007/978-3-030-87602-9_10) contains supplementary material, which is available to authorized users.

Keywords: Graph neural networks · Reproducibility · Brain connectivity graphs · Predictive medicine · Generative learning

1 Introduction

With the resurgence of artificial intelligence techniques in many medical fields such as network neuroscience, the adoption of deep learning models in neurological diagnosis and prognosis has increased dramatically [1–3]. In fact, deep learning models have demonstrated their power in representing complex and nonlinear relationships within data samples and tackle a wide variety of tasks such as classification and regression. This important breakthrough has led to their proliferation and generalization beyond Euclidean data structures and extension to other non-usual data types such as graphs. In particular, recent graph neural network (GNN) based architectures have been proposed to investigate the brain connectomes which are modeled as graphs where nodes and edges represent the different regions of interest (ROIs) and the connectivities between them, respectively [4–7]. However, despite the important progress in the last two years to boost the accuracy performance of GNN models [8,9], researchers have raised concerns about the *explainability* of these frameworks [10,11]. Examples include the lack of interpretability and *reproducibility* of the implemented architectures which makes it hard to draw generalizable clinical conclusions. To fill this gap, we need to tap into the learning process of these models and quantify the *reproducibility across distribution shifts and fractures between training and testing sets and the reliability of the clinical biomarkers across various experiments* (Fig. 1).

With the continual improvement and development of *predictive intelligence* in medicine, fundamentally designed to generate and predict data from minimal resources, one still needs to evaluate the reproducibility of biomarkers identified using generated clinical datasets in relation to those identified using ground truth datasets. In recent years, there has been a large stream of works developed to synthesize data, especially in the medical imaging field. This is motivated by their impact on alleviating the high costs of medical data acquisition. Most of the state-of-the-art works aiming to generate medical data from another rely on deep learning frameworks [12–14]. However, conventional deep learning methods do not achieve competitive performance when applied to non-Euclidian structures such as graphs. To mitigate this issue, [15] has proposed topoGAN which predicts a target graph from a graph in the source domain while ensuring the topological soundness of the predicted graphs in the target domain. Although this method outperformed a lot of previous tasks, its robustness in terms of biomarkers which are considered as discriminative by classifiers is not addressed. Consequently, there is a significant motivation for quantifying robustness of generative learning in terms of biomarkers reproducibility in comparison with ground-truth dataset. In fact, while known by their ability in capturing complex patterns in rich input data domain, GNNs can also be used to assess *how likely could predictive medicine preserve discriminativeness of biomarkers.* As a result, GNN models *can go beyond conventional classification boundaries and open novel angles to generalizable clinical interpretations of brain disorders.*

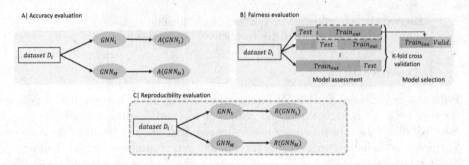

Fig. 1. *Diagram of the 3 evaluation settings used for classification models.* A| Given a set of M GNNs, typical comparison studies tend to select the model with the best performance in terms of accuracy without considering biomarker reproducibility. B| Given a dataset and a pool of GNNs, we perform the same data splits so that the models are trained, validated and tested on the same training, validation, test sets, respectively, to ensure fairness [16]. C| The proposed protocol for selection of the best GNN producing the most reproducible biomarkers while taking into account fairness.

Up to date, there is no such study in this domain. However, a limited number of studies have recently addressed the problem of reproducibility of traditional feature selection (FS) methods [17]. The generalization of this approach is not straightforward since GNNs have various architectures. Furthermore, unlike FS methods [18,19], GNNs do not directly generate features bijective to the set of edges and nodes in the original graph domain [10]. Moreover, since these methods are based on graph embedding, the dimensions and the shape of the data cannot be preserved throughout the whole forward propagation [20]. To circumvent these issues, by conceptualizing that FS methods extract the most discriminative features in the original data space, we can look at a weight learnt through a GNN as a discriminativeness measure of the proper original feature. To this aim, we are going to extract the learnt weights in the last embedding layer preserving the same original graph dimensions. Next, we construct a reproducibility graph modeling the relationship between GNNs. The weight of an edge corresponds to the overlap between the subsets of the most discriminative K nodes of a pair of GNNs. We identify the most reproducible GNN as the graph hub node.

To ensure the generalizability of the experiments and the trustworthiness of the selected biomarkers, we further evaluate our models with different settings. Specifically, we ran our models with two major learning mindsets. The first is conventional resourceful training where the model is trained on a whole set of samples with clear separation between validation and test sets [16]. The second method is frugal training which is based on few-shot learning approach [21]. In this paper, we propose the first graph-topology inspired framework to select the *most reproducible GNN model* in both *ground truth and synthesized connectomic datasets.* The main contributions of our work are:

1. Our framework unprecedentedly solves the problem of identifying the best GNN with the most reproducible biomarkers across multiple datasets having healthy and neurologically disordered subjects.

2. We assess reproducibility on two axis: ground-truth populations and predicted connectomes using generative learning.
3. We evaluate our models with different training strategies. We train our model in a resourceful manner using the conventional k-fold cross-validation. We also train our models in a frugal manner based on few-shot learning which is least investigated in graphs domain [21,22]. Based on this approach, we only perform the training on a limited number of samples.
4. It is able to quantify the reproducibility of clinical biomarkers through a deep learning process leading to better interpretability and trustworthiness assessment of such high-performance frameworks.
5. We perform our experiments in a fair manner leading to an unbiased comparison of the models as detailed in [16]. In more detail, the models are trained on the same data splits and each time the validation and the testing are performed in separate sets.

2   Proposed Method

In this section, we detail the main steps that constitute our proposed pipeline illustrated in (Fig. 2). We first train the different GNN architectures on a given brain connectivity dataset. After the training, we extract the learnt ROI weights respective to each GNN. Next, we identify the top discriminative biomarkers at different thresholds for each GNN. For each pair of GNNs, we define the reproducibility relationship as the overlap in their respective subset of most discriminative biomarkers. Next, we construct a reproducibility graph where a node represents a GNN architecture and an edge weight is quantified by the reproducibility calculated between a pair of GNNs. Finally, we identify the most reproducible GNN as the hub graph node.

Problem Statement. Let $\mathcal{D} = (\mathcal{G}, \mathcal{Y})$ denotes a dataset containing brain connectivity graphs with a set of brain neurological states to classify. Let $\mathcal{G} = \{\mathbf{G}_1, \mathbf{G}_2, \ldots, \mathbf{G}_n\}$ and $\mathcal{Y} = \{y_1, y_2, \ldots, y_n\}$ denote the set of the brain connectivity graphs and their labels, respectively. Each connectivity graph $\mathbf{G}_i$ is represented by a connectivity (adjacency) matrix $\mathbf{X}_i \in \mathbb{R}^{n_r \times n_r}$ and a label $y_i \in \{0, 1\}$ where n_r denotes the number of ROIs in the brain connectome.

Given a pool of M GNNs $\{GNN_1, GNN_2, \ldots GNN_M\}$, we are interested in training a GNN model $GNN_i : \mathcal{G} \rightarrow \mathcal{Y}$ using brain connectivity graph. We aim to identify the best GNN focusing on the most reproducible biomarkers differentiating between two brain states. To do so, we extract the weights vector $\mathbf{w}_i \in \mathbb{R}^{n_r}$ learnt by the ith GNN model, where $i \in \{1, 2, \ldots, M\}$.

Model Selection and Evaluation. To ensure fairness in the assessment of the models, we conducted separate model selection and evaluation steps as illustrated in (Fig. 1–**B**) and following the protocol detailed in [16]. The purpose of the model selection is to tune the hyperparameters based on the performance on a validation set. To do so, we make a holdout training where we separate a subset from the training set and exclude it from training and use it as a validation set

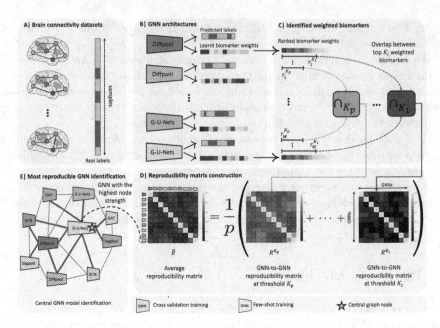

Fig. 2. *Overview of the proposed framework for identifying the most reproducible GNN in biomarker discovery.* **(A) Brain connectivity datasets.** We split our dataset into training, validation and testing sets with equal class proportions across subsets. **(B) GNN architectures.** We classify the dataset using five different GNN architectures. We use two training modes for each GNN. For each model, we extract the learnt weights respective of the biomarkers. **(C) Identified weighted biomarkers.** We extract the top-ranked biomarkers at different thresholds. **(D) Reproducibility matrix construction.** We calculate the overlap between the biomarkers weights vectors across models at each threshold. Next, we generate the average of the reproducibility matrices across the different thresholds. **(E) Most reproducible GNN identification.** We select the most reproducible GNN as the node having the highest strength in the reproducibility graph.

for the model selection. Next, we select the hyperparameters combination that brought the best results and use it in the model evaluation step where each model is assessed on multiple test sets via $k - fold$ cross-validation (CV). $k - fold$ CV uses k different training/test splits to evaluate the performance of the model. For each split, the model is tested on a subset of samples never used in the model selection step. This protocol is motivated by the fact that there are some issues in the clearness of the separation between both steps in different state-of-the-art deployed GNN implementations leading to biased and ambiguous results [16]. We also ensure label stratification in the different data partitions so that the class proportions are preserved across all the training/test/validation splits.

GNN Training Modes. To ensure the generalizability of the experiments we conduct two different training types: *resourceful* and *frugal*. The resourceful training is the conventional training approach based on $k - fold$ cross-validation protocol. It trains each GNN on all the training samples and follows the same fairness diagram illustrated in (Fig. 1). The model selection is performed via an independent holdout training. Next, the model is assessed on a separate test set never used in the model selection step. The frugal training is based on few-shot learning approach where only few samples are used for training. Afterwards, the model is then tested on all the remaining samples in the whole dataset.

Dataset Prediction. To generate brain connectivity datasets, we predict graphs in the target domain from subjects in the source domain. In our work, the different views of brain connectomes are considered as domains. Consequently, we predict a view from another. For this purpose, we adapt the method proposed in [15] to create the synthesized brain network views. This method consists of three major steps. First, we extract the source views and the target views from brain connectivity graphs. Next, the model creates an affinity matrix for each view from all the population graphs based on MKML [23]. Second, the model learns embeddings of population graphs using a GCN [24] based encoder and then cluster them into different groups to disentangle heterogeneity. In other terms, this step lies in mapping subjects into low-dimensional embeddings encoded in an affinity matrix that comprises all the population. Finally, using a set of cluster-specific generators and discriminators, the model reconstructs the brain graphs in the target domain. The generators and discriminator are regularized by local and global loss functions which preserve the topological properties of target domains. Next, we will identify the most reproducible GNN and therefore biomarkers in both ground truth and generated multi-view brain networks. Ideally, the predicted dataset will share similar biomarkers with the ground truth one by the most reproducible GNN architecture.

Biomarker Selection. In conventional classification GNNs, we extract the predicted labels to measure the classifier performance. Here, in addition to the accuracy, we extract the most relevant biomarkers in the original graphs to quantify the reproducibility of GNNs. Unlike traditional FS methods, the extraction of the most important biomarkers is not straightforward. Moreover, each GNN has its own architecture which makes the extraction of such information not generalizable. Furthermore, most of the GNNs use node and edge embeddings which affect the shape of the graph. To overcome these limitations, we aim to extract the most weighted biomarkers in the last embedding layer preserving the original graph dimensions. Given n_r biomarkers in each connectivity graph, we rank these biomarkers based on the absolute value of the respective embedding weight learnt throughout the neural network as illustrated in Algorithm 1. Based on that ranking, we extract $\mathbf{r}_i^{K_h}$ vector, containing the top K_h biomarkers weights learnt by the ith GNN where K_h denotes the threshold representing the top-ranked ROIs and $h \in \{1, \ldots p\}$. In this work, in addition to the prediction output of the classification GNNs, we also extract the region-wise weights learnt by a GNN (Fig. 2–**C**).

Definition 1. *Let $\mathbf{w}_i, \mathbf{w}_j \in \mathbb{R}^{n_r}$ denote two weight vectors learnt by GNN_i and GNN_j, respectively. We denote r_i^K and r_j^K as the two sets containing the* *regions corresponding to the top K elements in* $\mathbf{w}_i, \mathbf{w}_j$, *respectively. We define the reproducibility* $\mathbf{R}^K : \mathbb{R}^{n_r} \times \mathbb{R}^{n_r} \to \mathbb{R}$ *at threshold K as:* $\mathbf{R}^K(\mathbf{w}_i, \mathbf{w}_j) = \frac{|r_i^K \cap r_j^K|}{K}$.

Algorithm 1. GNN-to-GNN reproducibility matrix construction

1: **INPUTS:**
   Set of GNN specific weight absolute value vectors $\{\mathbf{w}_1, \mathbf{w}_2, \ldots, \mathbf{w}_M\}$ learnt on
   a training dataset
      of interest $\mathcal{D}$, threshold: K_h
2: **for** each pair of trained GNNs i and j in $\{1, \ldots, M\}$ **do**
3:     $r_i^{K_h} \leftarrow$ Top K_h regions from $\mathbf{w}_i$
4:     $r_j^{K_h} \leftarrow$ Top K_h regions from $\mathbf{w}_j$
5:     $\mathbf{R}_{ij}^{K_h} \leftarrow \frac{|r_i^{K_h} \cap r_j^{K_h}|}{K_h}$.
6: **OUTPUT:** GNN-to-GNN reproducibility matrix $\mathbf{R}^{K_h}$ at threshold K_h on the
   dataset $\mathcal{D}$

GNN-to-GNN Reproducibility Matrix. Given M GNN models and a threshold index h, we quantify the reproducibility between each pair of GNNs using the top K_h discriminative features. The reproducibility measure needs to reflect the commonalities between two models given their learnt biomarkers' weights. In other terms, this measure quantifies the similarities between two ranked weight vectors of two models. We define the reproducibility between two GNNs i and j as the overlap between their respective K_h top biomarkers as follows: $\mathbf{R}_{ij}^{K_h} = \frac{|r_i^{K_h} \cap r_j^{K_h}|}{K_h} \in \mathbb{R}$. The final reproducibility matrix $\mathbf{R}$ is of size $M \times M$, where M is the number of GNN models (Algorithm 1). For more generalizability, we introduce $\{\mathbf{R}_{ij}^{K_1}, \mathbf{R}_{ij}^{K_2}, \ldots \mathbf{R}_{ij}^{K_p}\}$ reproducibility matrices using different thresholds $\{K_1, K_2, \ldots K_p\}$ of the ranked weight vectors. Next, we generate the average reproducibility matrix $\bar{\mathbf{R}}$ by merging all the reproducibility matrices across the different p thresholds (Fig. 2–**D**).

Most Reproducible GNN Selection. To select the most reproducible GNN, we follow a graph theory based approach. The generated overall reproducibility matrix can be represented as a graph where its nodes are the different GNN architectures and the edges represent the reproducibility between pairs of models. Next, we rank the GNNs based on the node strength. We identify the most reproducible model as the node having the greatest strength in the reproducibility graph as follows: $GNN^* = \text{argmax}_i(s(i)) = \text{argmax}_i(\sum_{j=1}^{M} \bar{\mathbf{R}}_{ij})$, where $s(i)$ is the strength of the ith GNN node in the overall reproducibility graph (Fig. 2–**E**).

3   Results and Discussion

Evaluation Datasets. We evaluated our reproducibility framework on two large-scale datasets. The dataset (ASD/NC) includes 300 subjects equally partitioned between autism spectral disorder (ASD) and normal control (NC) states extracted from Autism Brain Imaging Data Exchange ABIDE I public dataset [25]. Each subject in each dataset is represented by two cortical brain networks (i.e., views). Their connectivities are respectively derived using the maximum principal curvature and cortical thickness network. For each dataset, we have brain connectivities from both right hemisphere (RH) and left hemisphere (LH). Each hemisphere in both datasets is labelled into 35 ROIs which are represented by the nodes of the graphs using Desikan-Killiany Atlas [26] and FreeSurfer [27]. For each ROI_i and for each cortical attribute, we calculate the average cortical measurement $\bar{a}_i$ across all its vertices. The connectivity value between ROI_i and ROI_j is the absolute distance between their average cortical attributes: $|\bar{a}_i - \bar{a}_j|$. These T1-w MRI processing steps to generate cortical morphological networks have been widely used in the network neuroscience literature to examine brain disorders, gender differences as well as healthy brain mapping [28–30].

Pool of GNNs. For our reproducibility framework, we used 5 state-of-the-art GNNs: DiffPool [31], GAT [32], GCN [24], SAGPool [33] and Graph-U-Net [34]. DiffPool learns a differential pooling to generate a hierarchical representation of an input graph. At each pooling layer, DiffPool learns a soft assignment of nodes into clusters which will be the nodes of the following layer [31]. GAT is originally designed for node classification. Here, we adapt it to perform graph classification task. GAT assigns different weights to the neighborhood to perform the aggregation in the following layer. GCN is also originally implemented for node classification task. Here, we adapt it to solve a graph classification problem by adding a linear layer that projects the node scores into one single score for the whole graph. GCN sequentially performs convolution operations using learnable weights that encode local graph structure and neighborhood features [24]. SAGPool uses graph pooling and unpooling depending on self-attention performed by graph convolutions [33]. Graph-U-Nets is based on a U-shape combining encoders and decoders to perform pooling and unpooling of the graph, respectively [34].

Training Settings. To generalize our experiments, beside the resourceful training approach based on the k- fold cross-validation, we also evaluated the GNNs with a frugal training approach based on few-shot learning. We used 5-fold cross-validation and performed 100 runs for the few-shot learning training selecting 4 samples (2 samples from each class) each run. Specifically, for each run, we used a different randomization for the training samples selection. We also used 3 thresholds for the top biomarkers extraction which are 5, 10 and 20.

Datasets Prediction. To evaluate the biomarker reproducibility in the generated brain connectomic dataset in comparison to ground truth dataset, we used topoGAN [15] to predict a target cortical view from a source one. Specifically, since we have two brain network views, each time, we select one to predict the other. Although our main purpose is to assess the reproducibility of a generative model, we also provide the efficiency of the trained model. Table 1 displays the topoGAN prediction results in terms of mean absolute error (MAE), Pearson

Table 1. *Prediction results of a target view from source view using different evaluation metrics.* View 1: maximum principal curvature. View 2: mean cortical thickness. MAE: mean absolute error. PCC: Pearson correlation coefficient. EC: eigenvector centrality. PC: PageRank centrality.

Dataset	View	MAE	PCC	MAE (EC)	MAE (PC)
ASD/NC (RH)	View 1	0.0803	0.0167	0.0267	0.0037
	View 2	0.0287	−0.0204	0.0149	0.0022
ASD/NC (LH)	View 1	0.0768	−0.0088	0.0267	0.0036
	View 2	0.0261	0.0107	0.0144	0.0026

correlation coefficient, MAE in eigenvector centralities and MAE in PageRank centralities.

Most Reproducible GNN. Both Fig. 3-**A** and **Fig** 4-**A** display the reproducibility scores for GNN models with ground-truth and generated ASD/NC datasets. For instance, Fig. 3-**A** shows that the most reproducible methods are SAGPool trained with few-shot learning and GAT for ground-truth and generated ASD/NC datasets using the maximum principal curvature measure, respectively. SAGPool and DiffPool trained with few-shot learning were identified as the most reproducible methods while using the mean cortical thickness for ASD/NC ground-truth and generated datasets, respectively. Figure 4-**A** displays the reproducibility scores for both views of ASD/NC left hemisphere datasets. For the first view, SAGPool trained with both few shots and cross-validation is the most reproducible method for both the real and predicted datasets, respectively. Using the second view, SAGPool and DiffPool trained with few shots were identified as the most reproducible methods using ground-truth and generated datasets. The most reproducible model is not necessarily the most accurate model. Since it reflects biomarker reliability, reproducibility is one aspect of model performance in addition to accuracy.

Most Reproducible Biomarkers. Both Fig 3-**B** and Fig 4-**B** display the weights respective to cortical regions (i.e., biomarkers) learnt by the most reproducible model using ground-truth and generated ASD/NC datasets. For the ASD/NC ground-truth RH and LH datasets, the inferior parietal cortex and unmeasured corpus callosum were identified as the most reproducible biomarkers using the first maximum principle curvature view. [35] found that the inferior parietal cortex is among the regions having similarity in deactivation behavior during some tasks between autism patients and their unaffected siblings. The unmeasured corpus callosum has been identified as the most reproducible biomarker for the ASD/NC RH dataset using the mean cortical thickness view. There have been many studies on the impact of this biomarker in relation to autism such as [36] who discussed the relationship between the size of corpus callosum and ASD. The transverse temporal cortex was also identified as the most reproducible biomarker for the ASD/NC LH using the second view. [37]

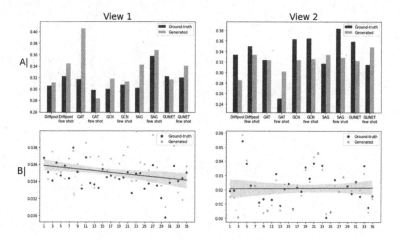

Fig. 3. *Reproducibility scores of models and weights of most reproducible model on ASD/NC right hemisphere dataset.* A| Reproducibility scores of models for both view 1 and view 2. B| Learnt weights of cortical regions with the most reproducible model for both views. ASD: autism spectrum disorder. NC: normal control. View 1: maximum principal curvature. View 2: mean cortical thickness.

suggested that abnormalities in the temporal cortex region reflect a fundamental early neural developmental pathology in autism.

Biomarker Reproducibility in Generated Connectomic Datasets. Figure 3-**B** and Fig. 4-**B** display a comparison between the learnt weights using the ground-truth and generated datasets. We can see that the distributions of weights using real and predicted data are close except for the first view of the ASD/NC left hemisphere dataset (Fig. 4-**B**). Moreover, the majority of the most discriminative biomarkers have changed with the generative learning except for ASD/NC right hemisphere using the second view. Both figures show that although the most discriminative biomarkers are not preserved through predictive medicine, the majority of the biomarkers have a close range of weights with generative learning. To sum up, generative learning can succeed partially in preserving the same distribution of the learnt feature maps using the most reproducible model. However, the most discriminative biomarkers have notably been affected by generative learning which is reflected by the discrepancy between real and predicted learned biomarker weights.

Limitations and Future Directions. Although our work has covered variations of different parameters and datasets to accurately evaluate reproducibility, it has some limitations. First, it does not cover all the information obtained from the cortical measurements. In our future work, it can be extended to other connectivity measures and other neurological disorders datasets to ensure more generalizability. Second, in this work, we only focused on reproducibility. We aim to jointly evaluate reproducibility and discriminativeness of biomarkers through generative learning. Finally, although we used 5 GNNs for the classification task, we have only used one model for the generative learning step. In our future work,

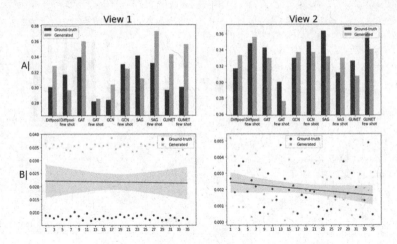

Fig. 4. *Reproducibility scores of models and weights of most reproducible model on ASD/NC left hemisphere dataset.* A| Reproducibility scores of models for both view 1 and view 2. B| Learnt weights of cortical regions with the most reproducible model for both views. ASD: autism spectrum disorder. NC: normal control. View 1: maximum principal curvature. View 2: mean cortical thickness.

we intend to encompass other generative learning models. This will also pave the way to comparison studies of predictive medicine models based on reproducibility.

4   Conclusion

In this paper, we evaluated the emerging GNN models from two angles: their reproducibility when using ground truth and synthesized datasets and in resourceful and frugal training settings. We proposed the first framework to identify and quantify the reproducibility of GNNs for neurological biomarker discovery in both real and generated datasets of neurological disorders. Our method takes advantage of the efficiency of GNNs which are known to be powerful on datasets of graphs in two training settings: few-shot and usual cross-validation training settings. We also used the node strength concept to identify the most reproducible GNN for a particular dataset. We used this evaluation to investigate the reproducibility of GNNs and the reliability of data prediction in biomarker reproduciblity. This is useful to evaluate the reliability of the models through inter-domain medical data predictions. Our framework is generic and can be used with different neurological disorder datasets for various tasks. Therefore, in our future work, we will generalize our framework to other disorders with various connectivity measures and new predictive medicine models.

Acknowledgements. This work was funded by generous grants from the European H2020 Marie Sklodowska-Curie action (grant no. 101003403, http://basira-lab.com/normnets/) to I.R. and the Scientific and Technological Research Council of Turkey to

I.R. under the TUBITAK 2232 Fellowship for Outstanding Researchers (no. 118C288, http://basira-lab.com/reprime/). However, all scientific contributions made in this project are owned and approved solely by the authors. M.A.G is supported by the same TUBITAK 2232 Fellowship.

References

1. Marblestone, A.H., Wayne, G., Kording, K.P.: Toward an integration of deep learning and neuroscience. Front. Comput. Neurosci. **10**, 94 (2016)
2. Richards, B.A., et al.: A deep learning framework for neuroscience. Nat. Neurosci. **22**, 1761–1770 (2019)
3. Storrs, K.R., Kriegeskorte, N.: Deep learning for cognitive neuroscience. arXiv preprint arXiv:1903.01458 (2019)
4. Bessadok, A., Mahjoub, M.A., Rekik, I.: Graph neural networks in network neuroscience. arXiv preprint arXiv:2106.03535 (2021)
5. Guye, M., Bettus, G., Bartolomei, F., Cozzone, P.J.: Graph theoretical analysis of structural and functional connectivity MRI in normal and pathological brain networks. Magn. Reson. Mater. Phys. Biol. Med. **23**, 409–421 (2010)
6. van den Heuvel, M.P., Sporns, O.: A cross-disorder connectome landscape of brain dysconnectivity. Nat. Rev. Neurosci. **20**, 435–446 (2019)
7. Corps, J., Rekik, I.: Morphological brain age prediction using multi-view brain networks derived from cortical morphology in healthy and disordered participants. Sci. Rep. **9**, 1–10 (2019)
8. Xu, K., Hu, W., Leskovec, J., Jegelka, S.: How powerful are graph neural networks? arXiv preprint arXiv:1810.00826 (2018)
9. Dwivedi, V.P., Joshi, C.K., Laurent, T., Bengio, Y., Bresson, X.: Benchmarking graph neural networks. arXiv preprint arXiv:2003.00982 (2020)
10. Pope, P.E., Kolouri, S., Rostami, M., Martin, C.E., Hoffmann, H.: Explainability methods for graph convolutional neural networks. In: Proceedings of the IEEE/CVF Conference on Computer Vision and Pattern Recognition, pp. 10772–10781 (2019)
11. Yuan, H., Yu, H., Gui, S., Ji, S.: Explainability in graph neural networks: a taxonomic survey. arXiv preprint arXiv:2012.15445 (2020)
12. Li, R., et al.: Deep learning based imaging data completion for improved brain disease diagnosis. In: Golland, P., Hata, N., Barillot, C., Hornegger, J., Howe, R. (eds.) MICCAI 2014. LNCS, vol. 8675, pp. 305–312. Springer, Cham (2014). https://doi.org/10.1007/978-3-319-10443-0_39
13. Ben-Cohen, A., et al.: Cross-modality synthesis from CT to pet using FCN and GAN networks for improved automated lesion detection. Eng. Appl. Artif. Intell. **78**, 186–194 (2019)
14. Goodfellow, I.J., et al.: Generative adversarial networks. arXiv preprint arXiv:1406.2661 (2014)
15. Bessadok, A., Mahjoub, M.A., Rekik, I.: Brain multigraph prediction using topology-aware adversarial graph neural network. Med. Image Anal. **72**, 102090 (2021)
16. Errica, F., Podda, M., Bacciu, D., Micheli, A.: A fair comparison of graph neural networks for graph classification. arXiv preprint arXiv:1912.09893 (2019)
17. Georges, N., Mhiri, I., Rekik, I., Initiative, A.D.N., et al.: Identifying the best data-driven feature selection method for boosting reproducibility in classification tasks. Pattern Recogn. **101**, 107183 (2020)

18. Líu, H., Motoda, H., Setiono, R., Zhao, Z.: Feature selection: an ever evolving frontier in data mining. In: Feature Selection in Data Mining, pp. 4–13 (2010)
19. He, Z., Yu, W.: Stable feature selection for biomarker discovery. Comput. Biol. Chem. **34**, 215–225 (2010)
20. Zhang, C., Song, D., Huang, C., Swami, A., Chawla, N.V.: Heterogeneous graph neural network. In: Proceedings of the 25th ACM SIGKDD International Conference on Knowledge Discovery & Data Mining, pp. 793–803 (2019)
21. Garcia, V., Bruna, J.: Few-shot learning with graph neural networks. arXiv preprint arXiv:1711.04043 (2017)
22. Kim, J., Kim, T., Kim, S., Yoo, C.D.: Edge-labeling graph neural network for few-shot learning. In: Proceedings of the IEEE/CVF Conference on Computer Vision and Pattern Recognition, pp. 11–20 (2019)
23. Wang, B., Ramazzotti, D., De Sano, L., Zhu, J., Pierson, E., Batzoglou, S.: SIMLR: a tool for large-scale single-cell analysis by multi-kernel learning. bioRxiv, p. 118901 (2017)
24. Kipf, T.N., Welling, M.: Semi-supervised classification with graph convolutional networks. arXiv preprint arXiv:1609.02907 (2016)
25. Di Martino, A., et al.: The autism brain imaging data exchange: towards a large-scale evaluation of the intrinsic brain architecture in autism. Mol. Psychiatry **19**, 659–667 (2014)
26. Fischl, B., et al.: Sequence-independent segmentation of magnetic resonance images. Neuroimage **23**, S69–S84 (2004)
27. Fischl, B.: Freesurfer. Neuroimage **62**, 774–781 (2012)
28. Soussia, M., Rekik, I.: Unsupervised manifold learning using high-order morphological brain networks derived from T1-w MRI for autism diagnosis. Front. Neuroinform. **12**, 70 (2018)
29. Chaari, N., Akdağ, H.C., Rekik, I.: Estimation of gender-specific connectional brain templates using joint multi-view cortical morphological network integration. Brain Imaging Behav. 1–20 (2020)
30. Sserwadda, A., Rekik, I.: Topology-guided cyclic brain connectivity generation using geometric deep learning. J. Neurosci. Methods **353**, 108988 (2021)
31. Ying, R., You, J., Morris, C., Ren, X., Hamilton, W.L., Leskovec, J.: Hierarchical graph representation learning with differentiable pooling. arXiv preprint arXiv:1806.08804 (2018)
32. Veličković, P., Cucurull, G., Casanova, A., Romero, A., Lio, P., Bengio, Y.: Graph attention networks. arXiv preprint arXiv:1710.10903 (2017)
33. Lee, J., Lee, I., Kang, J.: Self-attention graph pooling. In: International Conference on Machine Learning, pp. 3734–3743 (2019)
34. Gao, H., Ji, S.: Graph U-nets. In: International conference on machine learning, pp. 2083–2092 (2019)
35. Spencer, M.D., et al.: Failure to deactivate the default mode network indicates a possible endophenotype of autism. Mol. Autism **3**, 1–9 (2012)
36. Lefebvre, A., Beggiato, A., Bourgeron, T., Toro, R.: Neuroanatomical diversity of corpus callosum and brain volume in autism: meta-analysis, analysis of the autism brain imaging data exchange project, and simulation. Biol. Psychiat. **78**, 126–134 (2015)
37. Eyler, L.T., Pierce, K., Courchesne, E.: A failure of left temporal cortex to specialize for language is an early emerging and fundamental property of autism. Brain **135**, 949–960 (2012)

Self Supervised Contrastive Learning on Multiple Breast Modalities Boosts Classification Performance

Shaked Perek[✉], Mika Amit, and Efrat Hexter

IBM Research, Haifa University, Mount Carmel, 31905 Haifa, Israel
{shaked.perek,mika.amit,efrathex}@il.ibm.com

Abstract. Medical imaging classification tasks require models that can provide high accuracy results. Training these models requires large annotated datasets. Such datasets are not openly available, are very costly, and annotations require professional knowledge in the medical domain. In the medical field specifically, datasets can also be inherently small. Self-supervised methods allow the construction of models that learn image representations on large unlabeled image sets; these models can then be fine-tuned on smaller datasets for various tasks. With breast cancer being a leading cause of death among women worldwide, precise lesion classification is crucial for detecting malignant cases. Through a set of experiments on 30K unlabeled mammography (MG) and ultrasound (US) breast images, we demonstrate a practical way to use self-supervised contrastive learning to improve breast cancer classification. Contrastive learning is a machine learning technique that teaches the model which data points are similar or different by using representations that force similar elements to be equal and dissimilar elements to be different. Our goal is to show the advantages of using self-supervised pre-training on a large unlabeled set, compared to training small sets from scratch. We compare training from scratch on small labeled MG and US datasets to using self-supervised contrastive methods and supervised pre-training. Our results demonstrate that the improvement in biopsy classification using self-supervision is consistent on both modalities. We show how to use self-supervised methods on medical data and propose a novel method of training contrastive learning on MG, which results in higher specificity classification.

Keywords: Self-supervised · Mammography · Ultrasound · Contrastive learning

1 Introduction

Deep learning methods have become the standard tool in solving common imaging problems such as segmentation and classification. However, these methods require a large amount of annotated data to perform well and train a robust network. In a real-world scenario, especially in the medical field, acquiring large

© Springer Nature Switzerland AG 2021
I. Rekik et al. (Eds.): PRIME 2021, LNCS 12928, pp. 117–127, 2021.
https://doi.org/10.1007/978-3-030-87602-9_11

quantities of labeled data is difficult. Datasets are usually small and annotations are expensive. In some cases, medical data are scarce, for example when it comes to rare diseases. In other cases, medical labels, whether they are global or local, may be subject to a professional's opinion. These labels can be noisy and degrade the quality of supervised training.

Recent developments in the field of deep learning have the potential to offer a smarter way to train networks, without requiring labels and annotations. Self-supervised networks can learn representations from unlabeled datasets, by learning various image tasks. Contrastive learning is a machine learning technique that learns the general features of a dataset without labels, by teaching the model which data points are similar or different. It aims to learn representations by forcing similar elements to be equal and dissimilar elements to be different. New methods such as SimCLR [4] and Momentum Contrast (MoCo) [5,8] both used contrastive learning on ImageNet data and show results that are in the same league as those for supervised training.

Previous works on medical imaging classification demonstrate the wide range of self-supervised method applications in this field. Jamaludin et al. [10] use contrastive loss on longitudinal spinal MRI data to recognize same patient scans in a self-supervised task, with an auxiliary task of predicting vertebral body level: their final task is classifying disk degradation. Another work by Zhou et al. [22] compares their self supervised framework, which is similar to MoCo and trained on chest X-rays, to other self supervised methods trained on ImageNet. In a recent work by Azizi et al. [1] on two medical datasets for chest X-ray and dermatology images, the authors adapted SimCLR with a multi-instance contrastive learning modification that compares two images of the same patient as two positive representations. Their approach improved disease classification and outperformed ImageNet pre-trained baselines.

Breast cancer is one of the most common forms of cancers among women worldwide [2]. Early detection of suspicious findings by breast screenings can improve the survival rate of patients [19]. If we can reduce the radiologists' workload by developing high-sensitivity models that support decision-making, such as whether to send a suspicious finding for biopsy, we can help improve the process for healthcare providers and patients.

Developing classification models for breast cancer screening requires large datasets, such as in state-of-the-art works [14,16], annotated with a biopsy proven results. Most healthcare providers use mammography (MG) in screening examinations; however, other modalities such as ultrasound (US) exams, are less frequently conducted as they are not part of the screening exam. They are used in cases where a patient has an irregular finding; the results are then forwarded for further examination. This makes the task of collecting a large dataset even more challenging.

Two prior works classify tumors as malignant or benign on a small US dataset [17,20] using supervised methods of CNN and deep polynomial networks. However, to achieve good results, they crop the image around a given mass. Such local annotations are not always available. Transfer learning for the binary classification of MG lesions to benign/malignant is also a common practice [11,21]. These works use pre-trained weights from ImageNet, an image set that is not focused on medical images. However, previous works have shown that a preferable approach is transfer learning from a similar domain [7,16].

In this work, we explore the scenario of having many unlabeled breast images and a small set of labeled data. To the best of our knowledge, this is the first time self-supervised contrastive learning with MoCo has been used on MG and US images. We show its advantages in a common medical research use-case, where a larger image set with no outcome is more easily attainable. Our contribution is threefold: (i) We demonstrate the redundancy in large annotated sets, by conducting three biopsy classification experiments on a small set of data: training from scratch using randomly initialized weights, fine-tuning a supervised task trained on the same modality, and fine-tuning the self-supervised network. (ii) We propose an upgrade to MoCo that provides higher specificity for biopsy classification in MG images. (iii) We modify MoCo in a way that enables training on medical data.

2   Methods

We briefly describe Momentum Contrast for unsupervised learning [8], which is used in this paper. Given an input image X, which undergoes two random augmentations X_q, X_k, a query representation is defined as $q = f_q(X_q)$, where f is an encoder network. Similarly, a key representation is defined as $k = f_k(X_k)$. Contrastive learning brings representations of similar elements, (q, k) closer, and set dissimilar examples apart. Similar pairs are defined as "positive examples" whereas other pairs are defined as "negative examples". MoCo employs a queue, Q, of keys representations, detached from the minibatch size. The size of Q is large and it is filled with k at the end of each minibatch. Comparing q with Q gives the negative examples. MoCo initializes encoder f_k with f_q weights, however, during training f_q is updated using the gradients while f_k is updated with momentum update:

$$\theta_k = m\theta_k + (1 - m)\theta_q \tag{1}$$

Where θ_k, θ_q are the parameters of encoders f_k, f_q, respectively. Contrastive loss is used to measure similarity as defined by InfoNCE [15]

$$L_q = -\log \frac{\exp(q \cdot k_+/\tau)}{\sum_{i=0}^{K} \exp(q \cdot k_i/\tau)} \tag{2}$$

where K is the queue size including a single positive key, and k_+ is the positive of q. τ is a hyper-parameter that influences the spread of the distribution. The method, unlike SimCLR [4] enables training on a standard GPU by allowing the use of a smaller batch size.

We apply the self-supervised contrastive learning task, with a model backbone similar to MoCo-V2 [5]. We then define a three-class classification task of a full image, for two modalities, dividing it into negative-benign lesions, positive-malignant lesions, and normal healthy breast tissue that was not sent for biopsy. We used global biopsy proven labels per image. In ambiguous cases where two lesions were present, we used the more severe diagnosis for the classification, resulting in a single label per image.

2.1 Ultrasound

A breast US study consists of multiple slices, acquired in a single exam. Each slice may contain different information depending on the probe location on the breast. In suspicious cases where a lesion was sent for biopsy, some slices in the study may not even contain the lesion. However, all slices have a similar appearance. Our US dataset included a large in-house set of 30K unlabeled gray scale images used for self-supervised training. We used a smaller annotated set of roughly 2K images to fine-tune the MoCo backbone on the supervised task of biopsy classification. We also used the small annotated set to train a supervised ResNet50 [9] network from scratch. All of the classification tasks had three classes: negative, positive, and normal as previously defined. Examples of US images from each class are presented in Fig. 1.

In medical images, small subtle patterns can be important to the classification task and resizing can degrade results. Thus the input images were kept in their original resolution, but padded to a fixed size of 650 × 850.

2.2 Mammography

An MG study usually comprises four images: two for each view of Craniocaudal (CC) and Mediolateral-Oblique (MLO), for two breasts. The views share information regarding the breast. For example, if a lesion is present in one view it will be visible in the other. However, the general look of each view is very different as can be seen in Fig. 3. We define an adapted version of MoCo-V2, called MoCo-View, which is more suitable for dealing with the differences in each breast view. In the standard MoCo settings, each image in the query (q) batch is compared to all images accumulated in the queue, to set aside the

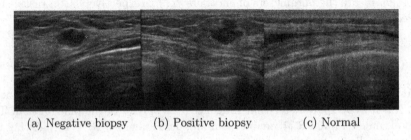

(a) Negative biopsy   (b) Positive biopsy   (c) Normal

Fig. 1. Ultrasound image examples. Labeled images from our in-house subset. Images are padded to the same shape of 650 × 850 if needed.

negative examples. However, in a two-view scenario, distinguishing between a CC and an MLO image is easy and there is no reason for the network to learn more complicated patterns. MoCo-View consists of two queues: Q_{CC} and Q_{MLO}. Each time a minibatch enters the queue, only images from the matching view are inserted in the respective queue. A detailed flow is depicted in Fig. 2.

For each one of the queues Q_{CC} and Q_{MLO}, we define the set of encoded keys $\{k_0^C, k_1^C, k_2^C, ...\} \in Q_{CC}$ and $\{k_0^M, k_1^M, k_2^M, ...\} \in Q_{MLO}$, respectively. These samples are considered the negative examples for comparing q when the viewpoint of q is CC or MLO, respectively. We use contrastive loss as in Eq. 2 with a minor modification:

$$L_q = \begin{cases} L_{q^{cc}} = -\log \frac{\exp(q^{cc} \cdot k_+/\tau)}{\sum_{i=0}^K \exp(q^{cc} \cdot k_i^*/\tau)}, & k_i^* \in Q_{cc} \\ L_{q^{mlo}} = -\log \frac{\exp(q^{mlo} \cdot k_+/\tau)}{\sum_{i=0}^K \exp(q^{mlo} \cdot k_i^*/\tau)}, & k_i^* \in Q_{mlo} \end{cases}, \quad (3)$$

where k_+ is the positive of q. Here, k_i^* is k_i^C when q's viewpoint is CC and k_i^M when its viewpoint is MLO. The sum K is over all negative samples and one positive.

The separation of the queues, enables a comparison of each view to similar views, thus making it more challenging for the network to distinguish between them, and thereby forcing the network to learn better representations. This extra label of an image view is easy to come by in MG, since images are in DICOM format, which holds extra information regarding the image. Examples of MG images from both views and each class are shown in Fig. 3.

The MG data used for training MoCo, consists of 30K images from 5 different providers. The encoder part is fine-tuned for the downstream classification task with ~5K sub-sections of the data, labeled as negative/positive/normal images. The same sub-section is also used to train a modified Inception-ResNet-V2 network [18] from scratch. The network is composed of 14 Inception-ResNet blocks that are concatenated to a global max pooling layer, followed by a convolutional classification head (defined in Sect. 3.2). Due to the very large size of MG images

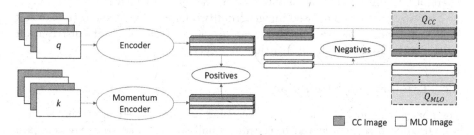

Fig. 2. MoCo-View pipeline. An input minibatch consists of images from two views, MLO in white and CC in blue. Two augmentations of the minibatch, q and k, are embedded using an encoder and momentum encoder, respectively. These representations are used to measure similarity between positive examples. Then, q representations are split into CC (blue) and MLO (white). Each of them is compared to its respective queue, and together they comprise the negative examples. (Color figure online)

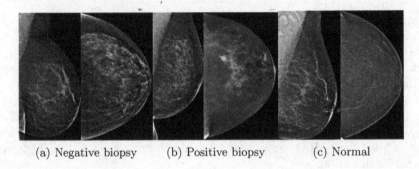

(a) Negative biopsy   (b) Positive biopsy   (c) Normal

Fig. 3. Mammography image examples. CC and MLO views examples for each class. MLO to the left and CC to the right. We use global labels with images resized to 224 × 224.

(400 × 3000), the input to both networks is resized to 224 × 224. Both the MoCo and MoCo-View encoder networks are also changed to the modified inception.

3   Experiments

3.1   Data

The data we used for both modalities is globally annotated with a biopsy proven outcome per image. We divided the MG data into two parts, with 80% train and 20% set aside for validation. All training data were used for the MoCo and MoCo-View training. The labeled subset used for fine-tuning was taken from the train set and we evaluated our model on the labeled part of the validation set. We trained the classification task for MG on 4 classes: negative, negative high-risk (HR), positive, and normal. Negative HR is a small class, containing 32 cases in the validation split. Although using this class helps the network learn better features, the end result is noisy and unreliable due to the set's small size. Therefore, the final result is combined with the larger negative class.

The US train set comprised images from two large providers and the labeled set for fine-tuning was from a third small provider. All labeled data distribution is detailed in Table 1 with total patient per class.

3.2   Training and Evaluation

Our goal is to show the advantages of using self-supervised pre-training on a large unlabeled set, compared to training small sets from scratch. We compare four models for MG modality with Inception-ResNet-V2 backbone model in each experiment: (1) Randomly initialized weights trained on a small labeled set. (2–3) Self-supervised using MoCo-View and MoCo-V2 trained on a large set without the use of labels and fine-tuned on the labeled small set. (4) Supervised pre-training of the large 30K MG set, on a binary classification task of normal cases versus cases sent for biopsy, then fine-tune using the small set. The last model is

Table 1. Data division for ultrasound and mammography fine tuning datasets.

	Train (# Patients/# Images)			Validation (# Patients/# Images)		
	Negative	Positive	Normal	Negative	Positive	Normal
US	119/320	124/372	387/1048	36/78	25/63	90/274
MG	1225/1404	1081/1720	1283/1602	252/371	278/433	327/400

used to set an upper-bound of performance for the task of 3-class classification using the small set.

The unsupervised training process comprises two steps: (1) Training an unsupervised encoder model, using a large unlabeled dataset. (2) Fine-tuning the encoder on a classification task using a smaller labeled set of the data.

We compare only two models for US, since US images are all taken from the same view and our larger set is completely unlabeled: (1) ResNet-50 with randomly initialized weights trained on a small set. (2) MoCo-V2, with ResNet-50 backbone model, trained on a large set without the use of labels, and fine-tuned on the small set.

We needed a few modifications to use MoCo-V2 on our medical dataset. In all our experiments, the MoCo hyperparameters were adjusted. The queue size K was adjusted to 1024, since our unlabeled sets have only approximately 30K images. τ was set to 0.2. The projection head of MoCo-V2 was as described in [5], consisting of a fully connected layer (FC), ReLU, and final FC. However, breast cancer classification is challenging and has more subtle features compared to natural images. Therefore, in the fine-tuning task, unlike MoCo's linear classification, we trained all layers without freezing and replaced the final FC layer with a convolutional classification head, consisting of 1×1 convolution with 256 filters, ReLU, dropout, and a final 1×1 convolution with the number of class channels. We used a minibatch of 4 for the US set to accommodate the use of the original resolution, and a minibatch of 64 for the resized MG images. An automated breast extraction procedure that zeros out the background and crops the image around the breast was applied to all inputs. For MG, we maintained the aspect ratio before resizing to 224×224 by zero-padding the smaller axis. We chose a model per modality according to the best performance each modality presented: ResNet-50 for US and Inception-ResNet for MG.

We also modified the augmentations to meet the limitations for the size of medical images. We resized the images to a constant shape, and used a random crop and scale as was done in the MoCo paper. We added random rotations with a degree range of $[-10, 10]$, random color jitter, random Gaussian blur, and random horizontal flip. We skipped the random gray scale since our input was already a single channel. Finally, we normalized the input range to $[-1, 1]$. The other parameters for MoCo and MoCo-View are the same as in [5]. For inception and ResNet-50 models trained from scratch, the learning rate is 10^{-4}. The fine-tuning has a lower learning rate of 10^{-5}. We used an Adam optimizer and trained using PyTorch 1.1 on 4 NVIDIA Tesla v100 16 GB GPUs.

We present our performance evaluation using the area under the receiver operating characteristics curve (AUROC) and the specificity achieved at the benchmark for radiologist-level sensitivity of 0.87 [12]. Confidence intervals of 95% were estimated by bootstrapping with 10,000 repetitions and p-value significance score according to the DeLong test [6].

4   Results

In this section we show the classification results for both modalities. We measured the AUROC for each class by doing one vs. all (e.g., positive biopsy is the positive class and negative biopsy with normals is the negative class). Figure 4 shows the ROC plots.

Table 2 aggregates all the AUC results. We can see the US results in the upper box of the table, where the ResNet50 training from scratch reached AUC of [0.81, 0.93, 0.91] and fine-tuning MoCo's ResNet50 encoder has an AUC of [0.91, 0.96, 0.95] for [negative, positive, normal], respectively. We can also see the specificity of ResNet50 [0.56, 0.83, 0.77] and MoCo [0.8, 0.92, 0.93] on US, for each class at a sensitivity of 0.87. The lower box results refers to the 4 MG

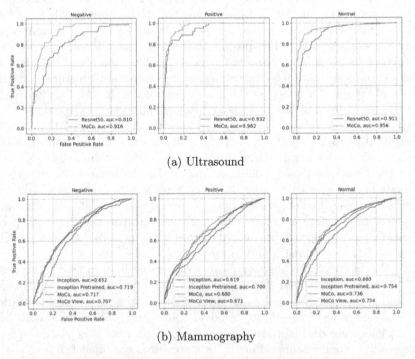

(a) Ultrasound

(b) Mammography

Fig. 4. US and MG 3-class classification AUROC curves. a) US two models, Resnet50 trained from scratch (blue) compared to self-supervised MoCo (orange) and b) MG 4 models results: inception trained from scratch (blue), Supervised pre-trained on a binary classification task (orange), self supervised pre-training of MoCo (green) and MoCo-View (red). (Color figure online)

Table 2. US and MG AUC classification results with 95% confidence interval. MG and US classified to: negative (benign)/positive (malignant)/normal (images not sent for biopsy). The best performance is depicted in bold. We use the DeLong test to show the significance. The lower box compares pairs of models and their matching p-values per class. As visible in US results, the improvement using MoCo is significant. For MG, MoCo and MoCo-View are significantly better compared to training from scratch. Comparing MoCo to MoCo-View, the negative and positive classes have a higher AUC for MoCo. However, this result is not significant, as seen in the lower p-values table. MoCo-View performs better in a significant way on the normal class. Supervised pre-training is also significantly better, as expected.

US models			
	Negative	Positive	Normal
ResNet50	0.810 [0.773–0.847]	0.932 [0.910–0.952]	0.911 [0.889–0.933]
MoCo	**0.916** [0.895–0.936]	**0.962** [0.950–0.974]	**0.956** [0.944–0.968]
p-value	2.01e−05	0.036	0.0008
MG models			
	Negative	Positive	Normal
Inception	0.652 [0.629–0.673]	0.619 [0.596–0.642]	0.680 [0.656–0.703]
MoCo	**0.717** [0.695–0.738]	**0.680** [0.658–0.703]	0.736 [0.714–0.757]
MoCo-View	**0.707** [0.683–0.728]	**0.671** [0.649–0.693]	**0.754** [0.734–0.774]
Supervised pre-train	0.719 [0.699–0.740]	0.700 [0.678–0.720]	0.754 [0.735–0.775]
MG p-values			
(Inception vs. MoCo)	3e−08	5.7e−05	6.6e−08
(Inception vs. MoCo-View)	8.1e−07	0.0005	3.6e−12
(MoCo vs. MoCo-View)	0.06	0.21	0.0018
(Inception vs. Supervised pre-train)	4.7e−08	1.5e−10	1.2e−11

models and their matching p-values. Each model is compared to a different model with their corresponding p-value significance per class.

5   Discussion

We successfully demonstrated how self-supervised contrastive learning can help overcome the challenge in medical imaging associated with training robust and accurate models from small annotated datasets. It can also reduce annotation costs by annotating only a small fraction of the data. Trained models will be much more robust and less sensitive to noise levels in the annotated data, since less annotations are required. In our experiments, unlike supervised learning, where larger sets are required, even if such sets can include annotation errors [3, 13], we can work with cleaner labeled sets, which are biopsy proven, even though this reduces the size of the dataset. We defined a set of experiments to support our claims. Our evaluations of US and MG show that the improvement achieved is consistent across the two modalities. We were able to achieve a significant improvement without the need for a large annotated dataset.

US data is especially hard to come by, since in most countries a US examination is only performed if a suspicious finding is present and further testing is necessary for the radiologist to make a decision. We show on the US modality, that fine-tuning a ResNet50 encoder from the MoCo model is significantly better ($p < 0.05$) than training from scratch. We also attain a specificity improvement of [42%, 11%, 21%] for each class at radiologists level sensitivity.

For MG data, we show that both self-supervised methods are significantly better than training from scratch on the biopsy classification task. In addition, the higher AUC for MoCo compared to our MoCo-View method is not significant for the negative and positive classes as shown by the p-value significance test. That said, the MoCo-View is significantly better for normal separation ($p < 0.001$). For the normal class, we also improved the specificity with MoCo-View by 10% compared to MoCo, and by 35% compared to training from scratch. This provides the potential of our model to reduce more unnecessary biopsies.

To summarize we show that: (i) MoCo and MoCo-View perform better than randomly initialized weights. (ii) MoCo-View performs better than MoCo for normal separation task. (iii) Using the sanity check of the upper bound performance for supervised pre-training on MG data, we show non-inferior results using a larger unlabeled set for MoCo/MoCo-View pre-training.

Due to memory limitations, we had to resize the MG images to allow training of the model. Lesions can be small with subtle patterns and resizing the image hinders our ability to classify correctly. Future work will focus on modeling a full resolution MG image. We also plan to test the generalization ability on held out datasets and carry out cross validation, train on different modalities and organs (such as liver CT), and test other tasks where labels are scarce.

References

1. Azizi, S., et al.: Big self-supervised models advance medical image classification. arXiv preprint arXiv:2101.05224 (2021)
2. Bray, F., Ferlay, J., Soerjomataram, I., Siegel, R.L., Torre, L.A., Jemal, A.: Global cancer statistics 2018: GLOBOCAN estimates of incidence and mortality worldwide for 36 cancers in 185 countries. CA: Cancer J. Clin. **68**(6), 394–424 (2018)
3. Calli, E., Sogancioglu, E., Scholten, E.T., Murphy, K., van Ginneken, B.: Handling label noise through model confidence and uncertainty: application to chest radiograph classification. In: Mori, K., Hahn, H.K. (eds.) Medical Imaging 2019: Computer-Aided Diagnosis, vol. 10950, pp. 289–296. International Society for Optics and Photonics, SPIE (2019)
4. Chen, T., Kornblith, S., Norouzi, M., Hinton, G.: A simple framework for contrastive learning of visual representations. In: International Conference on Machine Learning, pp. 1597–1607. PMLR (2020)
5. Chen, X., Fan, H., Girshick, R., He, K.: Improved baselines with momentum contrastive learning. arXiv preprint arXiv:2003.04297 (2020)

6. DeLong, E.R., DeLong, D.M., Clarke-Pearson, D.L.: Comparing the areas under two or more correlated receiver operating characteristic curves: a nonparametric approach. Biometrics 837–845 (1988)
7. Hadad, O., Bakalo, R., Ben-Ari, R., Hashoul, S., Amit, G.: Classification of breast lesions using cross-modal deep learning. In: 2017 IEEE 14th International Symposium on Biomedical Imaging (ISBI 2017), pp. 109–112. IEEE (2017)
8. He, K., Fan, H., Wu, Y., Xie, S., Girshick, R.: Momentum contrast for unsupervised visual representation learning. In: Proceedings of the IEEE/CVF Conference on Computer Vision and Pattern Recognition, pp. 9729–9738 (2020)
9. He, K., Zhang, X., Ren, S., Sun, J.: Deep residual learning for image recognition. In: Proceedings of the IEEE Conference on Computer Vision and Pattern Recognition, pp. 770–778 (2016)
10. Jamaludin, A., Kadir, T., Zisserman, A.: Self-supervised learning for spinal MRIs. In: Cardoso, M.J., et al. (eds.) DLMIA/ML-CDS -2017. LNCS, vol. 10553, pp. 294–302. Springer, Cham (2017). https://doi.org/10.1007/978-3-319-67558-9_34
11. Khan, S., Islam, N., Jan, Z., Din, I.U., Rodrigues, J.J.P.C.: A novel deep learning based framework for the detection and classification of breast cancer using transfer learning. Pattern Recogn. Lett. **125**, 1–6 (2019)
12. Lehman, C.D., Arao, R.F., et al.: National performance benchmarks for modern screening digital mammography: update from the breast cancer surveillance consortium. Radiology **283**(1), 49–58 (2017)
13. Li, Y., Zhang, Y., Zhu, Z.: Learning deep networks under noisy labels for remote sensing image scene classification. In: IGARSS 2019–2019 IEEE International Geoscience and Remote Sensing Symposium, pp. 3025–3028 (2019)
14. McKinney, S.M., et al.: International evaluation of an AI system for breast cancer screening. Nature **577**(7788), 89–94 (2020)
15. Oord, A.V.D., Li, Y., Vinyals, O.: Representation learning with contrastive predictive coding. arXiv preprint arXiv:1807.03748 (2018)
16. Schaffter, T., et al.: Evaluation of combined artificial intelligence and radiologist assessment to interpret screening mammograms. JAMA Netw. Open **3**(3), e200265–e200265 (2020)
17. Shi, J., Zhou, S., Liu, X., Zhang, Q., Lu, M., Wang, T.: Stacked deep polynomial network based representation learning for tumor classification with small ultrasound image dataset. Neurocomputing **194**, 87–94 (2016)
18. Szegedy, C., Ioffe, S., et al.: Inception-v4, Inception-ResNet and the impact of residual connections on learning (2016)
19. Tabár, L., et al.: The incidence of fatal breast cancer measures the increased effectiveness of therapy in women participating in mammography screening. Cancer **125**(4), 515–523 (2019)
20. Zeimarani, B., Costa, M.G.F., Nurani, N.Z., Bianco, S.R., De Albuquerque Pereira, W.C., Filho, C.F.F.C.: Breast lesion classification in ultrasound images using deep convolutional neural network. IEEE Access **8**, 133349–133359 (2020). https://doi.org/10.1109/ACCESS.2020.3010863
21. Zhang, X., et al.: Whole mammogram image classification with convolutional neural networks. In: 2017 IEEE International Conference on Bioinformatics and Biomedicine (BIBM), pp. 700–704 (2017)
22. Zhou, H.-Y., Yu, S., Bian, C., Hu, Y., Ma, K., Zheng, Y.: Comparing to learn: surpassing ImageNet pretraining on radiographs by comparing image representations. In: Martel, A.L., et al. (eds.) MICCAI 2020. LNCS, vol. 12261, pp. 398–407. Springer, Cham (2020). https://doi.org/10.1007/978-3-030-59710-8_39

Self-guided Multi-attention Network for Periventricular Leukomalacia Recognition

Zhuochen Wang[1], Tingting Huang[2,3], Bin Xiao[1], Jiayu Huo[1], Sheng Wang[1], Haoxiang Jiang[3], Heng Liu[3], Fan Wu[3], Xiang Zhou[4], Zhong Xue[4], Jian Yang[3(✉)], and Qian Wang[1(✉)]

[1] School of Biomedical Engineering, Shanghai Jiao Tong University, Shanghai, China
wang.qian@sjtu.edu.cn
[2] Department of Radiology,
The First Affiliated Hospital of Henan University of Chinese Medicine,
Zhengzhou, China
[3] Department of Radiology,
The First Affiliated Hospital of Xi'an Jiaotong University, Xi'an, Shanxi, China
yj1118@mail.xjtu.edu.cn
[4] Shanghai United Imaging Intelligence Co., Ltd., Shanghai, China

Abstract. Recognition of Periventricular Leukomalacia (PVL) from Magnetic Resonance Image (MRI) is essential for early diagnosis and intervention of cerebral palsy (CP). However, due to the subtle appearance difference of tissues between damaged and healthy brains, the performance of deep learning based PVL recognition has not been satisfactory. In this paper, we propose a self-guided multi-attention network to improve the performance for classification and recognition. In particular, we first conduct semantic segmentation to delineate four target regions and brain tissues as regions of interest (RoIs), which are pathologically related to PVL and should be focused in terms of the attention of the classification network. Then, the attention-based network is further designed to focus on the extracted PVL lesions when training the network. Moreover, the novel self-guided training strategy can provide comprehensive information for the classification network, and hence, optimize the generation of attention map then further improve the classification performance. Experimental results show that our method can effectively improve the precision of recognizing PVLs.

Keywords: Periventricular Leukomalacia · Attention mechanism · Self-guided training

1 Introduction

Cerebral palsy (CP) refers to a set of brain syndromes, such as movement disorders and postural abnormalities, which are attributed to various causes of

---

Z. Wang and T. Huang—have contributed equally.

© Springer Nature Switzerland AG 2021
I. Rekik et al. (Eds.): PRIME 2021, LNCS 12928, pp. 128–137, 2021.
https://doi.org/10.1007/978-3-030-87602-9_12

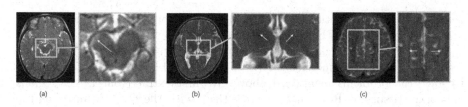

Fig. 1. Examples of PVL lesions on four essential RoIs related to CP: a) Cerebral peduncle; b) Thalamus and posterior limb of internal capsule (yellow arrows indicate lesion of posterior limb of internal capsule, and red arrows indicate lesion of thalamus); c) Centum semiovale. (Color figure online)

non-progressive brain injury. As a significant factor leading to disability of children, CP will affect a patient's entire life negatively, reducing the quality of life and incurring serious mental or economic burdens for their families. Accurate diagnosis of CP will effectively optimize neuroplasticity and functions of the patients [1], because it allows prompt access to early intervention during the brain development [2]. However, about half of infants with CP have no significant CP-related clinical manifestations in the early developmental period (i.e., the neonatal period) [3,4]. Therefore, a majority of them may miss the time window for early diagnosis and in-time intervention.

Since Periventricular leukomalacia (PVL) is identified as the most common cause of CP [5,6], it is vital to detect PVL lesions for early diagnosis and treatment. The presence of PVL is distinct in Magnetic Resonance Imaging (MRI) data for the majority of children with CP. From T2-MRI, as shown in Fig. 1, the main appearance of PVL includes high signals in four specific target regions of white matter, i.e., cerebral peduncle, thalamus, posterior limb of the internal capsule, and centrum semiovale [7]. However, these high signals are not always easy to be identified. The PVL lesions vary considerably in intensity and appearance, and the recognition of PVL lesion in MRI mainly relies on experiences of radiologists. Therefore, it is expected that deep learning-based computer-aided diagnosis can contribute to PVL recognition in clinical setting.

Unfortunately, it is hard to achieve satisfactory performance by simply applying state-of-the-art deep network in PVL recognition. The poor performance can be attributed to three major problems: 1) Neural networks do not take domain knowledge into consideration, which leads to inefficient feature extraction of the network; 2) It is difficult to localize and identify PVL lesions because of the low contrast between PVL lesions and surrounding normal tissues in MRI; 3) Training data can be limited, and data imbalance of healthy tissues and lesions greatly impacts the training procedure.

Therefore, we propose an attention-guided network to mitigate those issues. Related research [8,9] proves attention guidance can make the network to focus on target regions. For example, Li et al. [8] proposed the guided attention inference network to obtain more accurate attention maps under the segmentation-mask supervision. Ouyang et al. [9] utilized bounding boxes to guide the generation of the attention maps, so that the performance of thorax disease diagnosis in chest X-ray could be improved.

In this paper, we develop a self-guided multi-attention network for PVL lesion recognition from multi-modal MRI (i.e., T1-weighted and T2-weighted images) to solve the problems mentioned above. We first extract four target regions and brain tissues as RoIs, and combine them with the original image as the input of the classification network. Second, by establishing spatial consistency between the classification attention map and PVL lesion region in classification, the attention of the diagnosis network will be forced to focus on the lesion areas. Finally, we introduce a self-guided strategy to generate corresponding normal images from abnormal subjects, which acts as an efficient data augmentation method in training. Experiments clearly demonstrate the validity of our method.

2   Method

The overall structure of proposed method is shown in Fig. 2. The self-guided multi-attention network consists of three components: 1) RoI extraction; 2) attention-based PVL classification network; and 3) the self-guided training strategy. RoI extraction provides the position of potential lesions directly, and the attention loss is applied to constrain the attention map. In addition, we propose the self-guided strategy to provide more comprehensive information by the generated normal counterpart of the abnormal subject. The details of three components are introduced in the following sections.

2.1   RoI Extraction

We conduct RoI extraction as preprocessing of PVL classification as shown in the left of Fig. 2. The RoI contains two components: target regions and brain tissues.

We first build a slice selection network based on ResNet18 [10] to select the slices containing four target regions, which provide spatial information of potential PVL lesions. The selected slices are divided into four groups, each of which corresponds to one target region. Considering that T1-weighted MRI has higher contrast between the target regions and other tissues in brain, it is used for slice classification and following RoI extraction.

Next, we extract the RoIs for the classification network. For the target region segmentation, we employ the pointrend [11] which has been proved to be effective in semantic segmentation task. Moreover, as brain tissues provide healthy white matter reference for the PVL classification network, we also extract brain tissues from 3D MRI by using FSL [12]. By concatenating the extracted RoI maps with the T2-weighted image as the input of the downstream classification network, the network can pay attention to the area which is more likely to have PVL lesion.

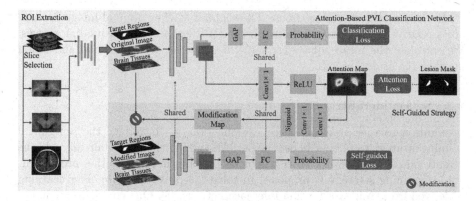

Fig. 2. The overall architecture of our network. 1×1 Conv refers to 1×1 convolution kernel. FC means fully-connected layer. GAP refers to global average pooling.

2.2 Attention-Based PVL Classification Network

Here we propose attention-based PVL classification network which can utilize the mask of PVL lesion as additional supervision. The architecture of the proposed network is shown in the upper half of Fig. 2. Due to low contrast between PVL and surrounding white matter, it is challenging for the conventional classification networks (e.g. ResNet) to localize the PVL without guidance. Consequently, we propose an attention-guidance pipeline, in which, we first extract the attention map of the network, and force the attention of the network to focus on PVL area by the attention loss between attention map and the mask of PVL lesion. With the guidance of the lesion mask, the generated attention map could be more accurate and tailored for PVL recognition.

To obtain the network's attention, the class activation map (CAM) is applied. We first collect the feature maps, then locate the high activation value in the feature maps as network's attention. We describe this process in detail as follows. The feature maps of backbone convolutional neural network are proved to contain essential high-level information that determines the final result of the network [13,14]. Also, the spatial information of the feature maps gives the position where the network should focus [15]. Therefore, the feature maps can be used to generate attention maps, and constrain the network optimization. The attention maps can be described as [16]:

$$I_{am} = \sum_{i=1}^{n} \omega_i \times I_f^i, \tag{1}$$

where I_f^i is ith feature map of the last convolution layer, ω_i is the weight of ith feature map. The weights are the weights of fully-connected layer, e.g. ω_3 is the weight of the 3th neuron in the fully-connected layer. Then, the weights of fully-connected layer are multiplied by feature maps, and the results are resized to the input image size so that the final attention maps are obtained.

In the training stage, we choose T2-weighted MRI as original image because of its relatively high contrast between the PVL lesion and surrounding tissues. We calculate the mean square error (MSE) between attention map and lesion mask as the attention loss.

2.3   Self-guided Strategy

In addition to the attention based classification, we introduce the self-guided training strategy to further boost the performance. Although we already add the attention loss, the network is still disturbed by information of other areas. Self-guided training strategy is proposed to solve the problem. Self-guided learning is derived from guided attention inference network [8]. Our goal is to guide the network to focus on the lesion by "un-focus" on irrelevant feature. To achieve that, we force the input PVL images to contain as little PVL-related features as possible. For example, by blocking PVL lesion area (high-responding area on attention map), regions beyond the focus area should include ideally not a single pixel that can trigger the network to recognize the input as a PVL case. Then the modified image is fed into the classification network, and the output should be opposite to the decision output from the input image before modification. The self-guided loss between the output and the normal label can thus be added to constrain the training. In this way, the self-guided strategy will optimize the generation of attention map.

Specifically, in our task, the intensity of PVL lesions in T2-weighted MRI is higher than normal white matter [17]. Therefore, we could convert PVL cases to normal by reducing the intensity of PVL lesions in T2-weighted MRI, to enforce the input to contain as little useful feature as possible. As shown in Fig. 2, we first obtain attention maps which indicate the area of PVL. To decrease intensity decrements of PVL, we attain modification maps from the attention maps through two convolution layers and an activation layer. The original PVL image is then subtracted by the modification maps to get "normal" images (denoted as modified image in Fig. 2). These "normal" images are fed into the classification network to calculate self-guided loss. Specifically, the ground-truth label of a modified "normal" image is negative. Binary cross entropy (BCE) loss is used for both classification loss and self-guided loss with opposite ground-truth labels.

3   Experiment

3.1   Data Preprocessing

We collected 163 cases of PVL patients and 103 normal cases. The age of subjects ranges from 6 months to 4 years. Every case contains both T1-weighted and T2-weighted images. Meanwhile, four target regions and PVL lesion were manually labeled by clinical doctor. Voxel size of images is $0.5 \times 0.5 \times 6\,\mathrm{mm}^3$. Skull stripping was operated on all images using FSL [12]. Moreover, we performed histogram matching to transform histogram of all images into that of

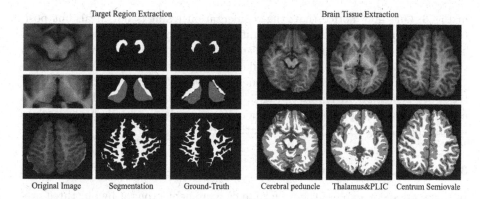

Fig. 3. Result of RoI extraction. In target region extraction, the left top column refers to extraction of cerebral peduncle, the second column refers to extraction of thalamus and posterior limb of internal capsule, the third column refers to extraction of centrum semiovale.

the template. Five fold cross-validation is used to evaluate all methods. Specifically, the data partition is subject-level. All reported result is evaluated on the validation set.

Table 1. Dice score of target region segmentation.

Target region	Dice
Cerebral peduncle	0.733
Thalamus	0.746
Posterior limb of internal capsule	0.702
Centrum semiovale	0.764

3.2   RoI Extraction

We exploit ResNet18 to train a slice classification network first to select slices containing RoIs. The classification accuracy is 0.941. In the selected slice, we crop the images according to the mean relative position of every target region. Furthermore, we exploit data augmentation methods, e.g., rotation and flipping. We trained segmentation networks to segment target regions. The Dice metrics are shown in Table 1. We can find that the performance of thalamus and centrum semiovale is relatively high because the shape of thalamus is simple and the contrast between centrum semiovale and other tissues is high. However, the Dice ratios of cerebral peduncle and posterior limb of internal capsule are relatively low due to the small size and fuzzy edge of cerebral peduncle and posterior limb of internal capsule. The visualization of target region segmentation is shown in Fig. 3. One may notice that, the segmentation result is not ideal, but position of segmentation is relatively accurately located in the corresponding region. Therefore, the coarse segmentation result can provide valuable information in the downstream tasks, i.e., PVL classification.

FSL [12] is exploited for brain tissue extraction. The visualization results of tissue segmentation are shown in Fig. 3. It can be found that the white matter is accurately segmented. Thus, the segmentation is able to provide healthy white matter reference for the PVL classification network.

3.3 PVL Classification

In this section, we explore the performance increment brought by our three proposed components. We conduct experiments in four settings: 1) vanilla ResNet34 [10], which refers to "Baseline" in Table 2. 2) ResNet34 with RoI extraction, which is annotated as "+RoI"; 3) ResNet34 with RoI extraction and attention-based method, which is denoted as "+RoI +attention"; 4) ResNet34 with RoI extraction, attention-based method, and self-guided training strategy, which is "+RoI +attention +self-guided" in Table 2. The quantitative comparison of PVL classification in different target regions with four aforementioned settings are shown in Table 2.

In the experiment of the baseline model, the ResNet34 works well in centrum semiovale (accuracy 0.900 ± 0.008) because of the high contrast between PVL lesion and other tissues. However, the result shows that the performance on cerebral peduncle is relatively low (accuracy 0.615 ± 0.012), because the abnormal position and size of lesion make it challenging to find.

Compared with the baseline, segmentation guided network ("+RoI") improves the performance of cerebral peduncle and thalamus by a large margin, in particular, accuracy increases from 0.615 to 0.758 for cerebral peduncle. It implies that the segmentation of RoIs help significantly to PVL lesion recognition. The performance boost in centrum semiovale is mere, indicating that the centrum semiovale is easily located in the classification network thus the external spatial guidance helps little.

As for the classification with attention-based method ("+RoI+attention"), all of four target regions get better accuracy compared with the performance of baseline and segmentation guided network. Specifically, the classification with attention-based method has average accuracy of 0.800, while baseline has average accuracy of 0.748 and "+RoI" has average accuracy of 0.793. This result demonstrates the proposed attention-guidance has positive impact to all four target regions.

In the last experiment, the performance of self-guided classification ("+RoI +attention +self-guided") improves a lot especially in cerebral peduncle and thalamus (accuracy increment of 0.019 and 0.018 respectively). In general, the average accuracy of classification of four target regions of our method is the best in all experiments with an average accuracy of 0.809. The visualization of attention map extracted from our network is shown in Fig. 4. It shows that the

Table 2. Result of PVL classification. CPe, PLIC, CS refers to cerebral peuncle, posterior limb of internal capsule, and centrum semiovale respectively. Avg. refers to average results of four target regions.

Method	Metrics	CPe	Thalamus	PLIC	CS	Avg.
Baseline	Precision	0.600 ± 0.024	0.647 ± 0.014	0.719 ± 0.008	0.864 ± 0.011	0.708
	Recall	0.692 ± 0.027	0.688 ± 0.019	0.742 ± 0.014	0.950 ± 0.006	0.768
	Accuracy	0.615 ± 0.012	0.750 ± 0.010	0.726 ± 0.021	0.900 ± 0.008	0.748
+RoI	Precision	0.746 ± 0.030	0.733 ± 0.023	0.714 ± 0.008	0.870 ± 0.011	0.779
	Recall	0.806 ± 0.023	0.688 ± 0.011	0.806 ± 0.021	0.925 ± 0.005	0.793
	Accuracy	0.758 ± 0.031	0.773 ± 0.006	0.742 ± 0.015	0.900 ± 0.008	0.793
+RoI+attention	Precision	0.706 ± 0.022	0.704 ± 0.028	0.728 ± 0.014	0.860 ± 0.018	0.749
	Recall	0.923 ± 0.005	0.718 ± 0.033	0.789 ± 0.011	0.920 ± 0.009	0.849
	Accuracy	0.769 ± 0.016	0.777 ± 0.010	0.747 ± 0.014	$\mathbf{0.905 \pm 0.007}$	0.800
+RoI +attention +self-guided	Precision	0.778 ± 0.019	0.769 ± 0.020	0.738 ± 0.013	0.855 ± 0.014	0.786
	Recall	0.808 ± 0.016	0.625 ± 0.010	0.777 ± 0.015	0.935 ± 0.004	0.782
	Accuracy	$\mathbf{0.788 \pm 0.009}$	$\mathbf{0.795 \pm 0.011}$	$\mathbf{0.751 \pm 0.013}$	0.905 ± 0.008	0.809

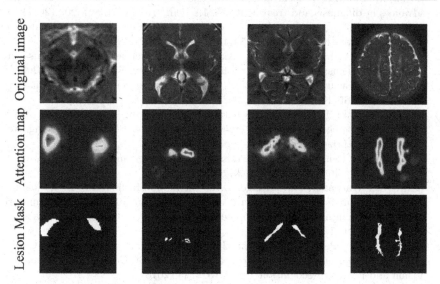

Fig. 4. Attention map extracted from the classification network.

attention of classification network focuses accurately on the PVL lesion, which demonstrates the validity of our network.

4   Conclusion and Discussion

In this paper, we present a specially designed network called self-guided multiattention network applied to PVL lesion recognition. The PVL lesions are localized in four fixed target regions in white matter according to clinical study. Based on this, the RoI extraction strategy is implemented to provide domain knowledge. And attention-based network is introduced to distinguish the slices with PVL lesion and slices without PVL lesion. Furthermore, the self-guided training strategy is adopted to enhance the classification performance. By integrating these components into the network, our proposed self-guided multi-attention network has achieved significant improvement in terms of accuracy.

It is deserved to be mentioned that the performance of centrum semiovale of our network improves a little through every step of experiment. It may indicate that the information of domain knowledge helps little to the training of network in the condition of high contrast of lesion. In the future, we will try to improve our network to offer better information for PVL lesion recognition and acquire higher performance.

References

1. Novak, I., et al.: Early, accurate diagnosis and early intervention in cerebral palsy: advances in diagnosis and treatment. JAMA Pediatr. **171**(9), 897–907 (2017)
2. Morgan, C., Fahey, M., Roy, B., Novak, I.: Diagnosing cerebral palsy in full-term infants. J. Paediatr. Child Health **54**(10), 1159–1164 (2018)
3. Drougia, A., et al.: Incidence and risk factors for cerebral palsy in infants with perinatal problems: a 15-year review. Early Human Dev. **83**(8), 541–547 (2007)
4. Deng, W., Pleasure, J., Pleasure, D.: Progress in periventricular leukomalacia. Arch. Neurol. **65**(10), 1291–1295 (2008)
5. Franki, I., et al.: The relationship between neuroimaging and motor outcome in children with cerebral palsy: a systematic review-part A. Structural imaging. Res. Dev. Disabil. **100**, 103606 (2020)
6. Novak, C.M., Ozen, M., Burd, I.: Perinatal brain injury: mechanisms, prevention, and outcomes. Clin. Perinatol. **45**(2), 357–375 (2018)
7. Ryll, U.C., Wagenaar, N., Verhage, C.H., Blennow, M., de Vries, L.S., Eliasson, A.-C.: Early prediction of unilateral cerebral palsy in infants with asymmetric perinatal brain injury-model development and internal validation. Eur. J. Paediatr. Neurol. **23**(4), 621–628 (2019)
8. Li, K., Wu, Z., Peng, K.-C., Ernst, J., Fu, Y.: Tell me where to look: Guided attention inference network. In: Proceedings of the IEEE Conference on Computer Vision and Pattern Recognition, pp. 9215–9223 (2018)
9. Ouyang, X., et al.: Learning hierarchical attention for weakly-supervised chest X-ray abnormality localization and diagnosis. IEEE Trans. Med. Imaging (2020)
10. He, K., Zhang, X., Ren, S., Sun, J.: Deep residual learning for image recognition. In: Proceedings of the IEEE Conference on Computer Vision and Pattern Recognition, pp. 770–778 (2016)
11. Kirillov, A., Wu, Y., He, K., Girshick, R.: PointRend: image segmentation as rendering. In: Proceedings of the IEEE/CVF Conference on Computer Vision and Pattern Recognition, pp. 9799–9808 (2020)
12. Smith, S.M., et al.: Advances in functional and structural MR image analysis and implementation as FSL. Neuroimage **23**, S208–S219 (2004)
13. Bengio, Y., Courville, A., Vincent, P.: Representation learning: a review and new perspectives. IEEE Trans. Pattern Anal. Mach. Intell. **35**(8), 1798–1828 (2013)
14. Mahendran, A., Vedaldi, A.: Visualizing deep convolutional neural networks using natural pre-images. Int. J. Comput. Vis. **120**(3), 233–255 (2016)

15. Selvaraju, R.R., Cogswell, M., Das, A., Vedantam, R., Parikh, D., Batra, D.: Grad-CAM: visual explanations from deep networks via gradient-based localization. In: Proceedings of the IEEE International Conference on Computer Vision, pp. 618–626 (2017)
16. Zhou, B., Khosla, A., Lapedriza, A., Oliva, A., Torralba, A.: Learning deep features for discriminative localization. In: Proceedings of the IEEE conference on computer vision and pattern recognition, pp. 2921–2929 (2016)
17. Khurana, R., Shyamsundar, K., Taank, P., Singh, A.: Periventricular leukomalacia: an ophthalmic perspective. Med. J. Armed Forces India **77**(2), 147–153 (2021)

Opportunistic Screening of Osteoporosis Using Plain Film Chest X-Ray

Fakai Wang[1,2(✉)], Kang Zheng[1], Yirui Wang[1], Xiaoyun Zhou[1], Le Lu[1],
Jing Xiao[4], Min Wu[2], Chang-Fu Kuo[3], and Shun Miao[1]

[1] PAII Inc., Bethesda, MD, USA
[2] University of Maryland College Park, College Park, MD, USA
[3] Chang Gung Memorial Hospital, Linkou, Taiwan ROC
[4] Ping An Technology, Shenzhen, China

Abstract. Osteoporosis is a common chronic metabolic bone disease that is often under-diagnosed and under-treated due to the limited access to bone mineral density (BMD) examinations, *e.g.*, Dual-energy X-ray Absorptiometry (DXA). In this paper, we propose a method to predict BMD from Chest X-ray (CXR), one of the most common, accessible and low-cost medical image examinations. Our method first automatically detects Regions of Interest (ROIs) of local and global bone structures from the CXR. Then a multi-ROI model is developed to exploit both local and global information in the chest X-ray image for accurate BMD estimation. Our method is evaluated on 1651 CXR cases with ground truth BMD measured by gold standard DXA. The model predicted BMD has a strong correlation with the ground truth (Pearson correlation coefficient *0.853*). When applied for osteoporosis screening, it achieves a high classification performance (AUC *0.928*). As the first effort in the field using CXR scans to predict the BMD, the proposed algorithm holds a strong potential in early osteoporosis screening and public health promotion.

Keywords: Bone mineral density estimation · Chest X-ray · Multi-ROI model

1 Introduction

Osteoporosis is the most common chronic metabolic bone disease, characterized by low bone mineral density (BMD) and decreased bone strength to resist pressure or squeezing force. With an aging population and longer life span, osteoporosis is becoming a global epidemic, affecting more than 200 million people worldwide [8]. Osteoporosis increases the risk of fragility fractures, which is associated with reduced life quality, disability, fatality, and financial burden to the family and the society. While with an early diagnosis and treatment, osteoporosis can be prevented or managed, osteoporosis is often under-diagnosed and

---

This work was done when Fakai Wang interned at PAII Inc.

© Springer Nature Switzerland AG 2021
I. Rekik et al. (Eds.): PRIME 2021, LNCS 12928, pp. 138–146, 2021.
https://doi.org/10.1007/978-3-030-87602-9_13

under-treated among the population at risk [4]. More than half of insufficiency fractures occur in individuals who have never been screened for osteoporosis [7]. The under-diagnosis and under-treatment of osteoporosis are mainly due to 1) low osteoporosis awareness and 2) limited accessibility of Dual-energy X-ray Absorptiometry (DXA) examination, which is the currently recommended modality to measure BMD.

Opportunistic screening of osteoporosis is an emerging research field in recent years [1–3,6]. It aims at using medical images originally for other indications to screen for osteoporosis, which offers an opportunity to increase the screening rate at no additional cost and time. Previous attempts mainly focused on using the CT attenuation (i.e., Housefield Unit) of the vertebrae to correlate with BMD and/or fracture risk. As the most common prescribed medical image scanning, plain films have much greater accessibility than CT scans. Its excellent spatial resolution permits the delineation of fine bony microstructure that may contain information that correlates well with the BMD. We hypothesize that specific regions of interest (ROI) in the standard chest x-rays (CXR) may help the opportunistic screening for osteoporosis.

In this work, we introduce a method to estimate the BMD from CXR and screen osteoporosis. Our method first locates anatomical landmarks of the patient's bone structures and extracts multiple ROIs that may provide imaging biomarkers for osteoporosis. We propose a novel network architecture that jointly processes the ROIs to accurately estimate the BMDs. We experiments on 1651 CXRs with paired DXA BMDs (ground truth). Our best model achieves a high correlation with the DXA BMD (*i.e.*, r-value 0.853), and an AUC area of 0.928 (using T-score of -2.5 as threshold for osteoporosis diagnosis).

In summary, our contributions are three-fold: 1) to our best knowledge, we are the first to develop models using CXR to estimate BMD, which could serve as a practical solution for opportunistic screening of osteoporosis; 2) we further on propose the anatomy-aware Multi-ROI model to combine global and local information in CXR for accurate BMD estimation. 3) we demonstrate that our method achieves clinically useful osteoporosis screening performance.

2   Methodology

2.1   Problem Overview

Our task is to use frontal-view chest X-ray scanning to predict the patient's lumbar BMDs. It is viable because of the distributed nature of osteoporosis and high bone density correlation of different areas. The task is formulated as a regression problem, with the model input being the CXR image, and the output being the predicted BMD value*(0.2–1.8)*. Each CXR image is paired with corresponding ground truth (DXA BMD on 4 lumbar vertebra, L1, L2, L3, L4) for the same patient. The baseline model directly utilize the part or the whole CXR scanning and make predictions (described in Sect. 2.2). However, since the lumbar spine is in a different area from the chest, a major challenge of the cross-region BMD prediction is to identify the highly correlated regions and

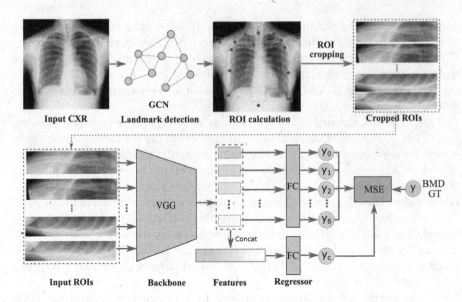

Fig. 1. An overview of the multi-ROI model pipeline (training), with landmark detection and ROI extraction in the upper part, and the multi-ROI model in the lower part. The landmark detector localizes 16 red dots on CXR images, to anchor local bones. Each dashed yellow box is normalized to the same orientation, height/width ratio. The cropped ROIs are used by all models. In the multi-ROI model, independent inputs include 6 local ROIs and the whole chest image. During training, Mean Square Error loss (MSE) is applied on 7 individual feature BMD predictions $y_0, .., y_6$, and on the concatenated feature BMD prediction y_c.

visual patterns. To solve this challenge, we further develop the Multi-ROI model, consisting of two steps: 1) extraction of Regions of Interest (ROIs) representing different chest bone regions (described in Sect. 2.3); 2) training a multi-ROI regression neural network that jointly analyzes the ROIs (described in Sect. 2.4).

2.2 BMD Estimation via a Single X-Ray Image

As a regression pipeline, our baseline model consists of a feature extractor and feature regressor. The input is the whole or part of the chest X-ray scanning. We employ general Convolutional Neural Network (CNN) as the backbone network to produce a feature map encoding of the input modality. Possible feature extractor includes backbones such as Resnet family, VGG family. We apply Global Average Pooling (GAP) to aggregate the feature map to a single feature vector. Although the inputs may have different resolutions, the average pooling layer reduce the spatial differences and produces the normalized feature vector. Fully connected layer is used to predict BMD. We use Mean Squared Error (MSE) loss to train the model.

2.3   Automatic ROI Extraction

There are many bone regions in the chest area, and it is unclear which local regions are more informative for BMD prediction. It is also unclear if certain pattern combination of different-area bones are more effective. In the baseline model, only a single input modality is utilized, without locality and scale information, which lacks the ability for fine-level exploitation. To enable the full potential, the desired model should be aware of local textures and global structures, to explore the correlations between different regions. Based on the clinical prior knowledge that osteoporosis is a metabolic bone disease affecting all bones in the human body, we hypothesize that all chest bone regions provide visual cues of the BMD. We need the ROIs from the following regions: left/right clavicle bones, cervical spine, left/right rib-cage area, T12 vertebra. We employ the graph-based landmark detector, Deep Adaptive Graph (DAG) [5] to automatically detect 16 landmarks which include 1) 3 points on the left/right clavicles, 2) 4 points along the left/right rib cages, 3) 1 point on the cervical spine and 1 point on the T12 vertebra. We manually labeled 1000 cases (16 landmarks on each CXR scan), for the DAG model training and selection. The resulting landmark detector can reliably extract all the keypoints. Once we get the landmark points for bone anchors, we crop the bone ROIs by a fixed size ($256 * 256$) and save all the cropped ROIs for the following stage. Examples of the CXR with 16 landmarks and extracted ROIs are shown in Fig. 1. Besides these 6 local ROIs, we also include the whole CXR ROI to provide global structural information.

2.4   BMD Estimation via Joint Analysis of the ROIs

Once we obtained the accurate ROIs of different bones in the CXR, we can exploit the BMD distribution in different regions. We proposed the multi-ROI model, which is a multi-input joint learning network architecture. By jointly utilizing the global and local ROIs in the chest, the network is capable of extracting visual patterns at different scales, which is vital for predicting BMDs. All the input ROIs go through the same feature extractor independently, generating 7 feature vectors in Fig. 1. These feature vectors are then decoded in two paths. In the first path, the individual feature vector can be directly processed by a fully connected (FC) layer to make predictions. In the second path, the 7 features form a joint feature vector, which is then decoded by an FC layer to predict BMD. During training, the mean squared error (MSE) loss is applied between ground truth and predicted BMDs, in both paths. During testing, the BMDs produced from the concatenated feature (in the second path) is taken as the prediction.

3   Experiments

3.1   Experiment Setup

Dataset. We collected 1651 frontal view CXR scans, with paired ground truth DXA BMD scores (on four lumbar vertebrae L1–L4). The data come from *anony-*

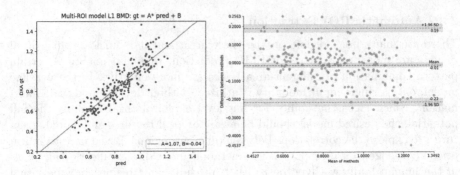

Fig. 2. The linear fitting curve (left) and BlandAltman plot (right) between Multi-ROI model predictions and ground truth (on L1 BMD). In the linear fitting curve, **A** represents the **slop**, while **B** represents the **y-intercept**. The horizontal axis represents the mean of prediction and DXA BMD, while the vertical axis represents the differences.

mous hospital after de-identification of the patient privacy. All experiments use the same data split, with 1057, 264 and 330 cases for training, validation and testing, respectively. There is no patient overlapping between data splits.

Metrics. Firstly, to get an intuitive measurement of data fitting ability, we draw the linear fitting curve and Bland-Altman plot. We use coefficient of determination in the linear fitting curve to measure model generalization. In the Bland-Altman plot, we use the error mean and variations to evaluate prediction quality. Secondly, to get a general conclusion towards clinical deployment, we calculate the *Pearson Correlation Coefficient* (R-value) and *Area Under Curve*. R-value measures the linear relationship between the predicted BMD and the ground truth BMD. It only considers the correlation between two sequences regardless their absolute values. While Area Under Curve (AUC) measures accumulated true positive (osteoporosis) rate under different judging thresholds for binary classification. BMD valuses are transformed into T-score values by checking the transforming table in DXA machine, and T-scores are used for the AUC calculation according to clinical practices. For osteoporosis diagnosis, we use T-score threshold of -2.5 (below -2.5 as positive osteoporosis, from the clinical definition of osteoporosis).

3.2 Implementation Details

We trained our model on a workstation with Intel Xeon CPU E5-2650 v4 CPU @ 2.2 GHz, 132 GB RAM, and 4 NVIDIA TITAN V GPUs. Our models are implemented with PyTorch. For both baseline model and multi-ROI model, all the input images/ROIs are resized to *(256, 256)* . During training, we apply augmentations such as scaling, rotation, translation, random flip. The SGD optimizer has a learning rate of 0.0001, a weight decay of 4e$-$4. All models are trained for 100 epochs.

Table 1. Comparison of baseline models (first 5 rows, Base+input modality) and multi-roi model (6th row, use all inputs modalities). VGG16 is used as feature extractor.

Model	L1		L2		L3		L4		Average	
	R-value	AUC	R-value	AUC	R-value	AUC	R-value	AUC	R-value	AUC
Base+cervi	0.823	0.915	0.817	0.930	0.802	0.937	0.738	0.841	0.795	0.906
Base+clavi	0.793	0.893	0.778	0.917	0.745	0.867	0.709	0.857	0.756	0.883
Base+lumb	0.750	0.829	0.791	0.889	0.754	0.859	0.711	0.841	0.751	0.855
Base+ribca	0.748	0.868	0.768	0.876	0.730	0.850	0.703	0.830	0.737	0.856
Base+chest	0.865	0.923	0.863	0.913	0.824	0.918	0.804	**0.901**	0.839	0.914
Milti-ROI	**0.882**	**0.939**	**0.874**	**0.95**	**0.851**	**0.926**	**0.807**	0.899	**0.853**	**0.928**

3.3   Input Modalities and Framework Comparison

Baseline Model and Input Modalities. Additional to the whole chest we have extracted 6 local ROIs, forming 5 input modalities (left and right clavicle ROIs form one, left and right ribcage ROIs form one.). Different from the multi-ROI model, the baseline model takes only one input modality. Therefore the baseline model is trained/evaluated 5 times separately on 5 input sources, while multi-ROI model takes all input modalities (6 local ROIs and the whole chest) in one experiment. Baseline experiments on different modalities are labeled as separate models, to investigate the modality effectiveness towards BMD prediction. These models are labeled as *Base+ROI*, and ROIs are from *cervical, clavicle, lumbar, ribcage, or the whole chest*, in Table 1. The *Base+chest* performs better than any other local ROIs due to global bone information of the whole chest modality.

Multi-ROI Model Architecture. As there are many bones in the chest, the one bone BMD may not be consistent with another. Baseline models lack the mechanism to solve conflicts of inconsistent BMD distributions. The baseline model also lack the ability to fuse local/global texture information. The key idea of *multi-ROI model* is to automatically identify the informative regions for estimating BMD. The multi-ROI model in Fig. 1 takes 7 input modalities all at once, which share the same encoding backbone. Each encoded feature vector is of size *512*, and the concatenated feature has a shape of *512 * 7*. Individual ROI features share the same fully connected layer, while the concatenated vector goes through a stand-along FC layer.

3.4   Multi-ROI Model Performance Analysis

Data Fitting Ability. To get an intuitive sense of model output quality, we draw the liner fitting curves in Fig. 2. The fitting parameters(A: slop, B: y-intercept) illustrate the overall fitting direction and modeling characteristics, indicating whether the predictions are compressed or translated around certain value. While the coefficient of determinant (R2) quantify the quality of data fitting. When both predicting on L1 BMD score, the baseline model get the linear fitting parameters of $A = 0.94$, $B = 0.07$ and R2 value of 0.721, while the multi-ROI model get $A = 1.07$, $B = -0.04$, $R2 = 0.778$. Although both models get good overall fitting directions, multi-ROI model shows better fitting goodness. The same comparison conclusions are found for predicting other bone (L1, L2, L3) BMDs.

To investigate the prediction errors, we draw the Bland-Altman plot in Fig. 2. When both predicting on L1 BMD score, the baseline model gets the error mean of -0.03 and error Standard Deviation of 0.117, while the multi-ROI model gets 0.01 and 0.102 respectively. Both models get nearly 0 error mean, but multi-ROI model have smaller standard deviation of prediction error. The same comparison conclusions are found for predicting other bone (L1, L2, L3) BMDs.

Osteoporosis Alarming Ability. To evaluate the model performance for osteoporosis prediction, clinic-orientated metrics are more meaningful than data fitting evaluation. We employ Pearson Correlation Coefficient (R-value) to measure sequential order correctness of prediction, and use AUC score to indicate disease judging ability. The multi-ROI model achieves an average R-value of 0.853 and an average AUC score of 0.928, surpassing the baseline models in Table 1. Compared with the *Base+chest* model, the *multi-ROI* model encodes additional 6 local ROI modalities, which allows for detailed local texture computation. The feature concatenation retains encoded information from both local ROIs and the global chest, enabling regional information interactions.

As an opportunistic screening method, the chest BMD prediction is mainly aimed at osteoporosis alarming. The model prediction would tell patients and doctors if the bone density falls below certain threshold in the T-score range. While the AUC reflects model alarming ability in the whole T-score range, the sensitivity and specificity at certain threshold reflect more specific performance. On L1 BMD prediction task with T-score threshold of -2.5, the baseline model (*Base+chest*) achieve sensitivity of 0.75 and specificity of 0.934, while the multi-ROI model achieves 0.8 and 0.912 respectively. It should be noticed that changing threshold would affect sensitivity and specificity in the opposite way. By calibrating the thresholds for each model, we can easily reach high specificity (>0.95) and applicable sensitivity (>0.6) at the same time. Similar conclusions apply to predicting other bone (L2, L2, L3) BMDs, though the sensitivity on L3, L4 prediction is lower.

Table 2. Bases: ensemble of 5 baseline models (same network architecture, but 5 different input modalities). All: ensemble of 5 baseline models + multi-ROI model.

Ensemble	L1		L2		L3		L4		Average	
	R-value	AUC	R-value	AUC	R-value	AUC	R-value	AUC	R-value	AUC
Bases	0.879	0.935	0.876	0.946	0.850	0.929	0.812	0.894	0.854	0.926
All	**0.888**	**0.947**	**0.886**	**0.954**	**0.869**	**0.939**	**0.815**	**0.907**	**0.864**	**0.937**

3.5 Performance of Model Ensemble

We observe overfitting during individual model training, due to the limited amount of labeled data. A common strategy to boost performance is model ensemble. We list individual model performance in Table 1, and the ensemble performance from these models in Table 2. *Bases* is the averaged predictions of all 5 baseline models (*Base+cervi, Base+clavi, Base+lumb, Base+ribca, Base+chest* models), and *All* is the averaged predictions of 5 baseline models and the multi-ROI model. *All* performs stronger in all the metrics. Although *Bases* has information from different input modalities, it still lacks the adaptive feature fusion process. In the *multi-ROI model*, the concatenated feature bears all regional information, and ROI importance is learned in the FC layer weights.

3.6 Network Backbone Comparison

Both the baseline and multi-ROI models use off-the-shelf Convolutional Neural Network (CNN) backbone as feature extractor. The CNN backbone encodes BMD information of input modality in a feature vector. Then a fully connected layer regresses the feature vector to make BMD prediction. We compare performance of different backbones in the Table 3. As we can see, VGG16 delivers the best performance, even better than Resnet. This is because bone texture encoding task involves only low-level pattern recognition, and a simpler architecture fits better on limited amount of labeled data.

Table 3. Comparison of different CNN backbones, under the same input modality (the whole chest) and network architecture (baseline model).

Backbone	L1		L2		L3		L4		Average	
	R-value	AUC	R-value	AUC	R-value	AUC	R-value	AUC	R-value	AUC
Resnet18	0.759	0.875	0.799	0.863	0.774	0.852	0.717	0.840	0.762	0.858
Resnet34	0.787	0.903	0.787	0.866	0.768	0.887	0.723	0.861	0.766	0.879
Resnet50	0.799	0.894	0.823	0.888	0.795	0.880	0.741	0.858	0.789	0.880
Vgg11	0.843	0.919	0.838	**0.916**	**0.832**	0.917	0.772	0.881	0.821	0.908
Vgg16	**0.865**	**0.923**	**0.863**	0.913	0.824	**0.918**	**0.804**	**0.901**	**0.839**	**0.914**

3.7 Evaluation of CXR BMD Prediction Task

Adjacent bones have more consistent BMD scores, while the correlation declines as distances increase. In the lumbar DXA BMD (ground truth), L1 has a *R-value* of *0.907* to L2, a *R-value* of *0.883* to L3, but only a *R-value* of *0.825* to L4. So the cross-bone model predictions may not exceed the neighboring bone BMD *R-values* (such as L1 and L3). The *multi-ROI* model get a *R-value* of *0.882* and an *AUC* score of *0.939* on L1 BMD prediction, which is already close to the DXA BMD R-values between L1 and L3 (*0.883*). This further proves the promising applicability of our proposed algorithm.

Physiological differences between lumbar bones and chest bones could also contribute to performance bottle necks. The accuracy for cross-bone BMD prediction vary from region to region, which can be observed in *Base+cervi* model and *Base+ribca* model in Table 1. The cervical vertebra has a more similar role to the lumbar in spine activity, while bones in the rib cage stay more static during body activities.

4   Conclusion

In this paper, we studied the task of BMD prediction from chest X-ray images. We develop the baseline model and compare different input modalities. We further on introduced a multi-ROI model to jointly analyze both local and global CXR regions. Experimental results and analysis show that by incorporating multiple ROIs, our model is able to produce significant improved performance compared to baseline models. Our method shows great potential to be applied to opportunistic screening of osteoporosis.

References

1. Cheng, X., et al.: Opportunistic screening using low-dose ct and the prevalence of osteoporosis in China: a nationwide, multicenter study. J. Bone Mineral Res. **36**(3), 427–435 (2020)
2. Dagan, N., Elnekave, E., Barda, N., Bregman-Amitai, O., Bar, A., Orlovsky, M., Bachmat, E., Balicer, R.D.: Automated opportunistic osteoporotic fracture risk assessment using computed tomography scans to aid in FRAX underutilization. Nat. Med. **26**(1), 77–82 (2020)
3. Jang, S., Graffy, P.M., Ziemlewicz, T.J., Lee, S.J., Summers, R.M., Pickhardt, P.J.: Opportunistic osteoporosis screening at routine abdominal and thoracic CT: normative L1 trabecular attenuation values in more than 20 000 adults. Radiology **291**(2), 360–367 (2019)
4. Lewiecki, E.M., Leader, D., Weiss, R., Williams, S.A.: Challenges in osteoporosis awareness and management: results from a survey of US postmenopausal women. J. Drug Assess. **8**(1), 25–31 (2019)
5. Li, W., et al.: Structured landmark detection via topology-adapting deep graph learning. arXiv preprint arXiv:2004.08190 (2020)
6. Pickhardt, P.J., et al.: Automated abdominal CT imaging biomarkers for opportunistic prediction of future major osteoporotic fractures in asymptomatic adults. Radiology **297**(1), 64–72 (2020)
7. Smith, A.D.: Screening of bone density at CT: an overlooked opportunity (2019)
8. Sözen, T., Özışık, L., Başaran, N.Ç.: An overview and management of osteoporosis. Eur. J. Rheumatol. **4**(1), 46 (2017)

Multi-task Deep Segmentation and Radiomics for Automatic Prognosis in Head and Neck Cancer

Vincent Andrearczyk[1]([⊠]), Pierre Fontaine[1,2], Valentin Oreiller[1,3],
Joel Castelli[4], Mario Jreige[3], John O. Prior[3], and Adrien Depeursinge[1,3]

[1] University of Applied Sciences and Arts Western Switzerland, Sierre, Switzerland
vincent.andrearczyk@hevs.ch
[2] Univ Rennes, CLCC Eugene Marquis, INSERM, LTSI - UMR 1099,
35000 Rennes, France
[3] Centre Hospitalier Universitaire Vaudois (CHUV), Lausanne, Switzerland
[4] Radiotherapy Department, Cancer Institute Eugène Marquis, Rennes, France

Abstract. We propose a novel method for the prediction of patient prognosis with Head and Neck cancer (H&N) from FDG-PET/CT images. In particular, we aim at automatically predicting Disease-Free Survival (DFS) for patients treated with radiotherapy or both radiotherapy and chemotherapy. We design a multi-task deep UNet to learn both the segmentation of the primary Gross Tumor Volume (GTVt) and the outcome of the patient from PET and CT images. The motivation for this approach lies in the complementarity of the two tasks and the shared visual features relevant to both tasks. A multi-modal (PET and CT) 3D UNet is trained with a combination of survival and Dice losses to jointly learn the two tasks. The model is evaluated on the HECKTOR 2020 dataset consisting of 239 H&N patients with PET, CT, GTVt contours and DFS data (five centers). The results are compared with a standard Cox PET/CT radiomics model. The proposed multi-task CNN reaches a C-index of 0.723, outperforming both the deep radiomics model without segmentation (C-index of 0.650) and the standard radiomics model (C-index of 0.695). Besides the improved performance in outcome prediction, the main advantage of the proposed multi-task approach is that it can predict patient prognosis without a manual delineation of the GTVt, a tedious and time-consuming process that hinders the validation of large-scale radiomics studies. The code will be shared for reproducibility on our GitHub repository.

Keywords: Head and neck cancer · Radiomics · Automatic segmentation · Deep learning

1 Introduction

Radiomics is the quantitative analysis of radiological images to obtain prognostic patient information [1]. The standard approach for radiomics involves the extraction of hand-crafted visual features from radiology images followed by prognostic

© Springer Nature Switzerland AG 2021
I. Rekik et al. (Eds.): PRIME 2021, LNCS 12928, pp. 147–156, 2021.
https://doi.org/10.1007/978-3-030-87602-9_14

modeling. This standard approach generally requires manual annotations of Volumes Of Interest (VOIs), i.e. tumor region, to spatially localize the extraction of features.

In the past seven years, deep Convolutional Neural Networks (CNNs) (mostly variants of the UNet model [2]) have reached excellent results in medical image segmentation, including tumor segmentation [3,4]. The prediction of patient outcome (e.g. survival) using CNNs, however, has received less attention or success. For this task, the number of observations is smaller than for segmentation tasks, in which each pixel/voxel is an observation, when compared to patient-wise observations for radiomics. The level of abstraction required to predict the patient outcome is also higher than in a segmentation task. For survival tasks, moreover, the loss requires comparing several pairs of observations to estimate concordance, thereby requiring many observations as compared to other classical losses (e.g. cross-entropy and Dice). Survival losses are therefore not particularly suited for deep CNNs training with mini-batches. As a consequence, training a (3D) deep CNN from scratch or based on a pretrained network with a few hundred patients for the prediction of a complex outcome with censored data is generally not as efficient and successful as extracting hand-crafted radiomics features and training a simple survival model, e.g. the Cox Proportional Hazards (CPH) model.

Early works on Artificial Neural Networks (ANNs) for survival problems were proposed to learn nonlinear relationships between prognostic features (not imaging features as used in radiomics) and the risk for a given outcome [5]. More recently, deep ANNs [6,7] were shown to successfully model complex relationships between the radiomics features and their risk of failure. An extensive review is proposed in [8].

Radiomics studies that make use of CNNs have often focused on using automatic segmentations or deep features, obtained from a trained CNN, in a standard radiomics pipeline, e.g. in the context of brain tumor [9]. In lung cancer, CNNs trained to perform tumor segmentation in FluoroDeoxyGlucose-Positron Emission Tomography (FDG-PET) and Computed Tomography (CT) were shown to identify a rich set of survival-related features with remarkable prognostic value [10]. CNNs have also been trained end-to-end for classification radiomics tasks, e.g. in Head and Neck cancer (H&N) [11]; see [12] for a review on deep learning-based radiomics. A few recent works proposed the use of CNNs trained with a Cox loss for survival tasks. Zheng et al. [13] proposed to train a CNN with a Cox loss for the prediction of patient survival in the context of pancreatic ductal adenocarcinoma. A survival CNN was used in [14] to predict the outcome of patients with brain cancer from histopathology images and genomic biomarkers, as well as in [15] for a survival analysis of rectal cancer based on PET/CT images.

Multi-task learning is a well-studied sub-field, notably in robotics and autonomous driving [16,17]. It was shown that learning related tasks in parallel while using a shared representation can improve the individual learning efficiency and prediction accuracy, as features learned for each task can help the learning of other tasks [16]. In medical imaging, multi-task training was recently used for several tasks and modalities. A multi-task method was proposed in [18] for brain tumor segmentation in Magnetic Resonance Imaging (MRI), leveraging fully annotated images (segmentation task) and weakly annotated images (classification of presence or absence of a tumor). Again for brain tumor segmentation, Weninger *et al.* [19] combined three tasks that share the encoder, namely tumor segmentation, image reconstruction (auto-encoder), and the classification of presence/absence of enhancing tumor. Another multi-task deep learning model was developed in [20] to jointly identify COVID-19 patients and segment the lesions from chest CT images. In histopathology, multi-task was used to improve tumor tissue classification by combining the prediction of auxiliary clinically relevant features as well as a domain adversarial task [21]. We are, however, not aware of previous works that make use of multi-task deep learning combining segmentation and survival losses in a model that can be trained end-to-end, neither multi-task learning applied to PET-CT radiomics nor to H&N cancer.

We base our work on the idea that deep features learned for tumor segmentation may also be useful for the prediction of patient outcome. In particular, neuron activations triggered by the segmentation task are expected to spatially guide the network to extract prognostically relevant patterns in the tumor area. In addition, training the model for tumor segmentation as an auxiliary task to outcome prediction allows the network to better exploit the scarce and precious radiomics training data where one observation is one patient. In this work, we show the potential of our multi-task approach for the prediction of outcome for patients with H&N cancer from PET/CT images. The main task is the prediction of Disease-Free Survival (DFS), to which we combine the auxiliary task of segmentation of the primary Gross Tumor Volume (GTVt).

To summarize, the contribution of this work is to propose a fully automatic bi-modal 3D deep segmentation and prognostic model able to deal with survival data and to learn from relatively small datasets, thanks to the multi-task paradigm.

2   Methods

In this section, we present the different methods developed in our work and the dataset used for the experiments. The main contribution, i.e. multi-task deep radiomics, is presented in Sect. 2.2. All models are bi-modal as they take CT and PET images as input. An overview of the multi-task hybrid segmentation-deep radiomics architecture is depicted in Fig. 1. The code to reproduce the experiments will be shared on our GitHub repository[1].

---

[1] https://github.com/vandrearczyk.

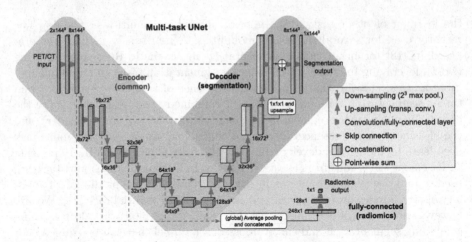

Fig. 1. 3D multi-modal (PET/CT) and multi-task architecture with a common down-sampling branch (green), an up-sampling segmentation branch (blue) and a radiomics branch (red). Residual convolutional layers are used in the down-sampling part. (Color figure online)

2.1 Segmentation

As a first model, we use a multi-modal 3D UNet for segmentation based on the model developed in [22], winner of the HECKTOR 2020 challenge[2] [4]. The architecture is presented in Fig. 1, including down-sampling (green) and up-sampling (blue) parts. The probabilities of the softmax activated outputs are thresholded at 0.5 to obtain a binary mask. More details on the implementation can be found in [22]. The model is trained with a Dice loss, computed as

$$\mathcal{L}_{Dice} = -2 \frac{\sum_k \hat{y}_k y_k}{\sum_k \hat{y}_k + \sum_k y_k},\qquad(1)$$

where $\hat{y}_k \in [0,1]$ is the softmax output for a voxel k, $y_k \in \{0,1\}$ is the value of this voxel in the 3D ground truth mask and the sum is computed over all voxels.

2.2 Multi-task Segmentation and Radiomics

The multi-task architecture is composed of the normal 3D segmentation with an additional radiomics branch (red in Fig. 1) at the bottleneck of the network. This radiomics branch is connected to multiple layers of the downsampling path using skip connections to gather information at multiple scales and complexity. It is composed of a global average pooling layer to aggregate the spatial information, a densely connected layer (ReLU activated) with 128 neurons and a prediction layer with a single neuron. Dropout with 0.5 probability is added before the

---

[2] The main difference is the reduced number of filters to be able to use a larger batch-size for the survival loss in Sect. 2.2.

dense layers for regularization. For the segmentation task, we use the Dice loss defined in Eq. (1). For the radiomics task, we use a Cox loss [7] computed as

$$\mathcal{L}_{Cox} = -\frac{1}{N_{E=1}} \sum_{i:E_i=1} \left(\hat{h}(x_i) - \log \sum_{j \in \mathcal{H}(T_i)} e^{\hat{h}(x_j)} \right), \qquad (2)$$

where $\hat{h}(x_i)$ is the estimated log-hazard for input x_i. E_i, T_i, x_i are the event indicator (0 for censored data), the observed duration, and the baseline covariate data in the ith observation respectively. $\mathcal{H}(T_i)$ is the hazard indicator for individuals that have not experienced the event of interest before time T_i. The Cox loss is motivated by clinical needs such as group stratification based on risk. It alleviates the need to model the underlying hazard, a strength of the Cox model in this scenario.

For learning the radiomics task only, we use the same network as the multi-task one, without the up-sampling part of the UNet (i.e. only green and red parts in Fig. 1). The model is trained with the Cox loss defined in Eq. (2). Note that in this third method, the manually annotated contours are used neither during training nor testing.

2.3   Training Scheme

All networks are trained with an Adam optimizer based on a cosine decay learning rate (initial 10^{-3}) and a batch size of 10, for 60 epochs. The three training scenarios are obtained by modifying the pair of loss weights (w_{rad}, w_{seg}) in the combined loss $\mathcal{L} = w_{rad}\mathcal{L}_{Cox} + w_{seg}\mathcal{L}_{Dice}$: $(0, 1)$ for the segmentation, $(1, 1)$ for the multi-task, and $(1, 0)$ for the radiomics. For each of the three settings, the best model based on the combined loss obtained on the validation data is used for testing. The training data is resampled to balance the proportion of censored data for the Cox loss (82% of censored data on average before resampling and 50% after). Data augmentation is applied to the training data including random shifts (maximum 30 voxels), mirroring (sagittal axis) and rotations (maximum five degrees). The models are implemented in TensorFlow 2 and trained on a Nvidia V-100 GPU (32 GB). The average training time of the multi-task model for one fold of the Cross-Validation (CV, see Sect. 2.6) is 148 min.

2.4   Standard Radiomics

We compare the deep algorithms with a state-of-the-art standard radiomics models based on hand-crafted features. We implemented the feature extraction and survival model in Python 3.9 with the following libraries: SimpleITK (2.0.2), PyRadiomics (3.0.1), scikit-learn (0.22.2) and scikit-survival (0.12.0). The computation time of the feature extraction and training times for one fold on an AMD Ryzen 7 3700X with 8 cores are 13 min and 3 s respectively.

Feature Extraction. We extract features inside the GTVt from the CT and PET images using PyRadiomics [23]. A total of 274 features are extracted for each patient, including 18 first-order and 112 texture features extracted per modality[3] as well as 14 additional shape-based features. Note that this approach, as opposed to the deep learning counterparts presented above, requires test-time ground truth annotations of the primary tumors as VOIs to extract the features.

Survival Model. We first select features based on the univariate C-index estimated on the validation set (see Sect. 2.6 for the clarification of training, validation and test splits). This selection is recommended e.g. in [24] and is similar to an F1-score univariate selection. More precisely, we use a shifted version (i.e. |C-index − 0.5|) to account for both concordant and anti-concordant features. The resulting top 20 features are kept. Second, correlated features are removed when higher than a given threshold value $t \in [0.6, 0.65, 0.70, 0.75, 0.80]$, optimized on the validation set using grid-search, as recommended e.g. in [25]. The resulting feature set is used by a CPH model [26] to predict the hazard score of the DFS outcome. The best performing model on the validation data is kept and used to predict hazards on the test set.

2.5 Evaluation

The radiomics results, i.e. predictions of DFS, are evaluated using the Concordance index (C-index) ranging from zero to one [27]. A C-index of one corresponds to the best model prediction, while a value of 0.5 is equivalent to a random prediction. The segmentation results are evaluated using the Dice Similarity Coefficient (DSC) ranging from zero to one, with one reflecting a perfect similarity between predicted and ground-truth contours. The DSCs are averaged across multiples cases as specified in the results section.

2.6 Dataset and Experimental Setup

For the experiments, we use the HECKTOR 2020 data including 239 cases from five centers [4]. This dataset was used for tumor segmentation and we propose for the first time to use it for prediction of patient outcome. Each case includes a CT and a PET image (inputs), a ground truth annotation of the GTVt and the DFS patient outcome information, i.e. time to recurrence of the GTVt following the treatment (outputs). The number of events is 43, i.e. 18% of non-censored cases, whereas the remaining 82% cases did not encounter the event during the follow-up. The average follow-up time is 1182 days.

The PET and CT images are resampled to $1 \times 1 \times 1$ mm ($2 \times 2 \times 2$ mm for the standard radiomics pipeline) using trilnear interpolation. The volumes are cropped to $144 \times 144 \times 144$ voxels (after augmentation described in Sect. 2.3) using

---

[3] 24 GLCM, 16 GLRLM, and 16 GLDZM features. A Fixed Bin Number (FBN) of 64 and a Fixed Bin Size (FBS) of 50 are used for CT. A FBN of 8 and a FBS of 1 are used for PET.

the bounding boxes of the HECKTOR 2020 challenge [28]. The PET images are standardized individually to zero mean and unit variance. The CT images are clipped to $[-1024, 1024]$ and mapped to $[-1, 1]$. A 5-fold CV is used for all experiments. For each fold, 20% of the dataset used for testing and the remainder is split again randomly as 80% training, 20% validation.

3   Results

Performance results for radiomics, i.e. comparison of models on the DFS prediction task, and segmentation are reported in Tables 1 and 2, respectively. It is worth noting that the radiomics results of the segmentation-only model and the segmentation results of the radiomics-only model are only reported as sanity checks as they both achieve (expected) random performance on these tasks. There is no precedent work on this dataset nor task.

Regarding the comparison with state of the art, standard radiomics methods have been applied to other H&N datasets and other survival outcomes in e.g. [29], yet direct comparison is not possible. We therefore compare against the standard radiomics method described in Sect. 2.4.

Table 1. Performance comparison for the radiomics task. w_{rad} and w_{seg} are the radiomics and segmentation loss weights, respectively. We report the C-index for each fold of the CV as well as the average $\pm$ standard-error of the C-index.

model (w_{rad}, w_{seg})	fold-1	fold-2	fold-3	fold-4	fold-5	Mean
Deep radiomics (1, 0)	0.703	0.599	0.578	0.687	0.684	0.650 ±0.026
Deep multi-task (1, 1)	0.713	0.702	0.803	0.615	0.783	**0.723** ±0.033
Deep segmentation (0, 1)	0.413	0.377	0.473	0.348	0.570	0.416 ±0.024
Standard radiomics	0.827	0.710	0.627	0.687	0.624	0.695 ±0.0826

Table 2. Performance comparison for the segmentation task. w_{rad} and w_{seg} are the radiomics and segmentation loss weights, respectively. We report the average Dice score for each fold of the CV as well as the global average $\pm$ standard-error for the Dice.

model (w_{rad}, w_{seg})	fold-1	fold-2	fold-3	fold-4	fold-5	Mean
Deep radiomics (1, 0)	0.006	0.001	0.000	0.034	0.000	0.008 ± 0.007
Deep multi-task (1, 1)	0.700	0.589	0.361	0.595	0.677	0.584 ± 0.060
Deep segmentation (0, 1)	0.713	0.681	0.696	0.639	0.685	**0.683** ± 0.012

4   Discussion and Conclusions

The network design proposed in this work relies on the assumption that teaching to segment the tumoral volume will benefit the prediction of patient prognosis. The observed results seem to validate this hypothesis, as the most concordant DFS prediction is achieved when combined with the segmentation task (see Table 1) culminating to an average C-index of 0.723. The combination of Dice and survival losses allowed to spatially guide neuron activations with the former, and learn localized prognostically relevant patterns with the latter. As a reminder, only a large bounding box encompassing the spatially extended oropharyngeal region (automatically determined as of [28]) is provided to the network. Therefore, this approach does not require tumoral contours at test time, as opposed to the standard radiomics approach. Remarkably, this fully automatic approach outperformed the standard radiomics pipeline relying on manual contouring (C-index of 0.695). This result opens avenues for very large scale clinical studies to validate the prognostic models on patients for which we only have the outcomes but no manual annotations of the tumors.

We believe that this work contributes to the state of the art by proposing a fully automatic bi-modal 3D deep prognostic model able to deal with survival data and to learn from relatively small datasets. It can do so by optimally leveraging training data via the use of the highly related and observation-rich segmentation task.

A surprisingly high prognostic performance is achieved even without using the Dice loss (only deep radiomics), which is highlighted by a C-index of 0.650, 7.3% lower than the top result. This demonstrates the efficacy of the survival loss combined with an appropriate encoding architecture. Note that in this fully-radiomics method, the GTVt contours are provided neither during training nor testing.

The best-performing segmentation method achieved a Dice score of 0.683. For this GTVt segmentation, the segmentation task did not benefit from the radiomics task, where the Dice loss provided best results when used on its own (see Table 2). This observation may be due to an optimization issue, the PFS task adding noise to the gradients. Other loss weights could be used to favor the segmentation task, yet the main task of interest in this work is the outcome prediction and the segmentation is only used to boost the performance on the latter. Note that the segmentation performance is far from the winner of the HECKTOR 2020 challenge (average DSC of 0.759), in which the main and only task was tumor segmentation, and the model was based on a complex ensemble of UNets.

A limitation of the proposed work is the use of binary weights (w_{rad}, w_{seg}) given to the two losses in the multi-task model (i.e. unweighted sum of losses). In future work, we will explore other types of loss weighting such as geometric [30] and epistemic uncertainty losses [31]. As another interesting future work, one could also consider adding an auxiliary task of domain adversariability to the training with a branch similar to the radiomics one and with a gradient reversal to create domain invariant features and ensure good generalization to new scanners and image acquisition protocols [32]. We also plan to study activation maps to reveal the most prognostically relevant regions and patterns used by the deep radiomics model.

Acknowledgements. This work was partially supported by the Swiss National Science Foundation (SNSF, grant 205320_179069) and the Swiss Personalized Health Network (SPHN via the IMAGINE and QA4IQI projects).

References

1. Gillies, R.J., Kinahan, P.E., Hricak, H.: Radiomics: images are more than pictures, they are data. Radiology **278**(2), 563–577 (2016)
2. Ronneberger, O., Fischer, P., Brox, T.: U-net: convolutional networks for biomedical image segmentation. In: Navab, N., Hornegger, J., Wells, W.M., Frangi, A.F. (eds.) MICCAI 2015. LNCS, vol. 9351, pp. 234–241. Springer, Cham (2015). https://doi.org/10.1007/978-3-319-24574-4_28
3. Menze, B.H., et al.: The multimodal brain tumor image segmentation benchmark (BRATS). IEEE Trans. Med. Imaging **34**(10), 1993–2024 (2014)
4. Andrearczyk, V., et al.: Overview of the HECKTOR challenge at MICCAI 2020: automatic head and neck tumor segmentation in PET/CT. In: Andrearczyk, V., Oreiller, V., Depeursinge, A. (eds.) HECKTOR 2020. LNCS, vol. 12603, pp. 1–21. Springer, Cham (2021). https://doi.org/10.1007/978-3-030-67194-5_1
5. Faraggi, D., Simon, R.: A neural network model for survival data. Stat. Med. **14**(1), 73–82 (1995)
6. Ranganath, R., Perotte, A., Elhadad, N., Blei, D.: Deep survival analysis. In: Machine Learning for Healthcare Conference, pp. 101–114. PMLR (2016)
7. Katzman, J.L., Shaham, U., Cloninger, A., Bates, J., Jiang, T., Kluger, Y.: DeepSurv: personalized treatment recommender system using a cox proportional hazards deep neural network. BMC Med. Res. Methodol. **18**(1), 1–12 (2018)
8. Steingrimsson, J.A., Morrison, S.: Deep learning for survival outcomes. Stat. Med. **39**(17), 2339–2349 (2020)
9. Crimi, A., Bakas, S., Kuijf, H., Keyvan, F., Reyes, M., van Walsum, T.: Brainlesion: Glioma, Multiple Sclerosis, Stroke and Traumatic Brain Injuries: 4th International Workshop, BrainLes 2018, Held in Conjunction with MICCAI 2018, Granada, Spain, September 16, 2018, Revised Selected Papers, Part II, vol. 11384. Springer, Heidelberg (2019). https://doi.org/10.1007/978-3-030-11726-9
10. Baek, S., et al.: Deep segmentation networks predict survival of non-small cell lung cancer. Sci. Rep. **9**(1), 1–10 (2019)
11. Parekh, V.S., Jacobs, M.A.: Deep learning and radiomics in precision medicine. Expert Rev. Precis. Med. Drug Dev. **4**(2), 59–72 (2019)
12. Diamant, A., Chatterjee, A., Vallières, M., Shenouda, G., Seuntjens, J.: Deep learning in head & neck cancer outcome prediction. Sci. Rep. **9**(1), 1–10 (2019)
13. Zhang, Y., Lobo-Mueller, E.M., Karanicolas, P., Gallinger, S., Haider, M.A., Khalvati, F.: CNN-based survival model for pancreatic ductal adenocarcinoma in medical imaging. BMC Med. Imaging **20**(1), 1–8 (2020)
14. Mobadersany, P., et al.: Predicting cancer outcomes from histology and genomics using convolutional networks. Proc. Natl. Acad. Sci. **115**(13), E2970–E2979 (2018)
15. Li, H., et al.: Deep convolutional neural networks for imaging data based survival analysis of rectal cancer. In: 2019 IEEE 16th International Symposium on Biomedical Imaging (ISBI 2019), pp. 846–849. IEEE (2019)

16. Caruana, R.: Multitask learning. Mach. Learn. **28**(1), 41–75 (1997)
17. Standley, T., Zamir, A., Chen, D., Guibas, L., Malik, J., Savarese, S.: Which tasks should be learned together in multi-task learning? In: International Conference on Machine Learning, pp. 9120–9132. PMLR (2020)
18. Mlynarski, P., Delingette, H., Criminisi, A., Ayache, N.: Deep learning with mixed supervision for brain tumor segmentation. J. Med. Imaging **6**(3), 034002 (2019)
19. Weninger, L., Liu, Q., Merhof, D.: Multi-task learning for brain tumor segmentation. In: Crimi, A., Bakas, S. (eds.) BrainLes 2019. LNCS, vol. 11992, pp. 327–337. Springer, Cham (2020). https://doi.org/10.1007/978-3-030-46640-4_31
20. Multi-task deep learning based ct imaging analysis for covid-19 pneumonia: classification and segmentation
21. Graziani, M., Otálora, S., Muller, H., Andrearczyk, V.: Guiding CNNs towards relevant concepts by multi-task and adversarial learning. arXiv preprint arXiv:2008.01478 (2020)
22. Iantsen, A., Visvikis, D., Hatt, M.: Squeeze-and-excitation normalization for automated delineation of head and neck primary tumors in combined PET and CT images. In: Andrearczyk, V., Oreiller, V., Depeursinge, A. (eds.) HECKTOR 2020. LNCS, vol. 12603, pp. 37–43. Springer, Cham (2021). https://doi.org/10.1007/978-3-030-67194-5_4
23. Van Griethuysen, J.J.M., et al.: Computational radiomics system to decode the radiographic phenotype. Cancer Res. **77**(21), e104–e107 (2017)
24. Suter, Y., et al.: Radiomics for glioblastoma survival analysis in pre-operative MRI: exploring feature robustness, class boundaries, and machine learning techniques. Cancer Imaging **20**(1), 1–13 (2020)
25. Lambin, P., et al.: Radiomics: the bridge between medical imaging and personalized medicine. Nat. Rev. Clin. Oncol. **14**(12), 749–762 (2017)
26. David, C.R., et al.: Regression models and life tables (with discussion). J. Roy. Stat. Soc. **34**(2), 187–220 (1972)
27. Harrell, F.E., Lee, K.L., Mark, D.B.: Tutorial in biostatistics multivariable prognostic models. Stat. Med. **15**, 361–387 (1996)
28. Andrearczyk, V., Oreiller, V., Depeursinge, A.: Oropharynx detection in PET-CT for tumor segmentation. In: Irish Machine Vision and Image Processing (2020)
29. Vallieres, M., et al.: Radiomics strategies for risk assessment of tumour failure in head-and-neck cancer. Sci. Rep. **7**(1), 1–14 (2017)
30. Chennupati, S., Sistu, G., Yogamani, S., Rawashdeh, S.A.: MultiNet++: multi-stream feature aggregation and geometric loss strategy for multi-task learning. In: Proceedings of the IEEE/CVF Conference on Computer Vision and Pattern Recognition Workshops (2019)
31. Kendall, A., Gal, Y., Cipolla, R.: Multi-task learning using uncertainty to weigh losses for scene geometry and semantics. In: Proceedings of the IEEE Conference on Computer Vision and Pattern Recognition, pp. 7482–7491 (2018)
32. Andrearczyk, V., Depeursinge, A., Müller, H.: Neural network training for cross-protocol radiomic feature standardization in computed tomography. J. Med. Imaging **6**(3), 024008 (2019)

Integrating Multimodal MRIs for Adult ADHD Identification with Heterogeneous Graph Attention Convolutional Network

Dongren Yao[1,2], Erkun Yang[1], Li Sun[3], Jing Sui[4(✉)], and Mingxia Liu[1(✉)]

[1] Department of Radiology and BRIC, University of North Carolina at Chapel Hill, Chapel Hill, NC 27599, USA
mxliu@med.unc.edu
[2] Brainnetome Center and National Laboratory of Pattern Recognition, Institute of Automation, Chinese Academy of Sciences, Beijing 100190, China
[3] National Clinical Research Center for Mental Disorders and Key Laboratory of Mental Health, Ministry of Health, Peking University, Beijing 100191, China
[4] State Key Laboratory of Cognitive Neuroscience and Learning, Beijing Normal University, Beijing 100678, China
jsui@bnu.edu.cn

Abstract. Adult attention-deficit/hyperactivity disorder (ADHD) is a mental health disorder whose symptoms would change over time. Compared with subjective clinical diagnosis, objective neuroimaging biomarkers help us better understand the mechanism of brain between patients with brain disorders and age-matched healthy controls. In particular, different magnetic resonance imaging (MRI) techniques can depict the brain with complementary structural or functional information. Thus, effectively integrating multi-modal MRIs for ADHD identification has attracted increasing interest. Graph convolutional networks (GCNs) have been applied to model brain structural/functional connectivity patterns to discriminate mental disorder from healthy controls. However, existing studies usually focus on a specific type of MRI, and therefore cannot well handle heterogeneous multimodal MRIs. In this paper, we propose a heterogeneous graph attention convolutional network (HGACN) for ADHD identification, by integrating resting-state functional MRI (fMRI) and diffusion MRI (dMRI) for a comprehensive description of the brain. In the proposed HGACN, we first extract features from multimodal MRI, including functional connectivity for fMRI and fractional anisotropy for dMRI. We then integrate these features into a heterogeneous brain network via different types of metapaths, with each type of metapatch corresponding to a specific modality or functional/structural relationship between regions of interest. We leverage both intra-metaptth and inter-metapath attention to learn useful graph embeddings/representations of multimodal MRIs and finally predict subject's category label using these embeddings. Experimental results on 110 adult ADHD patients and 77 age-matched HCs suggest that our HGACN outperforms several state-of-the-art methods in ADHD identification based on resting-state functional MRI and diffusion MRI.

Keywords: Attention-deficit/hyperactivity disorder · Graph convolution network · Resting-state functional MRI · Diffusion MRI

© Springer Nature Switzerland AG 2021
I. Rekik et al. (Eds.): PRIME 2021, LNCS 12928, pp. 157–167, 2021.
https://doi.org/10.1007/978-3-030-87602-9_15

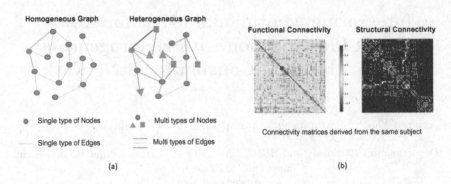

Fig. 1. Illustration of (a) homogeneous and heterogeneous graphs, and (b) functional and structural connectivity networks derived from resting-state fMRI and diffusion MRI of the same subject.

1 Introduction

Attention-Deficit/Hyperactivity Disorder (ADHD) is one of the most common neurodevelopmental disorders that occur in children, adolescents, and adults. Patients with ADHD often exhibit an ongoing pattern of inattention, hyperactivity and impulsive behavior [1].

Previous studies usually pay more attention to children with ADHD and have made substantial achievements in computer-aided diagnosis [2]. However, ADHD can last into adulthood, and symptoms along with adult patients can lead to difficulty at home or work and in social settings. The current diagnosis of ADHD in adults gathering multiple resources is mainly based on subjective observation or rating scale, such as diagnostic manual Diagnostic and Statistical Manual of Mental Disorders, Fifth Edition (DSM-5). Considering the fact that ADHD-associated symptoms would change over time, it is essential to employ objective imaging biomarkers to help us better understand the brain mechanism of adult ADHD for timely intervention and effective treatment [3–6].

In the past two decades, the use of non-invasive neuroimaging techniques such as electroencephalography (EEG), magnetic resonance imaging (MRI) and functional MRI to describe and understand the brain has progressed rapidly [7–9].

With the development of these brain imaging technologies, we have more direct ways to observe and measure those brain structural and functional alterations between healthy control and pathological conditions [10–13]. Currently, the etiological bases and neural substrates of adult ADHD are far from being fully understood. For instance, patients with ADHD suffer from inattention, hyperactivity, and impulsive behaviors that are not matched with their own age.

To better understand the neurobiological mechanisms underpinning this mental disorder, many fMRI-based studies have reported that large-scale brain dysfunctions may lie in ADHD patients [14,15]. Also, diffusion tensor imaging (DTI) based studies [16–18] have suggested that the underdevelopment of gray or white

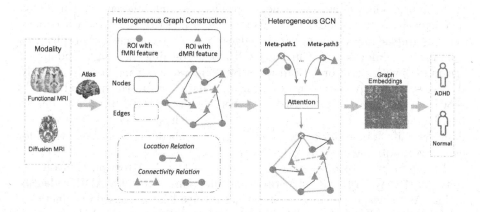

Fig. 2. Illustration of our Heterogeneous Graph Attention Convolutional Network (HGACN) for ADHD identification with multimodal MRI. The whole framework contains 3 parts: 1) Based on preprocessed functional and diffusion MRIs and specific atlas, we calculate Functional Connectivity (FC) from fMRI and Fractional Anisotropy (FA) from dMRI. 2) Based on the calculated functional and structural connectivity matrix, we construct a heterogeneous graph with two different nodes and three kinds of edges for each subject. 3) Based on the constructed graph, we employ a heterogeneous graph convolution network to learn a new graph embedding for each subject. With learned graph embedding, a fully-connected layer is leveraged for predicting the category label of a subject.

matter structure in the frontal lobe, striatum, and thalamus area may lead to the emergence of ADHD when they were in childhood. However, most of the existing studies focus on using a single MRI modality (*e.g.*, fMRI or dMRI) for investigating ADHD. Considering that these modalities can provide multi-view diverse and complementary information, it is highly desired to design a unified framework to effectively leverage multimodal MRIs for ADHD diagnosis.

Apart from advances in multimodal neuroimaging techniques, the development of graph convolution networks (GCNs) has allowed researchers to treat the human brain's structural and functional connectivity patterns as the "connectome" to learn rich graph embeddings/representations. Many previous works use graph theory information such as modularity and hierarchy. Based on a single modality, several previous studies [9, 19–21] employed Kipf or Chebyshev GCN and metric learning framework to measure the similarity between patients and healthy controls. With multiple modalities, Xing *et al.* proposed a DS-GCN model, consisting of GCN and Long term short Memory (LSTM) based on structural MRI (sMRI) and fMRI data, for automated diagnosis of mild cognitive decline [22]. Several studies employed GCN models to perform brain disorder analysis using DTI and fMRI data [23, 24]. These studies have shown that using GCN can achieve superior performance than traditional learning-based methods on brain image analysis. However, they generally treat graphs corresponding to different modalities as homogeneous graphs, ignoring inherent inter-modality heterogeneity of multimodal MRIs. As shown in Fig. 1 (a), compared with homogeneous graphs, multi-type nodes and edges comprise heterogeneous graphs, containing richer information. Current GCN models dealing with multimodal data (no matter with early or late fusion strategy) mine each modality separately, thus cannot effectively integrate multimodal data for subsequent analyses. As shown in Fig. 1 (b), after constructing heterogeneous fMRI-based and dMRI-based graphs (*i.e.*, connectivity matrices) for each subject, it is expected to design a unify framework to learn useful graph embeddings/representations for assisting in automated ADHD diagnosis.

In this paper, we propose a Heterogeneous Graph Attention Convolution Network (HGACN) to fuse structural and functional MRIs for ADHD identification. As shown in Fig. 2, we first extract diverse features based on functional MRI (fMRI) and diffusion MRI (dMRI) to depict the brain topology. For each subject, we employ the Automated Anatomical Labelling (AAL) template to generate a functional connectivity matrix/graph from resting-state fMRI and a fractional anisotropy (FA) matrix from dMRI. After pre-processing, we then model these features into one heterogeneous brain graph through different types of metapaths, with each type of meta-pathes corresponding to a specific modality or functional/structural relationship between paired regions of interest. We leverage both intra-metapth and inter-metapath attention to aggregate and learn useful graph embeddings for representing multimodal MRIs, and finally use them to identify adult ADHD from age-matched healthy controls (HCs). Experimental results on 110 adult ADHD and 77 age-matched HC subjects demonstrate the effectiveness of HGACN in integrating fMRI and dMRIs for ADHD identification. To the best of our knowledge, this is among the first attempts to fuse resting-state functional MRI and diffusion MRI data via an end-to-end heterogeneous GCN model for automated brain disorder analysis.

2   Materials and Method

2.1   Data and Image Pre-processing

The studied ADHD dataset contains 110 adult ADHD patients and 77 age-matched HCs recruited from a local hospital. Each subject is represented by three imaging modalities, including structural MRI (sMRI), rs-fMRI and dMRI. The demographic information about the studied subjects can be found in Table 1.

For rs-fMRI data, we leverage the Data Processing Assistant for Resting-State fMRI (DPARSF) to calculate functional connectivity matrix based on the AAL atlas. Specifically, we first discard the first ten time points, followed by slice timing correction, head motion correction, regression of nuisance co-variants of head motion parameters, white matter, and cerebrospinal fluid (CSF). Then, fMRI data are normalized with an EPI template in the MNI space, and resampled to the resolution of $3 \times 3 \times 3 \, mm^3$, followed by spatial smoothing using a 6 mm

Table 1. Demographic information of studied subjects in the Adult ADHD dataset. M: Male; F: Female; Std: Standard Deviation.

Category	#Subject	Site 1	Site 2	Gender (F/M)	Age (Mean±Std)
Adult ADHD	110	72	38	37/73	25.9 ± 4.9
Healthy Control (HC)	77	43	34	34/43	26.0 ± 3.9

full width half maximum Gaussian kernel. Finally, the AAL atlas, with 116 predefined regions-of-interest (ROIs) including cortical and subcortical areas, are nonlinearly aligned onto each scan to extract the mean time series for each ROI.

The dMRI data are preprocessed through FMRIB Software Library (FSL) with the following operations. 1) We first use the brain extraction tool (BET) to remove non-brain tissue; 2) then leverage nonuniform intensity normalization (N3) algorithm for filed bias correction on T1-weighted sMRIs; 3) further employ eddy current induced distortion to correct DWI scans; 4) also adjust the diffusion gradient strength accordingly, and 5) finally calculated the fractional anisotropy (FA) from DWI scans based on diffusion tensor fitting.

2.2 Method

As shown in Fig. 2, our HGACN model aims to combine multimodal MRIs into a unified heterogeneous graph learning framework for identifying ADHD patients from HCs. For each subject, we construct two heterogeneous graphs based on its functional and structural connectivity features derived from fMRI and dMRI, respectively. More details can be found in the following.

Construction of Heterogeneous Graph. The structure of a heterogeneous graph can be defined as $\mathbf{G} = (\mathbf{V}, \mathbf{E})$, where $\mathbf{V}$ and $\mathbf{E}$ denote the nodes and edges in the heterogeneous graph, respectively. The node set $\mathbf{V}$ has two different types of nodes (with each node corresponding to a specific ROI): 1) *functional nodes* $\mathbf{V}_f$, and 2) *structural nodes* $\mathbf{V}_d$. The edge set $\mathbf{E}$ consists of three types of pairwise ROI relationships: 1) *functionally connected edges* $\mathbf{E}_{ff}$ that represent functional uniformity between $\mathbf{V}_f$ nodes based on their Pearson correlation, 2) *structurally connected edges* $\mathbf{E}_{dd}$ that express structural connectivity coherence between $\mathbf{V}_d$ nodes based on their fractional anisotropy, and 3) *location-related edges* $\mathbf{E}_{df}$ that connect two different type of ROIs ($\mathbf{V}_d$ and $\mathbf{V}_f$) which are physically close in the predefined atlas. In this work, we employ AAL atlas containing 116 ROIs for brain ROI parcellation. Besides, for a heterogeneous graph, we define a meta-path to describe a composite relationship $(\mathbf{V}_1 \xrightarrow{E_1} \mathbf{V}_2 \xrightarrow{E_2} \cdots \xrightarrow{E_l} \mathbf{V}_{l+1})$ between a set of nodes (*e.g.*, $\mathbf{V}_1$ and $\mathbf{V}_{l+1}$).

After constructing heterogeneous graphs, we propose a novel Heterogeneous Graph Attention Convolutional Network (HGACN) to learn new graph embeddings/representations for input subjects. Denote I as an identity matrix, A as an adjacency matrix, and M as a degree matrix. Similar to standard GCNs [9,25] that are performed on homogeneous graphs $\mathcal{G}$, the aggregation operation with an adjacency matrix $A' = A + I$ is defined as follows:

$$H^{(l+1)} = \sigma(\widetilde{A} \cdot H^{(l)} \cdot W^{(l)}) \tag{1}$$

where $\sigma(\cdot)$ is an activation function such as ReLU, $\widetilde{A} = M^{-\frac{1}{2}} A' M^{-\frac{1}{2}}$ denotes the symmetric normalized adjacent matrix, and $W^{(l)}$ is trainable parameters for the l-th layer in GCN.

Through Eq. (1), we can only deal with nodes containing the same shared embedding space, therefore failing to handle heterogeneous graphs that contain different types of nodes and edges. To address this issue, we define a new graph convolution to explicitly take advantage of different types of node and edge information in a heterogeneous graph, as follows:

$$H^{(l+1)} = \sigma(\sum_{\tau \in \mathcal{T}} \widetilde{A}_\tau \cdot H_\tau^{(l)} \cdot W_\tau^{(l)}) \tag{2}$$

where $\mathcal{T}$ denotes all types of nodes ($\mathcal{T} = 2$ in this work) in heterogeneous graph, $\widetilde{A}_\tau$ denote the adjacency matrix corresponding to the τ-th type ($\tau = 1, \cdots, \mathcal{T}$). Based on Eq. (2), the new representation $H^{(l+1)}$ is generated by aggregating different types of edges with different transformation matrices $W_\tau^{(l)}$ from their neighboring nodes $H_\tau^{(l)}$.

Note that different types of neighboring nodes would have different impacts on a specific node, and the same type of nodes may be similarly affected by their neighbors. To model these characteristics, we propose to capture the intra- and inter-metapath attention information to learn more effective graph embeddings.

Intra-metapath Aggregation. For a specific node v, attention for intra-metapath aggregation will learn the weighs of the same types of neighboring nodes. Specifically, we denote $h_\tau = \sum_{v'} \widetilde{A}_{vv'} h_{v'}$ as the embedding of the τ-th type of nodes based on their neighbor node features $h_{v'}$. Then, we calculate the intra-metapath attention score as shown below:

$$\alpha_\tau = \sigma(\mu_\tau^T \cdot [h_v \oplus h_\tau]) \tag{3}$$

where μ_τ is the attention vector for the τ-th type τ and $\oplus$ denote the concatenation operation. After that, we compute the final node-level attention weight of the τ-th node type based on a softmax function as follows:

$$\alpha_\tau = \frac{\exp(\alpha_\tau)}{\sum_{\tau' \in \mathcal{T}} \exp(\alpha_{\tau'})} \tag{4}$$

Inter-metapath Aggregation. Besides the intra-metapath aggregation on a specific node type, we also propose to model the importance of different neighbor nodes between different types. Denote N_v as a node set, with each node denoting a neighbor of a center node v. Note that these neighbor nodes and the center node v may be defined in a different manner (*i.e.*, belonging to different types of node sets). Given a specific node v defined in the τ-th node type, we calculate the inter-metapath attention based on its neighbour nodes $v' \in N_v$ as follows:

$$\beta_{vv'} = \frac{\exp(\sigma(n^T \cdot \alpha_{\tau'}[h_v \oplus h_{v'}]))}{\sum_{i \in N_v} \exp(\sigma(n^T \cdot \alpha_i[h_v \oplus h_i]))} \qquad (5)$$

where h_v and $h_{v'}$ denote the embeddings of the node v and the node v', n is a to-be-learned attention vector, and $\alpha_{\tau'}$ and α_i are calculated from Eq. (3).

Based on intra- and inter-metapath aggregation with two attention operations, the final heterogeneous graph convolution can be written as follows:

$$H^{(l+1)} = \sigma(\sum_{\tau \in \mathcal{T}} \mathcal{B}_\tau \cdot H_\tau^{(l)} \cdot W_\tau^{(l)}) \qquad (6)$$

where the attention matrix $\mathcal{B}$ is a diagonal matrix whose v-th diagonal element is defined in Eq. (5).

Implementation. The whole HGACN framework stacks 3 heterogeneous graph convolution layers and one softmax layer for the final classification based on our learned new graph embedding for each subjects. We optimize the HGACN via the Adam algorithm, with the learning rate of 0.001, the number of epochs of 150, and the mini-batch size of 15. We implemented the HGACN based on Pytorch, and the model was trained with a single GPU (NVIDIA GeForce GTX TITAN with 12 GB memory).

3   Experiment

Experimental Setup. We evaluate the proposed HGACN and competing methods on the ADHD dataset through a 5-fold cross-validation strategy. The performance of MDD identification from age-matched HCs is measured by three metrics, including accuracy (ACC), sensitivity (SEN), and specificity (SPE).

Competing Method. We first compare the HGACN method with two conventional learning methods based on two connectivity matrices, including 1) Principal Component Analysis (PCA) + Support Vector Machine with Radial Basis Function kernel (**PS**), and 2) Clustering Coefficients (CC) with SVM (**CS**). The CC can measure the clustering degree of each node in a graph and can be treated as a feature selection algorithm. Hence, two different feature selection or extraction methods are used to reduce the dimensionality of original features. The number of components in PS is chosen from 3 to 50, with the step size of 1 via inner cross-validation. The CS method is associated with the degree of network sparsity, where the sparsity parameter is chosen from $\{0.10, 0.15, \cdots, 0.40\}$ according to cross-validation performance. The parameter C of SVM (with RBF kernel)

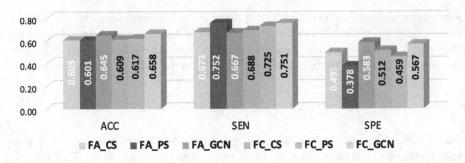

Fig. 3. Results of three methods (*i.e.*, CS, PS and GCN) with single modality data for ADHD vs. HC classification, *i.e.*, fractional anisotropy (FA) derived from dMRI and functional connectivity (FC) from resting-state fMRI.

used in CS and PS is chosen from $\{0.80, 0.85, \cdots, 3.00\}$ via cross-validation, and we use default values for the other parameters.

We also compare the proposed HGACN with the state-of-the-art deep learning method, *i.e.*, graph convolutional network (**GCN**) [25] with K-Nearest Neighbors graph generated from structural or functional connectivity matrix. The parameter k for constructing KNN graphs is chosen from $\{1, 2, \cdots, 30\}$. After modeling the node-centralized local topology (reflected by vertices and their connectivity) as the group-level shared by all subjects, the GCN method also contains 3 graph convolutional layers and one fully-connected layer.

Since two different data modalities are used in our study, we employ both early fusion (**EF**) and late fusion (**LF**) strategies for different methods to take advantage of these modalities. Specifically, for early fusion, we simply concatenate different types of features for the same subject as input features, followed by a specific machine learning algorithm (*e.g.*, CS). For late fusion, we will first perform classification (*e.g.*, via CS) based on each modality separately, and then combine/average their predictions to achieve the final diagnosis results.

Results Using Single Modality. As shown in Fig. 3, we report the disease classification results achieved by three machine learning methods based on single modality data. From Fig. 3, we have the following interesting observations. *First*, the GCN method is superior to other traditional methods when using functional connectivity (FC) features. It achieves at least 4% improvement compared with other methods. This demonstrates the necessity and effectiveness of exploiting graph topology on them. *Besides*, for two traditional learning methods and the GCN method, results on FC calculated from fMRI are slightly better than those on fractional anisotropy (FA) data derived from dMRI in most cases.

Results Using Multiple Modalities. To evaluate the efficiency of the proposed heterogeneous graph framework when dealing with multi-modality data, we compared our HGACN with three competing methods using two multi-

Table 2. Results of our HGACN and three competing methods (*i.e.*, CS, PS, and GCN) with early fusion and late fusion strategies in the task of ADHD vs. HC classification. The terms "EF_CS" and "LF_CS" denote CS with the early fusion and late fusion strategy, respectively.

Metric	EF_CS	LF_CS	EF_PS	LF_PS	EF_GCN	LF_GCN	HGACN
ACC	0.625(0.027)	0.631(0.031)	0.628(0.029)	0.622(0.049)	0.679(0.034)	0.667(0.042)	**0.701**(0.035)
SEN	0.583(0.030)	0.698(0.031)	0.743(0.052)	0.727(0.028)	0.669(0.038)	0.676(0.021)	**0.763**(0.045)
SPE	0.667(0.033)	0.574(0.029)	0.459(0.057)	0.467(0.048)	0.638(0.023)	**0.683**(0.031)	0.648(0.026)

modality fusion strategies, with results reported in Table 2. From Table 2, we can see that our HGACN method achieves at least 2% improvement in terms of ACC and SEN values, compared with the competing methods. This demonstrates the effectiveness of our proposed HGACN in integrating multimodal MRIs for ADHD diagnosis. Besides, with CS as the final discriminator, we can obtain better results using the late fusion strategy. In the remaining cases, early fusion is a better option. Recalling the results in Fig. 3, we can observe that one can obtain the overall better results using multi-modality data (no matter with early or late fusion).

These results imply that fMRI and sMRI contain complementary information that help boost the identification performance for ADHD patients.

4   Conclusion and Future Work

In this paper, we propose a heterogeneous graph attention convolution network (HGACN) to integrate different modalities data in a unified GCN framework. Specifically, we first calculate FC and FA features from fMRI and dMRI data. Then, each subject will construct a heterogeneous graph with two types of nodes and three different categories of edges. We further employ attention mechanisms to aggregate information between intra- and inter-metapaths. Finally, we use new graph embedding to predict the label of the subject. Thus, the proposed HGACN can effectively capture complementary information between different types of brain connectivity features. Experimental results on 187 subjects demonstrate that our method yields state-of-the-art performance in identifying adult ADHD patients from healthy controls.

In the current work, we only focus on functional/structural connectivity matrices to capture subject-level connectivity topology. Actually, information conveyed in structure MRIs such as tissue volume can help uncover the neuro-biological mechanisms of ADHD, which will be considered in our future work. Also, we simply combine multimodal MRIs acquired from two imaging site in the current work and ignore their inter-site differences. In the future, we will focus on alleviating inter-site data heterogeneity to further boost ADHD identification.

References

1. Swanson, J.M., et al.: Attention deficit hyperactivity disorder. In: Encyclopedia of Cognitive Science (2006)
2. Milham, M.P., Fair, D., Mennes, M., Mostofsky, S.H., et al.: The ADHD-200 consortium: a model to advance the translational potential of neuroimaging in clinical neuroscience. Front. Syst. Neurosci. **6**, 62 (2012)
3. Kessler, R.C., et al.: The prevalence and correlates of adult ADHD in the United States: results from the National Comorbidity Survey Replication. Am. J. Psychiatry **163**(4), 716–723 (2006)
4. Kooij, S.J., et al.: European consensus statement on diagnosis and treatment of adult ADHD: the European Network Adult ADHD. BMC Psychiatry **10**(1), 1–24 (2010)
5. Tenev, A., Markovska-Simoska, S., Kocarev, L., Pop-Jordanov, J., Müller, A., Candrian, G.: Machine learning approach for classification of ADHD adults. Int. J. Psychophysiol. **93**(1), 162–166 (2014)
6. Yao, D., Sun, H., Guo, X., Calhoun, V.D., Sun, L., Sui, J.: ADHD classification within and cross cohort using an ensembled feature selection framework. In: 2019 IEEE 16th International Symposium on Biomedical Imaging (ISBI 2019), pp. 1265–1269. IEEE (2019)
7. Zhang, L., Wang, M., Liu, M., Zhang, D.: A survey on deep learning for neuroimaging-based brain disorder analysis. Front. Neurosci. **14**, 779 (2020)
8. Yang, E., et al.: Deep Bayesian hashing with center prior for multi-modal neuroimage retrieval. IEEE Trans. Med. Imaging **40**, 503–513 (2020)
9. Yao, D., et al.: A mutual multi-scale triplet graph convolutional network for classification of brain disorders using functional or structural connectivity. IEEE Trans. Med. Imaging **40**(4), 1279–1289 (2021)
10. Liu, M., Zhang, J., Yap, P.T., Shen, D.: View-aligned hypergraph learning for Alzheimer's disease diagnosis with incomplete multi-modality data. Med. Image Anal. **36**, 123–134 (2017)
11. Yao, D., et al.: Discriminating ADHD from healthy controls using a novel feature selection method based on relative importance and ensemble learning. In: 2018 40th Annual International Conference of the IEEE Engineering in Medicine and Biology Society (EMBC), pp. 4632–4635. IEEE (2018)
12. Soussia, M., Rekik, I.: Unsupervised manifold learning using high-order morphological brain networks derived from T1-w MRI for autism diagnosis. Front. Neuroinform. **12**, 70 (2018)
13. Mahjoub, I., Mahjoub, M.A., Rekik, I.: Brain multiplexes reveal morphological connectional biomarkers fingerprinting late brain dementia states. Sci. Rep. **8**(1), 1–14 (2018)
14. Cortese, S., et al.: Toward systems neuroscience of ADHD: a meta-analysis of 55 fMRI studies. Am. J. Psychiatry **169**(10), 1038–1055 (2012)
15. Guo, X., et al.: Shared and distinct resting functional connectivity in children and adults with attention-deficit/hyperactivity disorder. Transl. Psychiatry **10**(1), 1–12 (2020)
16. Ellison-Wright, I., Ellison-Wright, Z., Bullmore, E.: Structural brain change in attention deficit hyperactivity disorder identified by meta-analysis. BMC Psychiatry **8**(1), 51 (2008)
17. Xia, S., Li, X., Kimball, A.E., Kelly, M.S., Lesser, I., Branch, C.: Thalamic shape and connectivity abnormalities in children with attention-deficit/hyperactivity disorder. Psychiatry Res.: Neuroimaging **204**(2–3), 161–167 (2012)

18. Luo, Y., Alvarez, T.L., Halperin, J.M., Li, X.: Multimodal neuroimaging-based prediction of adult outcomes in childhood-onset ADHD using ensemble learning techniques. NeuroImage: Clin **26**, 102238 (2020)
19. Ktena, S.I., et al.: Metric learning with spectral graph convolutions on brain connectivity networks. Neuroimage **169**, 431–442 (2018)
20. Ma, G., et al.: Deep graph similarity learning for brain data analysis. In: Proceedings of the 28th ACM International Conference on Information and Knowledge Management, pp. 2743–2751 (2019)
21. Bessadok, A., Mahjoub, M.A., Rekik, I.: Graph neural networks in network neuroscience. arXiv preprint arXiv:2106.03535 (2021)
22. Xing, X., et al.: Dynamic spectral graph convolution networks with assistant task training for early MCI diagnosis. In: Shen, D., et al. (eds.) MICCAI 2019. LNCS, vol. 11767, pp. 639–646. Springer, Cham (2019). https://doi.org/10.1007/978-3-030-32251-9_70
23. Song, X., Frangi, A., Xiao, X., Cao, J., Wang, T., Lei, B.: Integrating similarity awareness and adaptive calibration in graph convolution network to predict disease. In: Martel, A.L., et al. (eds.) MICCAI 2020. LNCS, vol. 12267, pp. 124–133. Springer, Cham (2020). https://doi.org/10.1007/978-3-030-59728-3_13
24. Yu, S., et al.: Multi-scale enhanced graph convolutional network for early mild cognitive impairment detection. In: Martel, A.L., et al. (eds.) MICCAI 2020. LNCS, vol. 12267, pp. 228–237. Springer, Cham (2020). https://doi.org/10.1007/978-3-030-59728-3_23
25. Kipf, T.N., Welling, M.: Semi-supervised classification with graph convolutional networks. arXiv preprint arXiv:1609.02907 (2016)

Probabilistic Deep Learning with Adversarial Training and Volume Interval Estimation - Better Ways to Perform and Evaluate Predictive Models for White Matter Hyperintensities Evolution

Muhammad Febrian Rachmadi[1]([⊠]) [iD], Maria del C. Valdés-Hernández[2] [iD],
Rizal Maulana[3], Joanna Wardlaw[2], Stephen Makin[4], and Henrik Skibbe[1]

[1] Brain Image Analysis Unit, RIKEN Center for Brain Science, Wako, Japan
`febrian.rachmadi@riken..jp`
[2] Centre for Clinical Brain Sciences, University of Edinburgh, Edinburgh, UK
[3] Faculty of Computer Science, Universitas Indonesia, Depok, Indonesia
[4] University of Aberdeen, Aberdeen, UK

Abstract. Predicting disease progression always involves a high degree of uncertainty. White matter hyperintensities (WMHs) are the main neuroradiological feature of small vessel disease and a common finding in brain scans of dementia patients and older adults. In predicting their progression previous studies have identified two main challenges: 1) uncertainty in predicting the areas/boundaries of shrinking and growing WMHs and 2) uncertainty in the estimation of future WMHs volume. This study proposes the use of a probabilistic deep learning model called Probabilistic U-Net trained with adversarial loss for capturing and modelling spatial uncertainty in brain MR images. This study also proposes an evaluation procedure named volume interval estimation (VIE) for improving the interpretation of and confidence in the predictive deep learning model. Our experiments show that the Probabilistic U-Net with adversarial training improved the performance of non-probabilistic U-Net in Dice similarity coefficient for predicting the areas of shrinking WMHs, growing WMHs, stable WMHs, and their average by up to 3.35%, 2.94%, 0.47%, and 1.03% respectively. It also improved the volume estimation by 11.84% in the "Correct Prediction in Estimated Volume Interval" metric as per the newly proposed VIE evaluation procedure.

Keywords: Progression prediction · White matter hyperintensities · Volume interval estimation

Electronic supplementary material The online version of this chapter (https://doi.org/10.1007/978-3-030-87602-9_16) contains supplementary material, which is available to authorized users.

I. Rekik et al. (Eds.): PRIME 2021, LNCS 12928, pp. 168–180, 2021.
https://doi.org/10.1007/978-3-030-87602-9_16

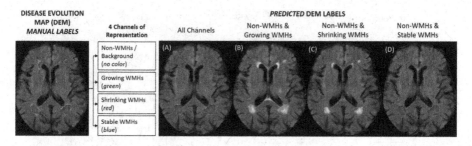

Fig. 1. (Left) Example of Disease evolution map (DEM) produced by subtracting manually generated labels of WMHs at baseline (t0) from manually generated labels of WMHs at follow-up (t1). Green regions are for growing WMHs, red regions are for shrinking WMHs, and blue regions are for stable WMHs (i.e., no changes from t0 to t1). Note there is another channel used to represent the non-WMHs/background in the supervised deep learning model. **(Right)** Different visualizations can be produced based on which channels are used in the testing/inference. *From left to right*: (A) All predicted channels are used to visualize the whole segmentation, (B) only the predicted non-WMHs and growing WMHs channels are used to visualize the segmentation of growing WMHs, (C) only the predicted non-WMHs and shrinking WMHs channels are used to visualize the segmentation of shrinking WMHs, and (D) only the predicted non-WMHs and stable WMHs are used to visualize the segmentation of stable WMHs.

1   Introduction

White matter hyperintensities (WMHs) are neuroradiological features often seen in T2-FLAIR brain MRI, characteristic of small vessel disease (SVD), which are associated with stroke and dementia progression [12]. Clinical studies indicate that the volume of WMHs on a patient may decrease (i.e., regress), stay the same, or increase (i.e., progress) over time [2,12].

Previous studies have proposed various unsupervised and supervised deep learning models to predict the progression (i.e., evolution) of WMHs [8,9]. In the supervised approaches, a deep learning model learns to perform multi-class segmentation of non-WMHs, shrinking WMHs, growing WMHs, and stable WMHs from the namely *disease evolution map* (DEM). The DEM is produced by subtracting manually generated labels of WMHs at baseline (t0) from manually generated labels of WMHs at follow-up (t1) (see Fig. 1).

One study [8] exposed two big challenges in predicting the progression of WMHs: 1) spatial uncertainty in predicting regions of WMHs dynamic changes and their boundaries (i.e., voxels of growth and shrinkage), and 2) uncertainty in the estimation of future WMHs volume (i.e., closeness between the predicted volume of WMHs and the true future volume of WMHs). In relation to the first challenge, it was observed that it is difficult to distinguish the intensities/textures of shrinking and growing WMHs in the MRI sequence used by the study (i.e., T2-FLAIR). This type of uncertainty is commonly known as *aleatoric uncertainty* [4]. In relation to the second challenge, the study showed that different predictive

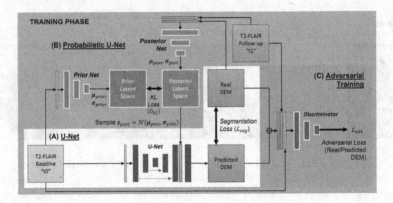

Fig. 2. Illustration of the deep learning models' training phase used in this study. We investigate three different training schemes, which are (A) deterministic training using U-Net [10], (B) probabilistic training using probabilistic U-Net [5], and (C) adversarial training using a GAN discriminator [3,7], all of which can be combined together. Symbol ⊕ stands for OR operation. Full schematics (i.e., figures) of all networks are available in the Supplementary Materials.

models produced similar error and correlation values in estimating the future volume of WMHs, making it harder to determine the best predictive model.

Our main contributions are listed as follows. Firstly, we propose a combination of probabilistic deep learning model with adversarial training to capture spatial uncertainties to predict WMHs evolution. Secondly, we propose a new evaluation procedure, which we name volume interval estimation (VIE), for achieving better interpretation and higher confidence in our predictive models in estimating the future volume of WMHs. The codes and trained model are available on our GitHub page (https://github.com/febrianrachmadi/probunet-gan-vie).

2 Proposed Approach

2.1 Probabilistic Model for Capturing Spatial Uncertainty

Uncertainties are unavoidable when predicting the progression of WMHs, and a previous study showed that incorporating uncertainties into a deep learning model produced the best prediction results [8]. However, the models evaluated in [8] only incorporate external uncertainties (i.e., non-image factors of stroke lesions' volume and unrelated Gaussian noise) and not primary/secondary information coming from brain MRI scans (e.g. statistical spatial maps showing the association of specific WMHs voxels with clinical variables like smoking status).

In this study, we propose the use of the Probabilistic U-Net [5] to capture uncertainties from the brain MR images when predicting the progression of WMHs. The Probabilistic U-Net combines a U-Net [10] with an auxiliary decoder

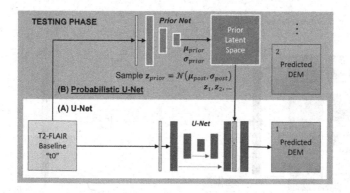

Fig. 3. Illustration of the testing/inference phase of the deep learning model used in this study. In this study, we perform two types of inference (based on the training phase previously performed): (A) deterministic inference using U-Net and (B) probabilistic inference using probabilistic U-Net.

network called Prior Net. The Prior Net models uncertainty in the data as a multivariate Gaussian distribution called prior latent space. The Prior Net learns the prior latent space from another decoder network called Posterior Net that generates a posterior latent space from training data (Fig. 2(B)). The posterior latent space and the Posterior Net are only available during training. *Kullback-Leibler* Divergence ($\mathcal{D}_{KL}$) score is used during training to make the prior latent space similar to the posterior latent space. In testing/inference (Fig. 3), the learned prior latent space is used to sample z, which are broadcasted and concatenated to the original U-Net for generating some variations in the predicted segmentation for the same input image. While variations of prediction are inferred from a few samples from a low-dimensional latent space (i.e., sample z), most information used for predicting the evolution of WMHs in spatial space still comes from the U-Net (i.e., U-Net's feature maps that are concatenated with the samples).

2.2 Adversarial Training for the Predictive Deep Learning Model

A previous study [8] also showed that adversarial training can help producing good predictions by ensuring that each prediction (i.e., predicted DEM) "looks" similar to the real DEM. However, adversarial training was only used for a GAN-based model (i.e., without any manual DEM). In this study, we propose adding adversarial training/loss in the supervised approach where the GAN's discriminator tries to distinguish the "real" manual DEM from the "fake" predicted DEM produced by the U-Net/Probabilistic U-Net. Adding adversarial loss in the training phase is advantageous because it uses information from the entire image space (i.e., global context information) rather than local (i.e., pixelwise) information usually given by the traditional segmentation loss. Figure 2(C) shows how the GAN's discriminator is used in the training phase.

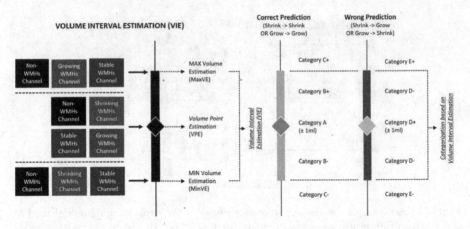

Fig. 4. (Left) Visualization of Volume Interval Estimation (VIE) produced by using subsets of predicted channels of non-WMHs, shrinking WMHs, growing WMHs, and stable WMHs. Note that the normal volume point estimation (VPE) is done by using all predicted channels. **(Right)** By using volume interval estimation, we can categorize prediction results more accurately (i.e., not only correct and wrong predictions). Detailed categorization scheme is shown in Table 1.

2.3   Volume Interval Estimation for Better Interpretation

One of the many challenges in predicting the progression of WMHs is to ascertain the quality of the prediction, especially when estimating the future volume of WMHs. Despite the existence of several metrics for quality control of an image estimation machine-learning algorithm [1], predictive deep learning models normally use the mean square error (MSE) to evaluate how close the predicted future volumes of WMHs are to the true future volumes after the training phase. However, how can we calculate the MSE in a real world scenario where the real future volume of WMHs is unknown?

For better interpretation and confidence in our prediction model, we propose using the Volume Interval Estimation (VIE). Instead of evaluating how close the predicted volume point estimation (VPE) is to the true volume of future WMHs at time point "1" (True time-point 1 Volume, or Tt1V), we evaluate where Tt1V lies within the VIE, i.e., the interval bounded by the maximum (MaxVE) and minimum (MinVE) volume estimations. VIE's interval is bounded by two extreme assumptions of WMHs progression: 1) there are no shrinking WMHs (which produces MaxVE) and 2) there are no growing WMHs (which produces MinVE). Note that the normal assumption for the WMHs progression (i.e., WMHs can be stable, growing, or shrinking) is located between these two extreme assumptions considering the stable WMHs to be regions of chronic damage (i.e., otherwise MinVE would be equal to zero). Thus, VPE is located between the MinVE and MaxVE. As illustrated in Fig. 4 (left), MinVE is produced by dropping the growing WMHs channel in the predicted DEM while MaxVE is produced by dropping the shrinking WMHs channel.

We can further categorize VIE according to 1) the location of Tt1V within VIE and 2) whether the volume estimation is correctly predicted or not (i.e., patient with growing WMHs is correctly predicted to have growing WMHs, and so on). Figure 4 (right) and Table 1 illustrate and describe each VIE's category.

Table 1. Categorization of the proposed volume interval estimation (VIE) based on the position of true future (follow-up) Total WMHs volume (Tt1V) in the predicted volume interval between maximum volume estimation (MaxVE), minimum volume estimation (MinVE), and volume point estimation (VPE). Visualization of the proposed volume interval estimation can be seen in Fig. 4. (For the dataset used in this study 1 ml is approximately 284 voxels, as 1 voxel represents a volume of 0.00351 ml.)

Category	Description
A	*Correct prediction* (VPE - 1 *ml* $<=$ Tt1V $<=$ VPE + 1 *ml*)
B+	*Correct prediction* (VPE + 1 *ml* $<=$ Tt1V $<=$ MaxVE)
B-	*Correct prediction* (MinVE $<=$ Tt1V $<=$ VPE - 1 *ml*)
C+	*Correct prediction* (Tt1V $>$ MaxVE)
C-	*Correct prediction* (Tt1V $<$ MinVE)
D+	*Wrong prediction*
	(VPE - 1 *ml* $<=$ Tt1V $<=$ VPE + 1 *ml*)
D-	(VPE + 1 *ml* $<=$ Tt1V $<=$ MaxVE **OR**
	VPE + 1 *ml* $<=$ Tt1V $<=$ MaxVE)
E+	*Wrong prediction* (Tt1V $>$ MaxVE)
E−	*Wrong prediction* (Tt1V $<$ MinVE)

3 Dataset and Experimental Setting

3.1 Dataset and Cross Validation

We use MRI data from all stroke patients ($n = 152$) enrolled in a study of stroke mechanisms [12], imaged at three time points (i.e., first time (baseline scan), at approximately 3 months, and a year after). This study uses the baseline (t0) and 1-year follow-up (t1) MRI data ($s = n \times 2 = 304$), both acquired at a GE 1.5T scanner following the same imaging protocol, explained in [11]. These data are pre-processed (co-registered, brain-extracted, filtered, and normalised) as explained in [8,9]. The spatial resolution of the images used in this study is $256 \times 256 \times 42$ with slice thickness of $0.9375 \times 0.9375 \times 4$ cubic mm. To make sure data from all patients are used in the testing and evaluation, we perform 4-fold cross validation where each fold uses 114 and 38 patients for training and testing respectively. Each model is trained for 64 epochs in one experiment.

3.2 Segmentation Loss ($\mathcal{L}_{seg}$)

In this study, we use the non-linear *softmax* function at the segmentation layer; see Eq. 1. The parameter s is the output of the segmentation layer. The network classifies each voxel either as non-WMHs, shrinking WMHs, growing WMHs, or stable WMHs. Thus, the number of output classes is set to $C = 4$.

$$p_i = \sigma(\boldsymbol{s})_i = \frac{e^{s_i}}{\sum_{j=1}^{C} e^{s_j}} \text{ for } i = 1, ..., C \tag{1}$$

We tested two different segmentation losses ($\mathcal{L}_{seg}$): 1) weighted cross entropy (WCE) (Eq. 2), and 2) *alpha* weighted focal loss (FL) [6] (Eq. 3). In both equations, tar_i is the true target class for each voxel and p_i is the probability of each voxel to be of the target class i. Whereas, w_i is the weight loss of class i in WCE and α_i is the weight loss of class i in FL. A larger weight loss for class i indicates that class i is predominant, contributing a larger loss value in total. Finally, γ is FL's hyperparameter, which is set to $\gamma = 2$ following the recommendation of the original paper [6]. Based on our preliminary experiments, the best weights for both WCE (i.e., $\boldsymbol{w} = (w_{i=1}, w_{i=2}, w_{i=3}, w_{i=4})$) and FL (i.e., $\boldsymbol{\alpha} = (\alpha_{i=1}, \alpha_{i=2}, \alpha_{i=3}, \alpha_{i=4})$) are 0.25, 0.75, 0.75, and 0.5 for non-WMHs ($i = 1$), shrinking WMHs ($i = 2$), growing WMHs ($i = 3$), or stable WMHs ($i = 4$) respectively.

$$\mathcal{L}_{seg}^{WCE} = WCE = -w_i \, tar_i \, log \, (p_i) \tag{2}$$

$$\mathcal{L}_{seg}^{FL} = FL = -\alpha_i \, tar_i \, (1 - p_i)^{\gamma} \, log \, (p_i) \tag{3}$$

3.3 *Kullback-Leibler* Divergence ($\mathcal{D}_{KL}$) for Probabilistic Loss

An additional *Kullback-Leibler* Divergence score ($\mathcal{D}_{KL}$) is used in the training if Probabilistic U-Net setting is used [5]. In this setting, Prior Net and Posterior Net are trained together with the generator (i.e., U-Net) for predicting the DEM. Let Q be the posterior distribution from the Posterior Net and P be the prior distribution from the Prior Net. The difference between the posterior distribution Q and the prior distribution P is penalized by Eq. 4 where X_{post} is the T2-FLAIR at t1, Y_{post} is the true DEM, and X_{prior} is the T2-FLAIR at t0. Following the original paper [5], the dimension for both z_{post} and z_{prior} is 6.

$$\mathcal{D}_{KL}(Q \parallel P) = \mathbb{E}_{z_{post} \sim Q, z_{prior} \sim P}[log \, Q(X_{post}, Y_{post}) - log \, P(X_{prior})] \tag{4}$$

In the training phase of the Probabilistic U-Net, each segmentation prediction is conditioned to $z_{post} \sim \mathcal{N}(\mu_{post}, \sigma_{post}) = Q(X_{post}, Y_{post})$ sampled from the Posterior Net. As per the original paper [5], the probabilistic segmentation loss $\mathcal{L}_{seg}^{prob}$ is defined by Eq. 5 with $\beta = 1$. Note that the segmentation loss of $\mathcal{L}_{seg}$ can be either WCE (Eq. 2) or FL (Eq. 3).

Table 2. Performance of U-Net and Probabilistic U-Net in Dice similarity coefficient (DSC) and volume point estimation (VPE). Note that higher DSC value is better ($\uparrow$), lower MSE value is better ($\downarrow$), and closer to 0 is better for Error ($\rightarrow 0$). The best result for each column is shown in bold and the second best is underlined. WCE stands for weighted cross entropy and while FL stands for focal loss.

Model	Cost function	DSC $\uparrow$				VPE	
		Shrink	Grow	Stable	Average	Error $\rightarrow 0$	MSE $\downarrow$
U-Net	WCE	0.1794 (0.072)	0.1970 (0.097)	0.6413 (0.159)	0.3393 (0.078)	-2.7127 (10.31)	112.87 (247.44)
	FL	0.1757 (0.077)	**0.2073** **(0.104)**	0.6483 (0.156)	0.3438 (0.076)	-2.7002 (10.08)	108.17 (256.61)
Prob. U-Net (t0 & DEM as inputs to Posterior Net)	WCE	0.1491 (0.061)	0.1524 (0.090)	0.6220 (0.171)	0.3079 (0.086)	-2.5095 (9.84)	102.44 (234.61)
	FL	0.1673 (0.074)	0.1858 (0.089)	0.6147 (0.184)	0.3226 (0.090)	-2.0297 (9.27)	89.56 (220.73)
Prob. U-Net (t1 & DEM as inputs to Posterior Net)	WCE	0.1964 (0.071)	0.2040 (0.091)	**0.6564** **(0.162)**	0.3522 (0.080)	-0.2953 **(8.33)**	69.05 (224.94)
	FL	**0.2092** **(0.082)**	0.2056 (0.092)	0.6507 (0.160)	**0.3552** **(0.080)**	-0.6650 (8.02)	**64.33** **(220.39)**

$$\mathcal{L}_{seg}^{prob} = \mathcal{L}_{seg}(P_i(p_i|X_{prior}, z_{post})) + \beta \cdot \mathcal{D}_{KL}(Q \parallel P) \qquad (5)$$

In the testing/inference phase, each segmentation prediction is conditioned to $z_{prior} \sim \mathcal{N}(\mu_{prior}, \sigma_{prior}) = P(X_{prior})$ sampled from the Prior Net. To get the final segmentation, we sampled 30 different z_{prior} from Prior Net to produce 30 different segmentation predictions for each patient and averaged all of them.

3.4 Adversarial Loss ($\mathcal{L}_{adv}$)

In this study, we modified the original adversarial loss [3] by adding a segmentation loss ($\mathcal{L}_{seg}$) for optimizing the generator to segment the DEM. Similar to the original paper [3], here the generator tries to minimize Eq. 6 while the discriminator tries to maximize it.

$$\mathbb{E}_{y \sim Y_{GAN}}[\log(D(y))] + \mathbb{E}_{x \sim X_{GAN}}[\log(1 - D(G(x))) + \mathcal{L}_{seg}(G(x))] \qquad (6)$$

In the Eq. 6, G is the generator, D is the discriminator, $x \sim X_{GAN}$ is the set of input images, $y \sim Y_{GAN}$ is the combination of true DEM and true images (i.e., T2-FLAIR for t0 and t1), $G(x)$ is the predicted DEM, $\mathbb{E}_y \sim Y_{GAN}$ is the expected value over Y_{GAN}, and $\mathbb{E}_x$ is the expected value over X_{GAN}. If G is U-Net then $X_{GAN} = X_{prior}$. Whereas, if G is probabilistic U-Net then $X_{GAN} = (X_{prior}, X_{post}, Y_{post})$. As in the previous section, $X_{prior}, X_{post}, Y_{post}$ correspond to the T2-FLAIR for t0, t1, and true DEM respectively.

In this study, we also evaluate three different combinations of Y_{GAN} to investigate which produces the best result. The tested combinations are 1) only the true DEM (DEM GAN), 2) true DEM and T2-FLAIR normalised values at t0

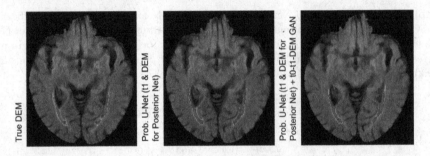

Fig. 5. Comparison of the true DEM (left) and predicted DEMs produced by using Probabilistic U-Net without adversarial training (middle) and Probabilistic U-Net with adversarial training with T2-FLAIR at t0 and true DEM (right).

(t0-DEM GAN), and 3) true DEM, T2-FLAIR normalised values at t0, and T2-FLAIR normalised values at t1 (t0-t1-DEM GAN). In these experiments, we used spectral normalization [7] for the discriminator network and trained it 5 times for each epoch.

4   Results

4.1   U-Net vs. Probabilistic U-Net

Table 2 shows the performances of U-Net and Probabilistic U-Net for predicting the spatial progression of WMHs (shown in Dice Similarity Coefficient (DSC)) and in volume point estimation (VPE). Following the original paper that proposed the Probabilistic U-Net [5], we first used T2-FLAIR at t0 and true DEM as inputs to Posterior Net. However, this approach was outperformed by the U-Net model. By, consequently, changing the input of Posterior Net to be T2-FLAIR at t1 and true DEM, the model using Probabilistic U-Net outperformed U-Net in our experiments. These show that the input data for the Posterior Net in the Probabilistic U-Net should differ from the input data for the other modules of this probabilistic architecture (i.e., U-Net and Prior Net). Table 2 also shows that the FL cost function produced better prediction results than the WCE in both DSC and VPE for all experimental settings.

4.2   Probabilistic U-Net with Adversarial Training

We investigated whether applying adversarial training with different input images can improve the performance of Probability U-Net. We evaluated these experiments using DSC, VPE, and the newly proposed VIE evaluation.

Table 3 shows that adversarial training with T2-FLAIR at t0 and true DEM slightly improved the prediction produced by Probabilistic U-Net in VPE (Error) and DSC (Stable). Figure 5 also shows that the predicted DEM produced by adversarial training more closely followed the true DEM by removing the small

Table 3. Performance of deep learning models trained with adversarial training for predicting the progression of WMHs in Dice similarity coefficient (DSC) and volume point estimation (VPE). Higher DSC value is better (↑), lower MSE value is better (↓), and closer to 0 is better for Error (→ 0). The best result for each column is shown in bold and the second best is underlined.

Model	DSC ↑				VPE	
	Shrink	Grow	Stable	Average	Error → 0	MSE ↓
Prob. U-Net (t1 & DEM for Posterior Net)	**0.2092** (**0.082**)	0.2056 (0.092)	<u>0.6507</u> (0.160)	**0.3552** (**0.080**)	−<u>0.6650</u> (8.02)	**64.33** (**220.39**)
Prob. U-Net (t1 & DEM for Posterior Net) + DEM GAN	0.1739 (0.083)	0.2083 (0.103)	0.6374 (0.172)	0.3399 (0.090)	2.0216 (9.32)	90.34 (180.32)
Prob. U-Net (t1 & DEM for Posterior Net) + t0-DEM GAN	<u>0.1911</u> (0.093)	0.2184 (0.103)	**0.6530** (**0.163**)	<u>0.3541</u> (0.089)	**0.3155** (**8.90**)	<u>78.83</u> (156.17)
Prob. U-Net (t1 & DEM for Posterior Net) + t1-DEM GAN	0.1737 (0.083)	**0.2367** (**0.100**)	0.6427 (0.169)	0.3511 (0.086)	−3.4385 (8.97)	91.70 (205.29)
Prob. U-Net (t1 & DEM for Posterior Net) + t0-t1-DEM GAN	0.1701 (0.083)	<u>0.2282</u> (0.102)	0.6425 (0.167)	0.3469 (0.083)	-3.3115 (8.83)	88.36 (220.39)
U-Net	0.1757 (0.077)	0.2073 (0.104)	0.6483 (0.156)	0.3438 (0.076)	−2.7002 (10.08)	108.17 (256.61)
U-Net + t0-DEM GAN	0.1849 (0.091)	0.2134 (0.099)	0.6468 (0.159)	0.3484 (0.079)	−1.1187 (9.58)	92.44 (191.44)

false positive clusters in the prediction results. These experiments show that, while Probabilistic U-Net without adversarial training consistently produced some of the best prediction results in terms of DSC, the Probabilistic U-Net with adversarial training predicted more realistic DEM, closer to the true DEM, and with better VPE values. Additionally, U-Net with adversarial training produced better prediction results than the original U-Net without adversarial training.

Table 4 shows the performances of the deep learning models evaluated using VIE. The percentage of patients with correctly predicted DEM (i.e., subjects with shrinking and growing WMHs correctly predicted as having shrinking and growing WMHs respectively) is given by the metric called "CP" (Correctly Predicted). We also calculated the percentage of patients having their true future volumes of WMHs (Tt1V) correctly estimated and located between MinVE and MaxVE, and expresses it under a metric named "CPinEVI" (Cor-

Table 4. Performance of deep learning models for predicting the future volume of WMHs evaluated in the newly proposed Volume Interval Estimation (VIE). The best result for each column is shown in bold and the second best is shown in underline. Symbol ($\uparrow$) indicates that higher values are better while symbol ($\rightarrow 0$) indicates that values closer to 0 are better. *Abbreviations*: "CP" stands for "Correct Prediction", "CPinEVI" stands for "Correct Prediction in Estimated Volume Interval", "(CP + WP)inEVI" stands for "Correct Prediction + Wrong Prediction but still in EVI", "VPE" stands for "Volume Point Estimation", "MaxVE" stands for "Maximum Volume Estimation, and "MinVE" stands for "Minimum Volume Estimation.

Model	CP $\uparrow$	CPinEVI $\uparrow$	(CP+WP) inEVI $\uparrow$	Distance to VPE (in *ml*)	
				MaxVE $\rightarrow 0$	MinVE $\rightarrow 0$
Prob. U-Net (t1 & DEM for Posterior Net)	**73.03%**	44.74%	51.32%	4.0862 (3.241)	−5.5700 (3.918)
Prob. U-Net (t1 & DEM for for Posterior Net) + DEM GAN	63.16%	30.26%	39.47%	2.5377 (3.0779)	−5.5978 (4.5046)
Prob. U-Net (t1 & DEM for for Posterior Net) + t0-DEM GAN	69.74%	39.47%	50.00%	2.6563 (3.0834)	−6.7103 (5.4319)
Prob. U-Net (t1 & DEM for for Posterior Net) + t1-DEM GAN	68.42%	44.74%	57.24%	2.8499 (2.7111)	−7.9550 (5.3201)
Prob. U-Net (t1 & DEM for for Posterior Net) + t0-t1-DEM GAN	**73.03%**	**48.68%**	<u>57.89%</u>	2.9383 (3.0793)	−7.6224 (5.5935)
U-Net	61.84%	36.84%	48.68%	2.9911 (3.3676)	-6.1355 (4.5706)
U-Net + t0-DEM GAN	<u>72.37%</u>	<u>46.71%</u>	**59.87%**	4.5915 (6.7208)	−6.2326 (4.7695)

rectly predicted in Estimated Volume Interval (EVI)). Based on the VIE categorization (Fig. 4 and Table 1), "CPinEVI" covers categories A, B+, and B-. Lastly, "(CP+WP)inEVI" shows the percentage of correctly and wrongly predicted patients with their Tt1V still located between MinVE and MaxVE. Based on Fig. 4 and Table 1, "(CP+WP)inEVI" covers categories A, B+, B-, D+, and D-.

Both "CPinEVI" and "(CP+WP)inEVI" are important for better interpretation and higher confidence in our predictive model. Metric "CPinEVI" is important not only in evaluation but also in real-word testing/inference. A predictive model with higher rate of "CPinEVI" in testing means that there is a high probability that the Tt1V lies between the predicted/estimated MinVE and MaxVE produced by the predictive model. On the other hand, "(CP+WP)inEVI" captures difficult cases where the future volume of WMHs is wrongly predicted by the predictive model but the Tt1V still lies between the predicted/estimated MinVE and MaxVE. These cases happen mostly when the WMHs volume change from t0 to t1 is very small. For example, a patient with WMHs volume of 5 ml at t0 and 5.5 ml at t1 (i.e., growing WMHs) is wrongly predicted by the model to have future WMHs volume of 4.5 ml (i.e., shrinkage in the total WMHs volume at t1) while having predicted MinVE and MaxVE of 4 ml and 6 ml respectively.

The results in Table 4, show that Probability U-Net with adversarial training using T2-FLAIR for t0, t1, and true DEM produced the best results in all metrics of VIE. While the rate of CP is the same with the Probabilistic U-Net without adversarial training, Probabilistic U-Net with adversarial training using T2-FLAIR for t0, t1, and true DEM produced better results than other probabilistic models in "CPinEVI" and "(CP+WP)inEVI" (48.68% and 57.89% respectively). It is worth to mention that the best result for "(CP+WP)inEVI" was produced by the U-Net with adversarial training using T2-FLAIR for t0 and true DEM (i.e., 59.87% respectively). However, as shown in Table 3, it did not outperform any Probabilistic U-Net settings in DSC and/or VPE.

Lastly, one can argue that higher rates of "CPinEVI" and "(CP+WP)inEVI" can be produced by expanding the VIE itself (i.e., smaller value of MinVE and larger value of MaxVE). However, as shown in Table 4, the predicted values of MinVE and MaxVE from different predictive models are relatively close to the predicted VPE in all settings (calculated by performing MinVE - VPE and MaxVE - VPE for the whole dataset).

5 Conclusion and Discussion

In this study, we propose the use of a probabilistic deep learning model (i.e., Probability U-Net) for capturing/modelling spatial uncertainty in the estimation of WMHs from brain MRI scans. The adversarial loss successfully improved the prediction results, ensuring the predicted DEM closely follows the global context of the true DEM by removing small clusters of false positives. Furthermore, we also propose a procedure to evaluate the predictive model called Volume Interval Estimation (VIE) for better evaluation, interpretation, and higher confidence in our predictive model. While the probability model with adversarial training produced some of the best results, VIE proved to be effective for interpreting and evaluating the predicted results. It is also worth to mention that there are still many useful evaluation metrics that can be derived from the VIE. Future works include incorporating VIE into the predictive model as a regularization term in the cost function. Preliminary results show an improvement in the prediction of WMHs evolution. Furthermore, to reduce aleatoric uncertainty, information from other MRI sequences (e.g. T1-weighted) and modalities (e.g. diffusion-weighted images) could be advantageous. Given the presence of WMHs in scans of older adults and dementia patients, re-training and testing the proposed schemes in a wider sample would be also beneficial.

Acknowledgements. Funds from JSPS (Kakenhi Grant-in-Aid for Research Activity Start-up, Project No. 20K23356) (MFR); Row Fogo Charitable Trust (Grant No. BRO-D.FID3668413) (MCVH); Wellcome Trust (patient recruitment, scanning, primary study Ref No. WT088134/Z/09/A); Fondation Leducq (Perivascular Spaces Transatlantic Network of Excellence); EU Horizon 2020 (SVDs@Target); and the MRC UK Dementia Research Institute at the University of Edinburgh (Wardlaw programme) are gratefully acknowledged. This research was also supported by the program for Brain

Mapping by Integrated Neurotechnologies for Disease Studies (Brain/MINDS) from the Japan Agency for Medical Research and Development AMED (JP21dm0207001).

References

1. Castorina, L.V., et al.: Metrics for quality control of results from super-resolution machine-learning algorithms-data extracted from publications in the period 2017-May 2021 [dataset] (2021). https://doi.org/10.7488/ds/3062
2. Chappell, F.M., et al.: Sample size considerations for trials using cerebral white matter hyperintensity progression as an intermediate outcome at 1 year after mild stroke: results of a prospective cohort study. Trials **18**(1), 1–10 (2017). https://doi.org/10.1186/s13063-017-1825-7
3. Goodfellow, I., et al.: Generative adversarial nets. In: Advances in Neural Information Processing Systems, vol. 27 (2014)
4. Hüllermeier, E., Waegeman, W.: Aleatoric and epistemic uncertainty in machine learning: an introduction to concepts and methods. Mach. Learn. **110**(3), 457–506 (2021). https://doi.org/10.1007/s10994-021-05946-3
5. Kohl, S., et al.: A probabilistic U-net for segmentation of ambiguous images. In: Advances in Neural Information Processing Systems, vol. 31 (2018)
6. Lin, T.Y., et al.: Focal loss for dense object detection. In: Proceedings of the IEEE International Conference on Computer Vision, pp. 2980–2988 (2017). https://doi.org/10.1109/ICCV.2017.324
7. Miyato, T., et al.: Spectral normalization for generative adversarial networks. In: International Conference on Learning Representations (2018)
8. Rachmadi, M.F., et al.: Automatic spatial estimation of white matter hyperintensities evolution in brain MRI using disease evolution predictor deep neural networks. Med. Image Anal. **63**, 101712 (2020). https://doi.org/10.1016/j.media.2020.101712
9. Rachmadi, M.F., del C. Valdés-Hernández, M., Makin, S., Wardlaw, J.M., Komura, T.: Predicting the evolution of white matter hyperintensities in brain MRI using generative adversarial networks and irregularity map. In: Shen, D., et al. (eds.) MICCAI 2019. LNCS, vol. 11766, pp. 146–154. Springer, Cham (2019). https://doi.org/10.1007/978-3-030-32248-9_17
10. Ronneberger, O., Fischer, P., Brox, T.: U-net: convolutional networks for biomedical image segmentation. In: Navab, N., Hornegger, J., Wells, W.M., Frangi, A.F. (eds.) MICCAI 2015. LNCS, vol. 9351, pp. 234–241. Springer, Cham (2015). https://doi.org/10.1007/978-3-319-24574-4_28
11. Valdés Hernández, M.D.C., et al.: Rationale, design and methodology of the image analysis protocol for studies of patients with cerebral small vessel disease and mild stroke. Brain Behav. **5**(12), e00415 (2015). https://doi.org/10.1002/brb3.415
12. Wardlaw, J.M., et al.: White matter hyperintensity reduction and outcomes after minor stroke. Neurology **89**(10), 1003–1010 (2017). https://doi.org/10.1212/WNL.0000000000004328

A Multi-scale Capsule Network for Improving Diagnostic Generalizability in Breast Cancer Diagnosis Using Ultrasonography

Chanho Kim[1], Won Hwa Kim[1,2], Hye Jung Kim[1,2], and Jaeil Kim[1(✉)]

[1] Kyungpook National University, 80 Daehak-ro, Buk-gu, Daegu, Republic of Korea
[2] Kyungpook National University Chilgok Hospital,
807 Hoguk-ro, Buk-gu, Daegu, Republic of Korea

Abstract. Recently, deep learning has shown promising results in medical image processing. However, computer-aided diagnosis (CAD) systems based on deep learning still struggle for the real-world deployment, due to its low generalizability and reliability. It is essential to improve the generalization performance to enable them to be used routinely in clinical practice. In this paper, we propose a capsule network with a multi-scale setting to achieve better generalization performance in the differential diagnosis of breast tumors using ultrasonography. The proposed network utilizes a Gaussian pyramid to learn multi-scale features of breast tumors and dynamic routing to improve its robustness against image quality with severe noises. To evaluate the generalizability of the proposed method, we collected breast ultrasound images from 4 different hospitals and used one dataset from 1 hospital as a train set and the rest as external validation sets. We compared the classification performance with other networks, which were employed for the ultrasound diagnosis in previous studies, on the external validation sets. We also conducted additional experiments: feature space visualization and robustness evaluation study with respect to the image noise. Our model showed better classification results than other networks, such as GoogLeNet and Inception-v3, in the external validation. Experimental results also indicate that the proposed network can learn more robust and noise-invariant features from breast ultrasound imaging.

Keywords: Computer-aided diagnosis · Capsule network · Generalizability · Breast cancer · Ultrasonography

1 Introduction

Breast cancer is one of the deadliest diseases in women worldwide. Early diagnosis of breast cancer generally increases the chances of successful treatment and effectively reduces death rates [17]. Various imaging modalities, such as

© Springer Nature Switzerland AG 2021
I. Rekik et al. (Eds.): PRIME 2021, LNCS 12928, pp. 181–191, 2021.
https://doi.org/10.1007/978-3-030-87602-9_17

magnetic resonance imaging (MRI), mammography, ultrasonography, and computer tomography (CT), are commonly used for screening and diagnosing breast cancers. The most widely-used screening modality is mammography, and ultrasonography is also a suitable tool for large-scale screening. Owing to the nature of ultrasonography using sound waves, with frequencies higher than the upper audible limit of human hearing, it is safe, non-invasive, and no risk of radiation [11,14] in the visualization of the internal structure of the breast. Despite such merits, breast cancer diagnosis with ultrasound imaging heavily depends on the expertise and experiences of operators, due to the shortcoming of breast ultrasound images, such as low contrast, speckle noise, and low spatial resolution [1,18]. Moreover, some factors like patients and modalities affect the quality of the ultrasound image. These limitations can lead to misdiagnosis causing unnecessary biopsies or surgeries.

In recent years, many studies have demonstrated that computer-aided diagnosis (CAD) can improve the diagnostic performance of radiologists [3,19]. CAD is a computer system that analyzes medical images and supports the diagnosis of medical images. Owing to the advances in deep learning techniques, many studies have reported that convolutional neural networks (CNN) can achieve good diagnostic performance using ultrasound images for breast cancer diagnosis [6,9,13]. However, the deep learning based methods still struggle for the real-world deployment, due to their limited generalizability in relation to data variability and noise. Considering the imaging parameters and operators' preferences to ultrasonography, the generalization is very difficult to be achieved without multi-center data acquisition and model robustness improvement. Especially, it is important to learn commonly-observed features of breast masses for benign or malignancy cases with less biases to training data sets for more reliable CAD systems.

For better generalization of CNNs, Geirhos, et al. have shown that ImageNet pre-trained networks have biased to texture features through experiments [5]. They utilize a style transfer network to remove image textures and to learn shape-based representation [4,8]. The network that learned shape-based representation showed improved object detection performance and robustness towards a wide range of image distortions. Sabour, et al. proposed a dynamic routing method to utilize capsule [12], which is a group of neurons forming a vector containing instantiation parameters, such as size, type, and position [7]. And the length of capsules represents the probability of the existence of an entity in an image. The most significant point of the capsule network is routing-by-agreement. Each capsule makes possible parent capsules through affine transformation, adds all predictions with coupling coefficient, and then predicts parent capsules. This approach can detect more robust features by dynamic routing.

In this paper, we propose a capsule network based method to improve generalizability and robustness in the breast cancer characterization. The proposed network utilizes a Gaussian pyramid of ultrasound images and routing-by-agreement for dynamic routing. To prove the effectiveness of our work, we collected ultrasound images from several hospitals and compared and analyzed the experimental results.

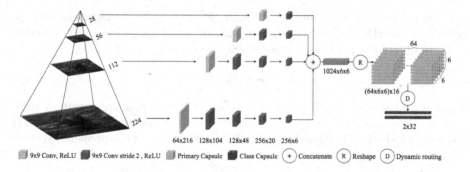

Fig. 1. Architecture of the proposed multi-scale capsule network utilizing Gaussian pyramid and routing-by-agreement.

2 Method

2.1 Gaussian Pyramid

Unlike Geirhos, *et al.* using style transfer network, we removed the texture of the ultrasound image through smoothing to train the network for shape information. In particular, we apply the Gaussian pyramid, a classical and efficient pyramid method, to extract multi-scale features. The image **I** of the l-th layer of the Gaussian pyramid is calculated as follows:

$$\mathbf{I}_l(i,j) = \sum_{m=-c}^{c} \sum_{n=-c}^{c} G(m,n) * \mathbf{I}_{l-1}(2i-1-m, 2j-1-n), \tag{1}$$

where $*$ is the convolution operation, (i,j) is spatial location of the image, and $G(m,n)$ is the Gaussian filter to be represented as follow:

$$G(m,n) = \frac{1}{2\pi\sigma^2} e^{-(m^2+n^2)/2\sigma^2}, \tag{2}$$

where σ refers to the variance related to the Gaussian filter. Specifically, the size of the Gaussian filter we use is 5×5. As its operation proceeds, the texture information progressively disappears by smoothing. In addition, the input image becomes smaller with its operation.

2.2 Dynamic Routing

First, capsules are composed of vectors by reshaping feature maps that have passed through the convolutional layers. Each capsule is a group of neurons that form a vector containing instantiation parameters, such as size, color, rotation, and position, of a particular object. In the capsule network, the direction of the vector represents the properties of the entity. And the length of the vector indicates the probability that the entity is present in the input image.

These capsules are acquired through the routing-by-arrangement mechanism [12]. Specifically, the low-level capsule $\mathbf{u}_i$ is multiplied by a weight matrix $\mathbf{W}_{ij}$ to obtain the predicted output capsule $\hat{\mathbf{u}}_{j|i}$.

$$\hat{\mathbf{u}}_{j|i} = \mathbf{W}_{ij}\mathbf{u}_i. \tag{3}$$

Like other parameters, $\mathbf{W}_{ij}$ is also trained through the backpropagation algorithm. These predicted vectors $\hat{\mathbf{u}}_{j|i}$ form high-level capsules but are not treated equally. The high-level capsules $\mathbf{s}_j$ are the weighted sum of the predicted vectors $\hat{\mathbf{u}}_{j|i}$ with coupling coefficients c_{ij}.

$$\mathbf{s}_j = \sum_i c_{ij}\hat{\mathbf{u}}_{j|i}. \tag{4}$$

The coupling coefficient c_{ij} is determined by the routing softmax and b_{ij}, the log prior probability that the low-level capsule should be coupled to the high-level capsule. b_{ij} is initialized to 0 and updated through the iterative process.

$$c_{ij} = \frac{\exp(b_{ij})}{\sum_k \exp(b_{ik})}. \tag{5}$$

During the processing, the length of each capsule must be less than 1 to represent the probability. Therefore, the length of each vector is adjusted through a nonlinear squash function while maintaining its direction. The squash function is defined as follows:

$$\mathbf{v}_j = \frac{\|\mathbf{s}_j\|^2}{1 + \|\mathbf{s}_j\|^2} \frac{\mathbf{s}_j}{\|\mathbf{s}_j\|}, \tag{6}$$

where $\mathbf{v}_j$ is the final output vector. This function shrinks short vectors to near 0 and long vectors close to 1 but not longer.

Through the squash function, the high-level capsules s_j become the final output capsules $\mathbf{v}_j$. Initially, all predicted vectors $\hat{\mathbf{u}}_{j|i}$ contribute equally to constructing the high-level capsules s_j, since b_{ij} is 0. However, the initial coupling coefficient c_{ij} is iteratively updated by measuring the arrangement that is the inner product of the final output vectors $\mathbf{v}_j$ and the predicted vectors $\hat{\mathbf{u}}_{j|i}$. This agreement is then added to b_{ij} to determine the coupling coefficient c_{ij} of the next iteration.

$$b_{ij} \longleftarrow \hat{\mathbf{u}}_{j|i} \cdot \mathbf{v}_j. \tag{7}$$

The coupling coefficient c_{ij} is updated by measuring the cosine similarity of the final output capsules $\mathbf{v}_j$ and the corresponding predicted capsules $\hat{\mathbf{u}}_{j|i}$ at each iteration. As a result, the high-level capsules s_j are aggregated from the more similar predicted capsules $\hat{\mathbf{u}}_{j|i}$. This whole process of predicting the final output capsules is dynamic routing.

To train the network for the classification task, we use margin loss, which is defined as follows:

$$L_k = T_k \max(0, m^+ - \|\mathbf{v}_k\|)^2 + \lambda(1 - T_k)\max(0, \|\mathbf{v}_k\| - m^-)^2, \tag{8}$$

where the λ means down-weighting of the loss for the absent class. In this paper, we set $m^+ = 0.9, m^- = 0.1$, and $\lambda = 0.5$. The total loss is simply the sum of the loss of each class capsule. This loss makes the class capsule have a long instantiation vector if and only if the corresponding breast tumor is present in the image. In contrast, it puts more loss on capsules with a long instance vector the absence of a corresponding breast tumor.

2.3   Network Architecture

The classification network consists of the feature encoder with 4 convolution pathways for the Gaussian pyramid and the capsule network with primary and class capsule layers. Figure 1 shows the entire architecture of the proposed network. The Gaussian pyramid reduces the size of the input image to gradually smooth out the texture information. Then, the resized images at each level of the Gaussian pyramid are given to the feature encoder with 4 pathways with different scales. Through the pathways, given images are transformed into feature maps of the same size. As a consequence, the last feature maps have a size of 6×6 and 256 channels. Then, the network concatenates all feature maps to aggregate information at each scale, which forms 1024 channels. By this aggregation process, the capsule network can learn primary capsules with more rich information about the shape and texture of relevant objects.

The primary capsule comprises $[64 \times 6 \times 6]$ vectors (each is a 16-D vector) by reshaping the feature maps and then passes them to the squash function for non-linearity and normalization from 0 to 1. Next, the network predicts the final class capsule through the dynamic routing process with these capsules. In this paper, the number of iteration of dynamic routing is 3. The final class capsule has two capsules (each is a 32-D vector), one for benign tumor and the other for malignant tumor. The magnitude of the class capsule vectors indicates the probabilities of each class (benign and malignant) for input image.

3   Experiments

3.1   Dataset

In this study, we acquired 2,252 ultrasound images of patients with malignant or benign breast tumors from 4 hospitals. Senior radiologists reviewed all breast images with associated radiological reports. Among the datasets, we chose the dataset from one hospital for network training. The data set is randomly divided into training, tuning, and internal validation sets (See Table 1). The data sets from three hospitals were used for the external validation to evaluate the generalizability of the proposed network and comparative models.

All breast ultrasound images were anonymized to erase personal information, such as patient name, patient ID, acquisition date, and manufacturer, using an in-house anonymization software. Then, we resized all images to a width of 224 pixels and a height of 224 pixels. The intensity of each picture has been normalized from 0 to 1 by dividing all intensity values with maximum intensity.

Table 1. Details of dataset composition.

Dataset		Benign	Malignant	Total
Train	Train	500	500	1000
	Tunning	100	100	200
Validation	Internal	168	200	368
	External-a	129	128	257
	External-b	75	172	247
	External-c	60	120	180
Total		1032	1220	2252

3.2 Implementation Details

For comparison, we employed ImageNet pre-trained networks, GoogLeNet [15] and Inception-v3 [16]. GoogLeNet used inception blocks that combined various filters in a layer and concatenated extracted information from each filter. Inception-v3 is an upgraded version of GoogLeNet. This network extended the inception block by replacing 7×7 and 5×5 convolution filters into 3×3 convolution filters, splitting a $n \times n$ convolution filter into a $1 \times n$ and $n \times 1$ convolution filter, and adjusting the position of pooling layers to avoid representational bottleneck. Both networks were initially trained using ImageNet data and fine-tuned with breast ultrasound images without using auxiliary classifiers [2].

Also, we implemented a dynamic routing capsule network, proposed by [12], without the feature encoder for the Gaussian pyramid to evaluate its effectiveness. Unlike the proposed network that received images of various sizes through Gaussian pyramids, this network takes an ultrasound image of 224×224 pixels as input and predicted a probability for each tumor class. The network consists of 4 convolution layers and the dynamic routing process. Each convolutional layer have 9×9 convolution filters with a stride of 2 and a ReLU activation function. The feature maps then form the primary capsule through the reshaping process and the squash function. Finally, the primary capsule predicts the class capsule through routing-by-agreement.

All these networks were implemented using the Pytorch library (Ver 1.6.1) with Python language (Ver 3.8.3) and trained on NVIDIA TITAN XP (11 GB memory). And, we trained them with a batch size of 32 using the backpropagation algorithm with the Adam optimizer, set a learning rate of 1e−4, and reduced it by 0.1% for each epoch [10]. Moreover, we trained the comparative models with blurred images using the Gaussian filter for smoothing. The size of the Gaussian filter was randomly selected to be less than 30 for every transform.

Table 2. Classification performance of each model in each dataset.

Dataset →	Internal			External-a		
Model ↓	Acc	Sens	Spec	Acc	Sens	Spec
GoogLeNet	92.66%	90.50%	95.24%	70.04%	92.97%	47.29%
GoogLeNet + Blur	93.75%	95.00%	92.26%	77.43%	86.72%	68.22%
Inception-v3	96.47%	97.50%	95.24%	75.49%	91.41%	59.69%
Inception-v3 + Blur	95.11%	95.00%	95.24%	78.99%	93.75%	64.34%
Capsule	98.64%	100.00%	97.02%	83.66%	94.53%	72.87%
Capsule + Blur	**99.18%**	99.50%	98.81%	87.94%	96.88%	79.07%
Proposed	97.01%	96.50%	97.62%	**88.72%**	92.19%	85.27%
Dataset →	External-b			External-c		
Model ↓	Acc	Sens	Spec	Acc	Sens	Spec
GoogLeNet	74.09%	75.00%	72.00%	73.33%	96.67%	26.67%
GoogLeNet + Blur	83.40%	80.23%	90.67%	80.56%	90.00%	61.67%
Inception-v3	81.78%	80.23%	85.33%	73.33%	82.50%	55.00%
Inception-v3 + Blur	83.81%	84.30%	82.67%	87.78%	96.67%	70.00%
Capsule	**85.83%**	82.56%	93.33%	86.67%	98.33%	63.33%
Capsule + Blur	85.02%	80.23%	96.00%	88.89%	96.67%	73.33%
Proposed	82.59%	76.16%	97.33%	**90.00%**	94.17%	81.67%

4   Experimental Results

4.1   Classification Performance

To evaluate the classification performance of each model, we calculated accuracy (Acc), sensitivity (Sens), and Specificity (Spec) on each dataset using a confusion matrix. The classification results are shown in Table 2. Capsule showed the best accuracy of 85.83% on External-b. On the other hand, Capsule trained with blurred image achieved the best accuracy of 99.18% on Internal. Meanwhile, our model achieved the best accuracy of 88.72% on External-a and 90.00% on External-c Also, it showed comparable performance on the others. Experimental results showed that the proposed model could achieve consistently high performance on more datasets.

4.2   Robustness Against Distortion

Speckle noise is one of the inherent problems with ultrasound. It destroys the resolution, brightness, and contrast of ultrasound images, which affects the diagnosis of breast tumors [1,18]. We decided on the variance of speckle noise as a hyperparameter and tested how the model accuracy decreases if the input image is distorted by speckle noise in the validation sets. As the variance of the noise increases, the noise of the image becomes denser. Examples of distortion at each

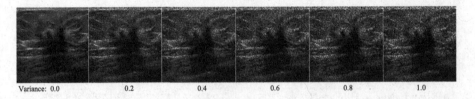

Variance: 0.0   0.2   0.4   0.6   0.8   1.0

Fig. 2. Examples of distortion on an ultrasound image at each variance.

value are shown in Fig. 2. And, Fig. 3 shows the results of the experiment. The models that trained on blurred images showed more robust performance than the model that did not. Also, capsule network-based models showed better performance than ImageNet pre-trained models. The proposed model showed the most robust performance regardless of noise in External-a and External-c, where it achieved the highest performance. Through the experiments, we demonstrated that routing-by-agreement, smoothing, and multi-scale features could improve the robustness of the network to noise.

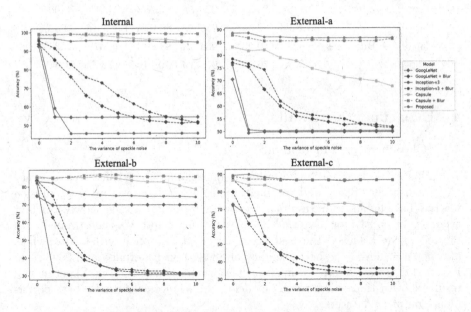

Fig. 3. Classification accuracy on noisy images.

4.3   Feature Space Visualization

To show how well each model distinguishes breast tumors, we visualized feature spaces using the t-SNE method, a parametric embedding technique for dimensionality reduction. In the case of ImageNet pre-trained models, the values before the Softmax function were visualized. And in the case of the capsule

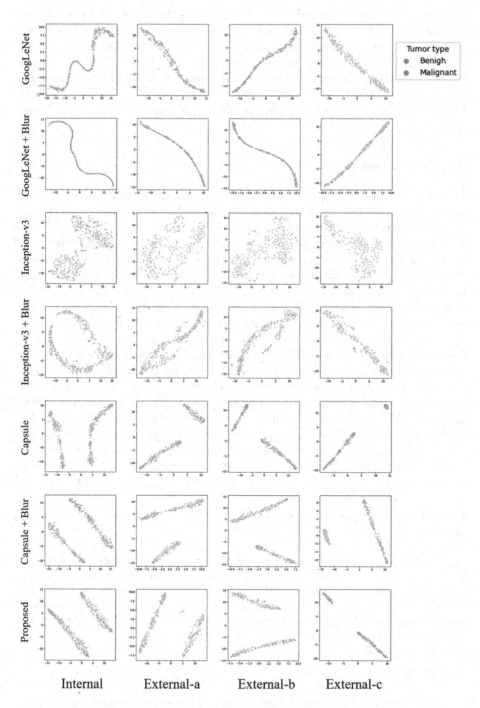

Fig. 4. Visualized feature space of each model on each dataset.

network-based models, including the proposed method, the final output vector was visualized. The Fig. 4 shows how well each model distinguishes breast tumors. Through the results, Inception-v3 was better at distinguishing breast tumors after training on the blurred images (see the third and fourth rows in Fig. 4). And, Capsule network-based models showed an ability to distinguish classes well regardless of the dataset (see the fifth, sixth, and seventh rows in Fig. 4). These results suggested that smoothing and routing-by-agreement could improve the network differentiation ability.

5   Conclusion

In this paper, we propose a capsule network-based architecture, which classify breast tumors as benign or malignancy classes. The proposed model showed consistent performance even with external data from multiple institutions. This network can extract multi-scale features and learned robust features by combining the Gaussian pyramid and capsule network. To prove the effectiveness of our work, we collected breast ultrasound images for external validation sets from 4 hospitals. First, we evaluated and compared the classification performance with other networks. The proposed network showed more generalizable performance on the external validation datasets. Next, we visualized the feature space using t-SNE, a method for dimensionality reduction. This experiment demonstrated that the capsule network-based architecture, including our work, could better differentiate between classes. Finally, we tested robustness against distortion by adding speckle noise to the ultrasound image. The experimental results also demonstrated that the proposed network showed more robustness to noise than other networks. Although the proposed network showed comparable performance, it showed lower performance than general capsule networks in some datasets. In the future, we will improve the generability and study the explainability of the network. The instant parameters of the class capsule represent the characteristics of the tumor. It is possible to create human-understandable explanations of breast tumors using these parameters. We plan to extend our work for explainable and generalizable CAD systems and apply them in real-world breast tumor diagnosis.

Acknowledgement. This research was supported by Basic Science Research Program through the National Research Foundation of Korea(NRF), funded by the Ministry of Education(2020R1I1A3074639) and the Ministry of Science and ICT (2020R1C1C1006453), and the Technology Innovation Program (Development of AI based diagnostic technology for medical imaging devices, 20011875), funded By the Ministry of Trade, Industry & Energy(MOTIE, Korea).

References

1. Abd-Elmoniem, K.Z., Youssef, A.B., Kadah, Y.M.: Real-time speckle reduction and coherence enhancement in ultrasound imaging via nonlinear anisotropic diffusion. IEEE Trans. Biomed. Eng. **49**(9), 997–1014 (2002)

2. Deng, J., Dong, W., Socher, R., Li, L.J., Li, K., Fei-Fei, L.: Imagenet: a large-scale hierarchical image database. In: 2009 IEEE Conference on Computer Vision and Pattern Recognition, pp. 248–255. IEEE (2009)
3. Drukker, K., Gruszauskas, N.P., Sennett, C.A., Giger, M.L.: Breast us computer-aided diagnosis workstation: performance with a large clinical diagnostic population. Radiology **248**(2), 392–397 (2008)
4. Gatys, L.A., Ecker, A.S., Bethge, M.: Image style transfer using convolutional neural networks. In: Proceedings of the IEEE Conference on Computer Vision and Pattern Recognition, pp. 2414–2423 (2016)
5. Geirhos, R., Rubisch, P., Michaelis, C., Bethge, M., Wichmann, F.A., Brendel, W.: Imagenet-trained cnns are biased towards texture; increasing shape bias improves accuracy and robustness. arXiv preprint arXiv:1811.12231 (2018)
6. Han, S., et al.: A deep learning framework for supporting the classification of breast lesions in ultrasound images. Phys. Med. Biol. **62**(19), 7714 (2017)
7. Hinton, G.E., Krizhevsky, A., Wang, S.D.: Transforming auto-encoders. In: Honkela, T., Duch, W., Girolami, M., Kaski, S. (eds.) ICANN 2011. LNCS, vol. 6791, pp. 44–51. Springer, Heidelberg (2011). https://doi.org/10.1007/978-3-642-21735-7_6
8. Huang, X., Belongie, S.: Arbitrary style transfer in real-time with adaptive instance normalization. In: Proceedings of the IEEE International Conference on Computer Vision, pp. 1501–1510 (2017)
9. Kim, C., Kim, W.H., Kim, H.J., Kim, J.: Weakly-supervised us breast tumor characterization and localization with a box convolution network. In: Medical Imaging 2020: Computer-Aided Diagnosis. vol. 11314, p. 1131419. International Society for Optics and Photonics (2020)
10. Kingma, D.P., Ba, J.: Adam: A method for stochastic optimization. arXiv preprint arXiv:1412.6980 (2014)
11. Moore, S.K.: Better breast cancer detection. IEEE Spectr. **38**(5), 50–54 (2001)
12. Sabour, S., Frosst, N., Hinton, G.E.: Dynamic routing between capsules. arXiv preprint arXiv:1710.09829 (2017)
13. Shin, S.Y., Lee, S., Yun, I.D., Kim, S.M., Lee, K.M.: Joint weakly and semi-supervised deep learning for localization and classification of masses in breast ultrasound images. IEEE Trans. Med. Imaging **38**(3), 762–774 (2018)
14. Stavros, A.T., Thickman, D., Rapp, C.L., Dennis, M.A., Parker, S.H., Sisney, G.A.: Solid breast nodules: use of sonography to distinguish between benign and malignant lesions. Radiology **196**(1), 123–134 (1995)
15. Szegedy, C., et al.: Going deeper with convolutions. In: Proceedings of the IEEE Conference on Computer Vision and Pattern Recognition, pp. 1–9 (2015)
16. Szegedy, C., Vanhoucke, V., Ioffe, S., Shlens, J., Wojna, Z.: Rethinking the inception architecture for computer vision. In: Proceedings of the IEEE Conference on Computer Vision and Pattern Recognition, pp. 2818–2826 (2016)
17. Wang, L.: Early diagnosis of breast cancer. Sensors **17**(7), 1572 (2017)
18. Wang, S., Huang, T.Z., Zhao, X.L., Mei, J.J., Huang, J.: Speckle noise removal in ultrasound images by first-and second-order total variation. Numer. Algorithms **78**(2), 513–533 (2018)
19. Yap, M.H., Edirisinghe, E., Bez, H.: Processed images in human perception: a case study in ultrasound breast imaging. Eur. J. Radiol. **73**(3), 682–687 (2010)

Prediction of Pathological Complete Response to Neoadjuvant Chemotherapy Using Multi-scale Patch Learning with Mammography

Ho Kyung Shin[1], Won Hwa Kim[1,2], Hye Jung Kim[1,2], Chanho Kim[1], and Jaeil Kim[1(✉)]

[1] Kyungpooksity, 80 Daehak-ro, Buk-gu, Daegu, Republic of Korea
[2] Kyungpook National University Chilgok Hospital,
807 Hoguk-ro, Buk-gu, Daegu, Republic of Korea

Abstract. Pathological complete response (pCR) indicates the absence of residual tumor in the breast and axillary nodes after neoadjuvant chemotherapy (NAC), which reduces cancerous tumor and improves the prognosis of breast-conserving surgery. To avoid eventual toxicities by NAC and improve the long-term survival outcome, the prediction of pCR using routine breast imaging is an important step to decide the patient treatment. In this paper, we propose a multi-scale patch learning method to predict pCR from pre-NAC mammography, which is widely used for early detection of breast cancer. We use two images (CC and MLO view) of mammography to integrate the texture and shape information of breast tumors in a form of image pyramid with multiple scales. We first extract fixed-sized patches from each pyramid level and concatenate them along the channel dimension to learn multi-scale features of the breast tumor and its surrounding regions. The proposed model achieved better prediction performance (0.803 AUC, 0.75 accuracy, 0.733 sensitivity, and 0.767 specificity) in pCR prediction task than other comparative methods which have been introduced for breast cancer characterization using mammography.

Keywords: Deep learning · Mammography · Multi-scale patch extraction · Breast cancer · pCR Prediction

1 Introduction

Breast cancer is the most common and dangerous type of cancer among women worldwide. Neoadjuvant chemotherapy (NAC) [14,24] is one of the widely used treatments for early-stage breast cancer to reduce tumor size and metastases and to improve the prognosis of breast-conserving surgery. Pathological complete response (pCR) [3,4,10], which indicates the absence of residual tumor in breast after NAC, is used to evaluate the efficacy of NAC. Achievement of pCR

© Springer Nature Switzerland AG 2021
I. Rekik et al. (Eds.): PRIME 2021, LNCS 12928, pp. 192–200, 2021.
https://doi.org/10.1007/978-3-030-87602-9_18

is the ideal outcome of NAC, and it is also used as a surrogate for overall survival (OS) and event-free survival (EFS). The proportion of patients achieving pCR ranges from 10–50%, depending on their cancer subtypes [17,22]. Therefore, predicting whether a patient will achieve pCR prior to NAC can reduce unnecessary toxicities by NAC and help to make a proper decision for alternative treatments.

Various imaging modalities such as MRI (Magnetic Resonance Imaging), mammography, and ultrasonography are commonly used for breast cancer diagnosis. Compared to other imaging modalities, MRI can quantify the breast tumor size most similarly to the pathological measurement [15,16,25]. For this reason, many studies have recently been conducted on the prediction of pCR using MRI. [5,19,21]. Ravichandran, et al. used images patches, extracted from tumor in DCE-MRI (Dynamic contrast enhanced Magnetic Resonance Imaging), and clinical receptor subtype for the pCR prediction. They achieved an area under the curve (AUC) of 0.77 using the patches only and 0.85 AUC by using clinical receptor subtype together [21]. Duanmu, et al. used MRI, molecular and demographic data, and they achieved 0.8 AUC [5]. Qu, et al. used pre-NAC and post-NAC MRI and they improved the pCR prediction by incorporating post-NAC MRI (0.97 AUC) [19]. Although MRI has the advantage of predicting pCR to NAC, it also has limitations such as higher cost, long acquisition time, lack of availability, and the need for intravenous gadolinium contrast administration. On the other hand, mammography has its own advantages of being more cost-effective and commonly used for screening breast cancer [6,7,18]. As a screening modality, mammography can detect abnormalities in breast before physical symptoms appear. Therefore, if pCR prediction is made through mammography, it will help determine effective treatment for patients at an earlier stage.

The goal of this study is to present a novel approach for predicting the response of NAC using mammography. Several deep learning studies using mammography have been introduced for breast mass characterization [2,12,20], but we cannot find any approach using mammography for the pCR prediction. In this paper, we proposed a neural network, named *Multi-scale Patch-Net*, to predict the pCR from mammography via multi-scale patch learning. In our approach, bilateral craniocaudal (CC) and mediolateral oblique (MLO) view images are used as inputs for the network. We first build the image pyramids of both images in various scales, then we extract fixed-size patches from the images at different scales and they are given individually as input to the network. *Multi-scale Patch-Net*, which consists of a feature extractor and a classifier of fully-connected layers, is trained in a end-to-end manner to learn the image texture and shape characteristics of breast tumor in mammography for the pCR prediction. To evaluate the pCR prediction performance of the proposed method, we performed a comparison study with previous deep learning methods using mammography for the breast mass characterization in terms of prediction accuracy, sensitivity, specificity, and AUC.

2   Materials and Methods

In this section, we first introduce a dataset used in this study, and then describe how to extract patches from multi-scale mammography images. Last, we introduce the architecture and details of *Multi-scale Patch-Net* as shown in Fig. 1.

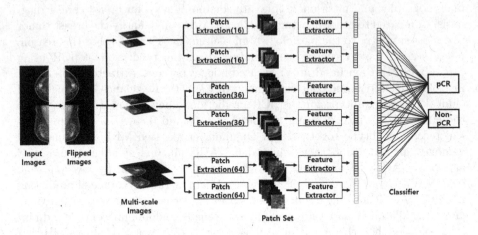

Fig. 1. Architecture of the *Multi-scale Patch-Net* for the pCR prediction

2.1   Materials

For this study, we collected a mammography dataset of 288 breast cancer patients at Kyungpook National University Chilgok Hospital. The mammography was taken prior to NAC for every patient. Images in this dataset have a resolution of 2560×3328 or 3328×4096. In this dataset, 144 patients achieved pCR after NAC (pCR group), while the remaining 144 did not (non-pCR group). We split the dataset into training and test sets via random selection. The train set consists of a total of 228 patients, 114 pCR and 114 non-pCR. And, the test set includes the mammography of 60 patients, 30 pCR and 30 non-pCR. Table 1 shows the summary for the train/test sets. For all patients, the mammography images of left and right breasts were acquired, but we excluded images without breast cancer in this study. For the consistency of inputs, we flipped MLO and CC images for left breast to right side as a preprocessing step.

2.2   Multi-scale Patch Extraction

Multi-scale patch extraction is a method of resizing mammography to various scales and then extracting patches from the resized images. First, MLO and CC view images of mammography are resized into various scales to build an image pyramid (Fig. 2) [1]. In this step, we resize the MLO and CC view images into

Table 1. Summary of the train/test set

	Train	Test	Total
pCR	114	30	144
non-pCR	114	30	144
	228	60	288

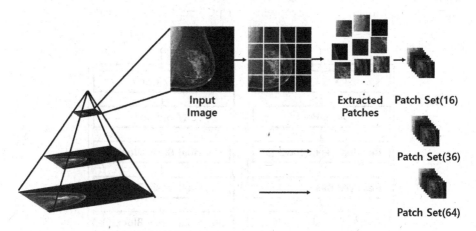

Fig. 2. Extracting patch sets from resized mammography

three scales, 1792×1792, 1356×1356, and 896×896, using bilinear interpolation. We extract fixed-size patches (224×224) without overlap between them from the resized images of the image pyramid. Then, the extracted patches are concatenated along channel dimension to create a patch set. Thus, the patch sets of the resized images at each pyramid level can have the different numbers of channels. At the most coarse level of the image pyramid, we extract 16 patches from MLO and CC view images separately. And, 32 and 64 patches are obtained at other finer levels of the pyramid. Through this multi-scale patch extraction process, the characteristics of the breast tumor and various surrounding areas can be shown to the prediction network.

2.3 Multi-scale Patch-Net

Multi-scale Patch-Net. consists of feature extractors and a classifier. For the MLO and CC view patch sets at three levels of the image pyramid, we build 6 feature extractor networks without weight sharing. The feature extractors learn the image texture, size, and shape characteristics of breast tumor at different scales. We performed an ablation study to determine the levels of the image pyramid. The backbone networks of the feature extractor (see Fig. 3(a)) is a residual network (ResNet-34) [8], which is randomly initialized for training. We remove the fully connected layer from ResNet-34 and change the kernel size of the first layer to 3 × 3. Reducing the size of the kernel is more beneficial to the patch sets including breast partially. We compare this feature extractor with the feature extractor without changing the kernel size (see Fig. 3(b)). The feature extractor takes a patch set as an input and provides a feature vector of 1×512 dimension. For the pCR prediction, the feature vectors of 6 feature extractors are concatenated together to be given to a classifier which consists of a fully-connected layer followed by sigmoid activation function. The entire network is trained using binary cross-entropy loss for pCR and non-pCR cases.

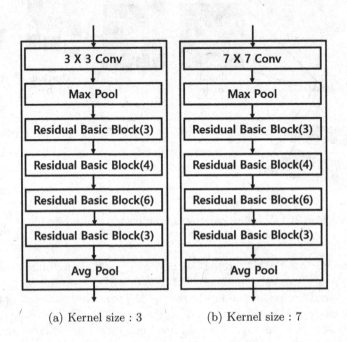

(a) Kernel size : 3   (b) Kernel size : 7

Fig. 3. Feature extractors with different kernel sizes in the first layer

3   Experiment

Since most of the studies on pCR prediction using deep learning were conducted based on MRI, we compared *Multi-scale Patch-Net* with mammography-based deep learning models, which performed well in the breast cancer characterization (benign vs. malignancy) [2,12,20]. Lévy, et al. introduced a GoogLeNet [23], of which two auxiliary classifier are removed, and they achieved an accuracy of 0.92 in breast cancer characterization [12]. Arevalo, et al. changed the last fully connected layer and activation function of AlexNet [11] and used batch normalization instead of Local Response Normalization (LRN) for every convolutional layer. Ensemble of the best three modified AlexNet achieved an accuracy of 0.8 [2]. Rampun, et al. proposed a model with two convolutional layers, two pooling layers, and fully connected layer with maxout activation function [20]. For fair comparison, these three models were trained using tumor regions only as their approaches. The tumor regions were manually delineated by two medical experts and the common regions between the manual segmentation were used as input for the comparative models.

In addition, we compared the proposed method with 3 single-scale CNN models, given input mammography images with different scales, to evaluate the effectiveness of our multi-scale approach. For the single-scale models, we resized the MLO and CC view images to 1792×1792, 1356×1356, and 896×896 respectively. We also evaluated the effect of receptive fields of the feature extractor by changing the kernel size of convolution layers into 7×7.

The proposed model and comparative models were implemented Pytorch 1.7.0. The experiments were performed on a workstation with NVidia TITAN RTX (24GB memory). All models were trained using Adam optimizer [9] with Cross Entropy loss. Learning rate was 7e-4, and it was scheduled with a cosine annealing scheduler [13].

4   Result

The performance of *Multi-scale Patch-Net* and comparative models is evaluated in terms of AUC, accuracy, sensitivity, and specificity. The prediction metrics of the five models are shown in Table 2. The proposed model with the kernel size 3 achieved an AUC of 0.803, accuracy of 0.75, sensitivity of 0.733 and specificity of 0.767. The proposed model outperformed the comparative models in AUC, accuracy, and sensitivity, and was relatively lower than the comparative models in specificity. This result indicate that using the whole mammography of two MLO and CC views can improve the pCR prediction performance compared to using only the tumor and surrounding regions (ROI-based approaches). The proposed model shows superior prediction performance than three single-scale Patch-Nets. This comparison proves that using multiple scales as input has an effect on prediction performance improvement rather than using a single scale. The proposed model without changing the kernel size achieved lower performance (0.661 AUC, 0.667 accuracy, 0.5 sensitivity, and 0.833 specificity) than the proposed model with kernel size 3×3 at the first layer. Comparison of these two models suggests that kernel size affects prediction performance when using extracted patches. Figure 4 shows Receiver Operating Characteristic(ROC) curves of the five models.

Table 2. Prediction performance of the proposed model and comparative models

	AUC	Accuracy	Sensitivity	Specificity
Lévy et al. [12]	0.4489	0.5667	0.6333	0.3333
Arevalo et al. [2]	0.5644	0.55	0.2667	0.9
Rampun et al. [20]	0.5355	0.6167	0.1333	**0.9333**
Single-scale Patch-Net(scale = 896 × 896)	0.6967	0.6833	0.7	0.6667
Single-scale Patch-Net(scale = 1356 × 1356)	0.6089	0.6667	0.5667	0.7667
Single-scale Patch-Net(scale = 1792 × 1792)	0.7267	0.7167	0.6	0.8333
Multi-scale Patch-Net(size = 7)	0.6611	0.6667	0.5	0.8333
Multi-scale Patch-Net(size = 3)	**0.8033**	**0.75**	**0.7333**	0.7667

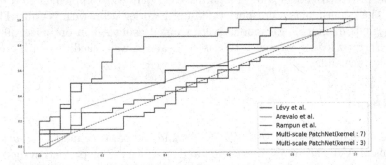

Fig. 4. Receiver Operating Characteristic (ROC) curve of the proposed model and comparative models.

5   Conclusion

In this paper, we proposed a mammography-based pCR prediction model, called *Multi-scale Patch-Net*. *Multi-scale Patch-Net* predicts whether a breast cancer patient will achieve pCR using screening mammography. The patch sets, which are extracted from different scales of the image pyramid, are fed into the individual feature extractors to obtain the deep learning features of breast at various scales. The classifier, the last part of our proposed model, predicts the achievement of pCR from the concatenated feature vectors of the feature extractors.

Most of the studies on the pCR prediction were conducted using MRI. So we compared our model with the other mammography based models [2,12,20] that show outstanding performance in the breast cancer characterization. The comparison result demonstrates that proposed model outperforms comparative models. With the proposed model, it is expected that pCR prediction will be performed cost-effectively at an early stage. For future work, We are planning to advance our patch extraction method towards extracting more influential patches from mammography. And the structure of feature extractors will also be improved.

Acknowledgement. This research was supported by Basic Science Research Program through the National Research Foundation of Korea(NRF), funded by the Ministry of Education(2020R1I1A3074639) and the Ministry of Science and ICT (2020R1C1C1006453), and the Technology Innovation Program (Development of AI based diagnostic technology for medical imaging devices, 20011875), funded By the Ministry of Trade, Industry & Energy(MOTIE, Korea).

References

1. Adelson, E.H., Anderson, C.H., Bergen, J.R., Burt, P.J., Ogden, J.M.: Pyramid methods in image processing. RCA Eng. **29**(6), 33–41 (1984)
2. Arevalo, J., González, F.A., Ramos-Pollán, R., Oliveira, J.L., Lopez, M.A.G.: Representation learning for mammography mass lesion classification with convolutional neural networks. Comput. Methods Programs Biomed. **127**, 248–257 (2016)
3. Cortazar, P., Geyer, C.E.: Pathological complete response in neoadjuvant treatment of breast cancer. Ann. Surg. Oncol. **22**(5), 1441–1446 (2015)
4. Cortazar, P., et al.: Pathological complete response and long-term clinical benefit in breast cancer: the ctneobc pooled analysis. Lancet **384**(9938), 164–172 (2014)
5. Duanmu, H., et al.: Prediction of pathological complete response to neoadjuvant chemotherapy in breast cancer using deep learning with integrative imaging, molecular and demographic data. In: Martel, A.L., et al. (eds.) MICCAI 2020. LNCS, vol. 12262, pp. 242–252. Springer, Cham (2020). https://doi.org/10.1007/978-3-030-59713-9_24
6. Elmore, J.G., Armstrong, K., Lehman, C.D., Fletcher, S.W.: Screening for breast cancer. JAMA **293**(10), 1245–1256 (2005)
7. Gøtzsche, P.C., Jørgensen, K.J.: Screening for breast cancer with mammography. Cochrane Database Syst. Rev. (6) (2013)
8. He, K., Zhang, X., Ren, S., Sun, J.: Deep residual learning for image recognition. In: Proceedings of the IEEE Conference on Computer Vision and Pattern Recognition, pp. 770–778 (2016)
9. Kingma, D.P., Ba, J.: Adam: A method for stochastic optimization. arXiv preprint arXiv:1412.6980 (2014)
10. Kong, X., Moran, M.S., Zhang, N., Haffty, B., Yang, Q.: Meta-analysis confirms achieving pathological complete response after neoadjuvant chemotherapy predicts favourable prognosis for breast cancer patients. Eur. J. Cancer **47**(14), 2084–2090 (2011)
11. Krizhevsky, A., Sutskever, I., Hinton, G.E.: Imagenet classification with deep convolutional neural networks. Adv. Neural. Inf. Process. Syst. **25**, 1097–1105 (2012)
12. Lévy, D., Jain, A.: Breast mass classification from mammograms using deep convolutional neural networks. arXiv preprint arXiv:1612.00542 (2016)
13. Loshchilov, I., Hutter, F.: Sgdr: Stochastic gradient descent with warm restarts. arXiv preprint arXiv:1608.03983 (2016)
14. Makhoul, I., Kiwan, E.: Neoadjuvant systemic treatment of breast cancer. J. Surg. Oncol. **103**(4), 348–357 (2011)
15. Marinovich, M.L., et al.: Meta-analysis of magnetic resonance imaging in detecting residual breast cancer after neoadjuvant therapy. J. Natl. Cancer Inst. **105**(5), 321–333 (2013)
16. Marinovich, M.L., et al.: Agreement between MRI and pathologic breast tumor size after neoadjuvant chemotherapy, and comparison with alternative tests: individual patient data meta-analysis. BMC Cancer **15**(1), 1–12 (2015)

17. Masuda, H., et al.: Differential response to neoadjuvant chemotherapy among 7 triple-negative breast cancer molecular subtypes. Clin. Cancer Res. **19**(19), 5533–5540 (2013)
18. Niell, B.L., Freer, P.E., Weinfurtner, R.J., Arleo, E.K., Drukteinis, J.S.: Screening for breast cancer. Radiol. Clin. **55**(6), 1145–1162 (2017)
19. Qu, Y.H., Zhu, H.T., Cao, K., Li, X.T., Ye, M., Sun, Y.S.: Prediction of pathological complete response to neoadjuvant chemotherapy in breast cancer using a deep learning (dl) method. Thoracic Cancer **11**(3), 651–658 (2020)
20. Rampun, A., Scotney, B.W., Morrow, P.J., Wang, H.: Breast mass classification in mammograms using ensemble convolutional neural networks. In: 2018 IEEE 20th International Conference on e-Health Networking, Applications and Services (Healthcom), pp. 1–6. IEEE (2018)
21. Ravichandran, K., Braman, N., Janowczyk, A., Madabhushi, A.: A deep learning classifier for prediction of pathological complete response to neoadjuvant chemotherapy from baseline breast dce-mri. In: Medical Imaging 2018: Computer-Aided Diagnosis. vol. 10575, p. 105750C. International Society for Optics and Photonics (2018)
22. Straver, M., et al.: The relevance of breast cancer subtypes in the outcome of neoadjuvant chemotherapy. Ann. Surg. Oncol. **17**(9), 2411–2418 (2010)
23. Szegedy, C., et al.: Going deeper with convolutions. In: Proceedings of the IEEE Conference on Computer Vision and Pattern Recognition, pp. 1–9 (2015)
24. Thompson, A., Moulder-Thompson, S.: Neoadjuvant treatment of breast cancer. Ann. Oncol. **23**, x231–x236 (2012)
25. Yuan, Y., Chen, X.S., Liu, S.Y., Shen, K.W.: Accuracy of MRI in prediction of pathologic complete remission in breast cancer after preoperative therapy: a meta-analysis. Am. J. Roentgenol. **195**(1), 260–268 (2010)

The Pitfalls of Sample Selection: A Case Study on Lung Nodule Classification

Vasileios Baltatzis[1,2]([✉]), Kyriaki-Margarita Bintsi[2], Loïc Le Folgoc[2],
Octavio E. Martinez Manzanera[1], Sam Ellis[1], Arjun Nair[3], Sujal Desai[4],
Ben Glocker[2], and Julia A. Schnabel[1,5,6]

[1] School of Biomedical Engineering and Imaging Sciences, King's College London,
London, UK
`vasileios.baltatzis@kcl.ac.uk`
[2] BioMedIA, Department of Computing, Imperial College London, London, UK
[3] Department of Radiology, University College London, London, UK
[4] The Royal Brompton and Harefield NHS Foundation Trust, London, UK
[5] Technical University of Munich, Munich, Germany
[6] Helmholtz Center Munich, Munich, Germany

Abstract. Using publicly available data to determine the performance of methodological contributions is important as it facilitates reproducibility and allows scrutiny of the published results. In lung nodule classification, for example, many works report results on the publicly available LIDC dataset. In theory, this should allow a direct comparison of the performance of proposed methods and assess the impact of individual contributions. When analyzing seven recent works, however, we find that each employs a different data selection process, leading to largely varying total number of samples and ratios between benign and malignant cases. As each subset will have different characteristics with varying difficulty for classification, a direct comparison between the proposed methods is thus not always possible, nor fair. We study the particular effect of truthing when aggregating labels from multiple experts. We show that specific choices can have severe impact on the data distribution where it may be possible to achieve superior performance on one sample distribution but not on another. While we show that we can further improve on the state-of-the-art on one sample selection, we also find that on a more challenging sample selection, on the same database, the more advanced models underperform with respect to very simple baseline methods, highlighting that the selected data distribution may play an even more important role than the model architecture. This raises concerns about the validity of claimed methodological contributions. We believe the community should be aware of these pitfalls and make recommendations on how these can be avoided in future work.

1 Introduction

Lung nodule characterization is the most difficult step in the pipeline of lung cancer diagnosis according to radiologists, which can be observed by a great

© Springer Nature Switzerland AG 2021
I. Rekik et al. (Eds.): PRIME 2021, LNCS 12928, pp. 201–211, 2021.
https://doi.org/10.1007/978-3-030-87602-9_19

inter-observer disagreement on the task [7,12]. A lung nodule is normally characterized with respect to texture, spiculation, lobulation, and its morphological appearance on a CT scan, and eventually it must be classified as either benign or malignant for patient management. The Lung imaging Reporting And Data System (Lung-RADS) [9] is a protocol that defines explicit guidelines for nodule management and follow up planning, and classifies pulmonary nodules in six categories, each of which has its own suggested follow up. Lung-RADS also integrates the PanCan Model [11], which provides a malignancy probability based on the morphology of a nodule and additional patient information. Certain diagnosis can only be made through biopsy, which, however, is invasive and not always feasible to have access to. While determining the malignancy of a nodule from its appearance on a CT scan is not a fail-proof method, it is still a very useful step of the lung cancer detection pipeline. It can have very important value to clinicians in conjunction with patient history and demographics.

Several deep learning methods have been proposed for automated nodule classification from CT. The publicly available Lung Image Database Consortium and Image Database Resource Initiative (LIDC) database [2,10] has been in the core of the majority of such efforts. The LIDC does not primarily contain pathology confirmed ground truths (besides a very small subset of cases), but rather radiologists' annotations. Nevertheless, it is still heavily used by the research community for the task of lung nodule classification. Interestingly, there are various design choices regarding sample selection that need to be considered, which can have severe impact on the reported results.

The contributions of this paper can be summarized as follows: 1) We analyze several published works reporting results on LIDC nodule classification and examine the different assumptions such as annotation aggregations methods, removal of cases based on clinical guidelines, and data augmentation, which all can affect the resulting sample selection process; 2) Through an extensive experimental analysis, we show that the selected data distribution can affect the difficulty of the task and may play an even more important role than the model architecture; 3) We demonstrate that reproducibility and direct model comparison is virtually impossible to achieve and provide suggestions towards making this feasible in future work, while also making our data selection publicly available to promote reproducibility. We illustrate the pitfalls of sample selection with a novel methodological approach of curriculum by smoothing for lung nodule classification. Our findings and insights will be of use to the community and aid in the design of future approaches for lung nodule classification.

2   State-of-the-Art in Lung Nodule Classification

The LIDC dataset contains more than 1000 scans. Each scan was reviewed by four radiologists who pinpointed lesion locations and assigned a variety of annotations including malignancy. For every nodule, each radiologist had to assign a malignancy rating from 1 (most likely benign) to 5 (most likely malignant). Nodules annotated with 3 were regarded as *indeterminate*.

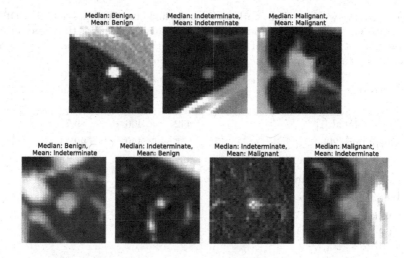

Fig. 1. Lung nodule examples from the LIDC. Top row: Nodules that have the same consensus regardless of the aggregation method used. Bottom row: Nodules that have different consensus depending on the aggregation method.

There are a number of preprocessing and data curation steps which are considered fixed when using the LIDC and almost all recent deep learning papers follow them. These include (1) retaining only nodules that have been annotated by at least three radiologists and (2) discarding nodules annotated as *indeterminate*. Subsequently, for each nodule a consensus annotation is extracted from the individual annotations through some form of aggregation or truthing (typically using mean, median, or majority voting). Example nodules from the LIDC with different consensus/aggregation combinations can be seen in Fig. 1. Given these relatively straightforward steps, it may be surprising to find that every paper we studied reports largely varying numbers for benign and malignant nodules and overall cases (see Table 1). Most studies report that they follow a procedure similar to previous work, however, rarely provide the exact details about either the sample selection process or the final dataset (e.g. by publishing a list of scan series IDs). Beside the differences in absolute numbers of benign and malignant cases, the characteristics of the underlying data distribution may change significantly. One of the most important characteristics is the size of a nodule (quantified by its diameter), as it plays an essential role in malignancy classification. Another discrepancy arises from the decision to remove cases that have a slice thickness >2.5 mm, which is based on clinical guidelines [5]. Images with thick slices are deemed unsuitable for lung cancer screening. This step was first suggested in the LUNA16 nodule detection challenge [15] and has also been adopted by other studies [20]. One of the few works that release their pre-processed data is by Al-Shabi et al. [1].

Here, we attempt to draw a direct comparison to their work with the dataset we have extracted from pre-processing LIDC (see Fig. 2). Something like this is

Table 1. Overview of previous work for lung nodule classification on LIDC-IDRI in terms of nodule counts and performance. Despite all papers using the same publicly available dataset, final numbers of benign and malignant cases vary largely making a direct comparison of the methods' performance impossible.

Method	Benign count	Malignant count	Accuracy (%)
Local-Global [1]	442	406	89.75
DeepLung [20]	554	450	90.44
Lightweight multi-CNN [14]	857	448	93.18
Interpretable hierarchical CNN [16]	3212	1040	84.20
NoduleX [3]	394	270	93.20
Multi-crop CNN [17]	880	495	87.14
Multi-task w/margin ranking loss [8]	972	450	93.50

not feasible for the other proposed methods which do not publicly release their sample selection. In this comparison, we want to highlight the important role that the aggregation method (mean vs median) plays in determining which samples are labeled as benign and malignant. When median aggregation is used, we see that a lot more nodules have an *indeterminate* consensus (i.e. median=3) and are therefore excluded, resulting in a smaller, more balanced dataset, which is much easier to separate based on the key characteristic of nodule diameter. Specifically, median aggregation leads to 442/406 benign/malignant nodules for [1] and 376/357 benign/malignant in our replicated pipeline, respectively. In contrast, mean aggregation results in 653/484 benign/malignant for [1] and 559/451 for us. A factor leading to a discrepancy between the two samples, even when the same aggregation method is used, is the fact that cases with a slice thickness >2.5 mm have been retained by [1]. These factors make reproducibility and direct comparison of methods nearly impossible.

3   Methodology

Here we present different methods and approaches, including our attempted contribution, which we considered for studying the impact of sample selection on lung nodule classification performance. We used several baselines and state-of-the-art deep learning approaches.

3.1   Diameter-Based Baselines

Diameter Threshold. The first baseline we set is not learning-based but a rather simplistic one. Specifically, given that the size of a nodule is a primary factor in determining whether a nodule is malignant or not (i.e. large nodules are most likely to be regarded by experts as malignant, while small nodules

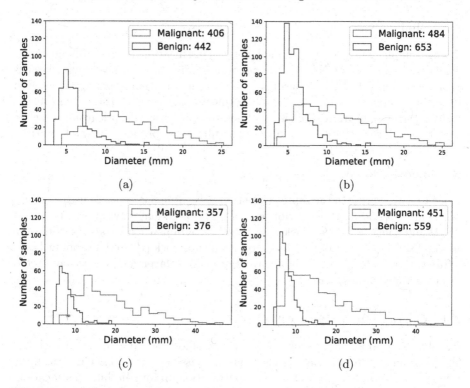

Fig. 2. Data distributions of benign and malignant samples over nodule diameter. (a) Median aggregation from [1], (b) Mean aggregation from [1], (c) Our median aggregation, (d) Our mean aggregation. Median aggregation produces fewer nodules in total (i.e. more nodules are classified as *indeterminate*) for both cases, and at the same time more balanced datasets.

as benign) we use the provided diameter annotation in LIDC and specify a threshold for classifying nodules into benign and malignant. This baseline is used as a surrogate to determine the difficulty of the classification, as the overall size difference between structures may be easily picked up by an image-based prediction model such as a convolutional neural network (CNN).

Regressed Diameter Threshold. Another baseline that we use is similar to the previous one but with a CNN that is trained to regress the diameter through a mean squared error loss. The classification is taking place by applying the threshold determined from the first baseline on the output of the CNN instead of the annotation. Again, if this baseline works well, one may conclude that the task given a specific dataset is not very difficult.

3.2   ShallowNet

We also implement a CNN for malignancy classification (termed ShallowNet), which is a bare-bones CNN comprising of four convolutional layers with kernels of shape 3x3 and ReLU activations, and corresponding max-pooling layers with kernels of shape 2x2, as well as a fully-connected layer with 1024 neurons at the end for the classification. This is a deliberately simplistic deep learning baseline used to compare with more complicated architectures proposed in the literature.

3.3   Local-Global

Since we have access to the sample selection of [1], it makes sense to use the state-of-the-art method on this distribution. The Local-Global network was proposed by [1] and consists of two blocks. Each block contains the following sequence: a residual sub-block [4] followed by a non-local sub-block [19] and a dropout layer. After the two blocks, there is an average pooling layer and a fully-connected layer for the classification.

3.4   Curriculum by Smoothing

Finally, we propose the use of curriculum by smoothing (CBS) [18], which has shown promising results on computer vision classification tasks. CBS plays the role of our attempted methodological contribution on lung nodule classification. The main idea behind CBS is to apply a Gaussian smoothing kernel to the output of each convolutional layer of a CNN. We use $\theta \circledast x$ to denote the convolution of a kernel θ with an input x. Typically, in a CNN, a convolution operation is followed by a non-linear *activation* function as described in Eq. 1:

$$z = activation(\theta_\omega \circledast x) \tag{1}$$

where θ_ω are the trainable parameters of a convolutional layer. The CBS formulation is presented in Eq. 2:

$$z = activation(\theta_G \circledast (\theta_\omega \circledast x)) \tag{2}$$

where θ_G is a predefined Gaussian kernel. The Gaussian kernel is deterministic and is not trained. During the early stages of training it has an initial standard deviation σ, which is annealed as training progresses. This way, high-frequency information is suppressed in the early training steps of the CNN and is only considered at later stages of the training process.

It is important to note that while we introduce CBS here as an approach that could enhance the performance of ShallowNet or Local-Global for the task of lung nodule classification, our purpose is not to propose a novel model architecture but rather to explore whether the selected sample distribution can play a more important role than the model architecture and highlight the pitfalls that occur in such a scenario.

4   Experimental Analysis

Following from the differences in the data distributions, we move to comparing some baseline models, as well as the proposed method from [1]. In this section we focus on two distributions to demonstrate the impact of sample selection and understand whether performance differences stem from the data or the methods. Specifically, we use the data produced with median aggregation (Fig. 2a) from [1] (henceforth denoted as $\mathcal{D}_1$), as this is the one the authors report results for, and mean aggregation (Fig. 2d) for our data (denoted as $\mathcal{D}_2$). We do not consider mean aggregation to be superior to median, but instead we want to study the differences in performance that are caused by this specific choice of truthing. Median aggregation leads to the two classes being more easily separated based on nodule diameter (Figs. 2a, 2c), even though 5–10 mm is considered the most difficult area to separate malignant from benign nodules. In both $\mathcal{D}_1$ and $\mathcal{D}_2$, a nodule is considered benign when the consensus has a value lower than 3 and malignant when it has a value greater than 3.

CT scans with a slice thickness greater than 2.5 mm are removed according to clinical guidelines [5] and every remaining scan is resampled to 1 mm isotropic resolution across all three dimensions and one 32x32 mm patch is extracted along each orthogonal plane at each nodule location. The final classification result for each nodule occurs from the averaging of the individual classification of each of its three planes. Some experiments include offline data augmentation (i.e. the size of the dataset itself is increased six-fold through the addition of nodule augmentations); these augmentations are the ones suggested by [1] and include rotations, horizontal flips and Gaussian smoothing. For the proposed methodological contribution of employing CBS we choose 3x3 sized kernels, with an initial standard deviation $\sigma = 1$ of the Gaussian smoothing kernel and an annealing of 0.5 every 5 epochs based on guidelines provided by the authors of [18] and our own validation performance. All models are evaluated using 10-fold cross validation and the reported results are the average of the performance across the 10 folds. The networks are trained using the Adam optimizer [6] with learning rate 10^{-3} and binary cross-entropy loss for 50 epochs and a batch size of 256 samples. We also deploy early stopping to avoid overfitting. All experiments were conducted using PyTorch [13].

The results of the comparison can be found on Table 2. First, we show that even separating the samples based on nodule diameter (i.e. thresholding) can achieve a quite high accuracy (85.02% for $\mathcal{D}_1$ and 83.46% for $\mathcal{D}_2$). In each case, we select the threshold that maximizes training accuracy. The threshold for the two cases is quite different (7.2 mm for $\mathcal{D}_1$ and 11.5 mm for $\mathcal{D}_2$) because of the different aggregation methods used and also because the equivalent diameter (i.e. the diameter of the sphere having the same volume as the nodule estimated volume) is the one used in [1]. Then we use a shallow CNN (ShallowNet) to regress the nodule diameter and use a threshold (7.7 mm for $\mathcal{D}_1$ and 11 mm for $\mathcal{D}_2$) on that, in order to classify the nodule. If we focus on $\mathcal{D}_1$, we see that a ShallowNet trained directly on malignancy can initially just outperform the diameter-based baselines (85.74%) but its performance gets better progressively

Table 2. Comparison of methods on the different data distribution settings. The reported results are averaged across the 10 folds. $\mathcal{D}_1$ is the data distribution used in [1], which has occurred from median aggregation, while $\mathcal{D}_2$ has been extracted from the LIDC by us using mean aggregation. We use accuracy (Acc), sensitivity (Sens) and specificity (Spec) to report the performance of each method and all the reported values are percentages (%). Even from the baselines, it is evident that $\mathcal{D}_1$ is an easier task to solve than $\mathcal{D}_2$. All methods perform better when augmented with CBS for $\mathcal{D}_1$. In $\mathcal{D}_2$ all configurations perform similarly to the diameter baseline, and there is no improvement from progressively increasing the complexity of the model by adding augmentations and/or CBS.

Method	$\mathcal{D}_1$			$\mathcal{D}_2$		
	Acc	Sens	Spec	Acc	Sens	Spec
Diameter threshold	85.02	90.14	80.31	83.46	69.62	**94.63**
CNN-regressed diameter threshold	84.43	84.23	84.61	81.58	68.95	91.77
ShallowNet	85.74	77.09	93.67	83.86	74.94	91.05
ShallowNet + CBS	86.80	78.57	**94.35**	82.77	71.17	92.12
ShallowNet (w/aug)	89.74	85.96	93.21	**84.35**	77.38	89.98
ShallowNet (w/aug) + CBS	**90.91**	89.40	92.30	82.37	73.61	89.44
Local-Global [1]	89.15	89.16	89.14	82.97	74.72	89.62
Local-Global + CBS	89.26	**91.40**	86.94	81.98	75.38	87.29
Local-Global (w/aug) [1]	89.75	90.17	88.17	82.57	**79.15**	85.33
Local-Global (w/aug) + CBS	**90.91**	90.64	91.17	81.88	70.06	91.41

when we use either CBS (86.80%) or offline augmentations (89.74%) and reaches up to 90.91% if we use both. We observe the same pattern for Local-Global [1] which starts from 89.15% when we do not use CBS or augmentations and eventually reaches 90.91% when we use both. The progressive gains from CBS and augmentations that are present in $\mathcal{D}_1$, however, are not replicated on $\mathcal{D}_2$. All the methods in that case perform very similar to the diameter-based baselines with the ShallowNet being the only one that surpasses them marginally in terms of accuracy (84.35% with augmentations).

5   Discussion

The LIDC dataset has been instrumental for the majority of recent works on lung nodule classification. Here, we take a critical look at the aspect of sample selection after discovering inconsistencies in the reported literature. We aimed to examine different factors that affect the performance of a model and thus the apparent value of its methodological contribution. Starting from the pre-processing steps that various studies have applied on the LIDC dataset, we observe that a number of different assumptions during the sample selection process can lead to very different resulting data distributions (Table 1). Such factors are the choice of the aggregation method (e.g. median or mean), in order to extract a consensus from the multiple annotations per nodule, or the removal of certain cases which are considered as unsuitable for the task due to clinical guidelines.

The aggregation method, in particular, plays a very important role. First, it is affecting the total number of nodules that are retained, since median aggregation leads to more nodules having an *indeterminate* consensus and consequently being removed, compared to mean aggregation. It is fair to say that these nodules, which are retained in the dataset with mean aggregation, are harder examples, and therefore, the classification task that occurs from mean aggregation is more difficult. Second, the prevalence of the two classes in the dataset changes substantially, since median aggregation leads to a more balanced, and potentially more favorable for classification, dataset.

It is easy to understand that these choices change the nature of the underlying data distribution and hence, of the classification task itself. The comparison of the performance of different methods applied on different distributions is thus complex and makes the objective assessment of the value of methodological contributions difficult, which we also demonstrate experimentally. We initially devise several baselines. The first one is a simple thresholding based on the nodule diameter annotation. A size-relevant annotation is usually a core part of a lung nodule dataset, including the LIDC, and therefore this baseline can be applicable in all future studies. In the second baseline we apply a threshold on the diameter predictions that have been regressed by a neural network. This can indicate the degree of bias that a neural network has towards associating large nodules with malignancy and small ones with a benign nature. Given the very similar performance of the ShallowNet trained on malignancy prediction itself with the ShallowNet that is trained to regress the diameter, we understand that this bias is actually quite severe. It is well documented [11] that the size of the nodule is an important factor in determining whether a nodule is benign, but from a clinical perspective there are also other indications such as texture or spiculation, which do not seem to be picked up by the neural network. The aforementioned baselines can describe the difficulty of the task, and we suggest their adaptation by the research community working on lung nodule classification. Additionally, we intend to publicly release our sample selection and we urge the research community to do the same to promote reproducibility.

The core argument of our paper is epitomized when we compare the performance of all methods on the two distributions. Overall, we see that on $\mathcal{D}_1$, adding data augmentation or increasing the complexity of the model (i.e. Local-Global instead of ShallowNet) consistently leads to a distinct increase in performance. The approach of using CBS during training results in a performance increase on every single method, outperforming marginally even the state-of-the-art (Local-Global w/augmentations) on $\mathcal{D}_1$. However, on $\mathcal{D}_2$, all methods are bounded by the diameter threshold baseline and even CBS is not having the impact it did on $\mathcal{D}_1$. This highlights the pitfalls of sample selection which may lead to incorrect conclusions about the methodological contributions. If we were to report only results on $\mathcal{D}_1$, we may have concluded that CBS is beneficial for lung nodule classification, and even outperforms previous works.

6 Conclusion

In this paper we have investigated the effect of sample selection in the context of lung nodule classification using deep learning. We have investigated different factors that cause the various published studies to report completely different number of nodules, and we show experimentally that these factors explicitly affect network performance. We have demonstrated that using progressively more and more complex methods systematically improves performance on the task, if and only if the assumptions regarding the data selection process allows for it. On the other hand, if the data distribution presents a more challenging classification task, as is the case when mean aggregation for the nodule annotations is used, then model complexity or data augmentation do not offer any kind of performance boost compared to even the simplest baseline.

Acknowledgments. This work is funded by the King's College London & Imperial College London EPSRC Centre for Doctoral Training in Medical Imaging (EP/L015226/1), EPSRC grant EP/023509/1, the Wellcome/EPSRC Centre for Medical Engineering (WT 203148/Z/16/Z), and the UKRI London Medical Imaging & Artificial Intelligence Centre for Value Based Healthcare. The Titan Xp GPU was donated by the NVIDIA Corporation.

References

1. Al-Shabi, M., Lan, B.L., Chan, W.Y., Ng, K.H., Tan, M.: Lung nodule classification using deep local-global networks. Int. J. Comput. Assist. Radiol. Surg. **14**(10), 1815–1819 (2019). https://doi.org/10.1007/s11548-019-01981-7
2. Armato, S.G., McLennan, G., Bidaut, L., et al.: The Lung Image Database Consortium (LIDC) and Image Database Resource Initiative (IDRI): a completed reference database of lung nodules on CT scans. Med. Phys. **38**(2), 915–931 (2011). https://doi.org/10.1118/1.3528204. http://www.ncbi.nlm.nih.gov/pubmed/21452728
3. Causey, J.L., Zhang, J., Ma, S., et al.: Highly accurate model for prediction of lung nodule malignancy with CT scans. Sci. Rep. **8**(1) (2018). https://doi.org/10.1038/s41598-018-27569-w
4. He, K., Zhang, X., Ren, S., Sun, J.: Deep residual learning for image recognition. In: Proceedings of the IEEE Computer Society Conference on Computer Vision and Pattern Recognition, vol. December 2016, pp. 770–778 (2016). https://doi.org/10.1109/CVPR.2016.90. http://image-net.org/challenges/LSVRC/2015/
5. Kazerooni, E.A., Austin, J.H., Black, W.C., et al.: ACR-STR practice parameter for the performance and reporting of lung cancer screening thoracic computed tomography (CT): 2014 (resolution 4). J. Thoracic Imaging **29**(5), 310–316 (2014)
6. Kingma, D.P., Ba, J.L.: Adam: a method for stochastic optimization. In: 3rd International Conference on Learning Representations, ICLR 2015 - Conference Track Proceedings (2015)
7. Lin, H., Huang, C., Wang, W., Luo, J., Yang, X., Liu, Y.: Measuring interobserver disagreement in rating diagnostic characteristics of pulmonary nodule using the lung imaging database consortium and image database resource initiative. Acad. Radiol. **24**(4), 401–410 (2017). https://doi.org/10.1016/j.acra.2016.11.022. http://www.ncbi.nlm.nih.gov/pubmed/28169141

8. Liu, L., Dou, Q., Chen, H., Qin, J., Heng, P.A.: Multi-task deep model with margin ranking loss for lung nodule analysis. IEEE Trans. Med. Imaging **39**(3), 718–728 (2020). https://doi.org/10.1109/TMI.2019.2934577

9. McKee, B.J., Regis, S.M., McKee, A.B., Flacke, S., Wald, C.: Performance of ACR lung-RADS in a clinical CT lung screening program. J. Am. Coll. Radiol. **13**(2), R25–R29 (2016). https://doi.org/10.1016/j.jacr.2015.12.009. http://www.ncbi.nlm.nih.gov/pubmed/25176499

10. McNitt-Gray, M.F., Armato, S.G., Meyer, C.R., et al.: The Lung Image Database Consortium (LIDC) data collection process for nodule detection and annotation. Acad. Radiol. **14**(12), 1464–1474 (2007). https://doi.org/10.1016/j.acra.2007.07.021. http://www.ncbi.nlm.nih.gov/pubmed/18035276

11. McWilliams, A., Tammemagi, M.C., Mayo, J.R., et al.: Probability of cancer in pulmonary nodules detected on first screening CT. New Engl. J. Med. **369**(10), 910–919 (2013). https://doi.org/10.1056/NEJMoa1214726. http://www.nejm.org/doi/10.1056/NEJMoa1214726

12. Nair, A., Bartlett, E.C., Walsh, S.L., et al.: Variable radiological lung nodule evaluation leads to divergent management recommendations. Eur. Respir. J. **52**(6), 1–12 (2018). https://doi.org/10.1183/13993003.01359-2018

13. Paszke, A., Gross, S., Chintala, S., Chanan, G., Yang, E., et al.: Automatic differentiation in Pytorch (2017). https://openreview.net/forum?id=BJJsrmfCZ

14. Sahu, P., Yu, D., Dasari, M., Hou, F., Qin, H.: A lightweight multi-section CNN for lung nodule classification and malignancy estimation. IEEE J. Biomed. Health Inform. **23**(3), 960–968 (2019). https://doi.org/10.1109/JBHI.2018.2879834

15. Setio, A.A.A., Traverso, A., de Bel, T., et al.: Validation, comparison, and combination of algorithms for automatic detection of pulmonary nodules in computed tomography images: the LUNA16 challenge. Med. Image Anal. **42**, 1–13 (2017). https://doi.org/10.1016/j.media.2017.06.015. http://www.ncbi.nlm.nih.gov/pubmed/28732268

16. Shen, S., Han, S.X., Aberle, D.R., Bui, A.A., Hsu, W.: An interpretable deep hierarchical semantic convolutional neural network for lung nodule malignancy classification. Expert Syst. Appl. **128**, 84–95 (2019). https://doi.org/10.1016/j.eswa.2019.01.048. https://linkinghub.elsevier.com/retrieve/pii/S0957417419300545

17. Shen, W., Zhou, M., Yang, F., et al.: Multi-crop convolutional neural networks for lung nodule malignancy suspiciousness classification. Pattern Recogn. **61**, 663–673 (2017). https://doi.org/10.1016/j.patcog.2016.05.029. https://linkinghub.elsevier.com/retrieve/pii/S0031320316301133

18. Sinha, S., Garg, A., Larochelle, H.: Curriculum by smoothing. In: Advances in Neural Information Processing Systems, vol. 33, pp. 21653–21664. Curran Associates, Inc. (2020). https://proceedings.neurips.cc/paper/2020/file/f6a673f09493afcd8b129a0bcf1cd5bc-Paper.pdf

19. Wang, X., Girshick, R., Gupta, A., He, K.: Non-local neural networks. In: Proceedings of the IEEE Computer Society Conference on Computer Vision and Pattern Recognition, pp. 7794–7803. IEEE Computer Society (2018). https://doi.org/10.1109/CVPR.2018.00813

20. Zhu, W., Liu, C., Fan, W., Xie, X.: DeepLung: deep 3D dual path nets for automated pulmonary nodule detection and classification. In: 2018 IEEE Winter Conference on Applications of Computer Vision (WACV) (2018). https://doi.org/10.1101/189928. http://arxiv.org/abs/1801.09555

Anatomical Structure-Aware Pulmonary Nodule Detection via Parallel Multi-task RoI Head

Haoyi Tao[1], Yuanfang Qiao[1], Lichi Zhang[1], Yiqiang Zhan[2], Zhong Xue[2], and Qian Wang[1(✉)]

[1] School of Biomedical Engineering, Shanghai Jiao Tong University, Shanghai, China
`wang.qian@sjtu.edu.cn`
[2] Shanghai United Imaging Intelligence Co., Ltd., Shanghai, China

Abstract. Automatic and accurate pulmonary nodule detection from Computed Tomography (CT) scans plays a vital role in efficient pulmonary cancer screening. Although recent anchor-based methods using Convolutional Neural Networks (CNNs) have achieved state-of-the-art performance in this task, they still have some limitations. First, they do not utilize any prior information such as blood vessel segmentation from images, which can effectively help the detection task. Second, they do not integrate enough context information in the nodule classification branch of the detection framework. Third, the detection is generally achieved by using one single model, which may be insufficient to produce satisfactory results. To overcome these limitations, here we first extract anatomical structures from CT scans and propose a weighted training patch sampling method based on the anatomical structure information. Besides, we propose a parallel multi-task region-of-interest (RoI) head for nodule classification. The proposed method is evaluated on the public Lung Nodule Analysis (LUNA16) challenge dataset. Our method achieves 98.1% in sensitivity at one false-positive per scan and 99.4% in sensitivity at two false-positives per scan.

Keywords: Lung nodule detection · CT · Deep learning · Multi-task

1 Introduction

Lung cancer is the most common cause of cancer-related death, which makes up almost 18% of all cancer deaths [13]. Low-dose lung CT screening provides an effective way for early diagnosis, which can greatly reduce the lung cancer mortality rate [14]. Advanced computer-aided diagnosis systems (CADs) are expected to automatically detect and segment pulmonary nodules with high sensitivity and low false-positive rates. In recent years, deep convolutional neural networks have developed greatly in the field of pulmonary nodule detection. The state-of-the-art methods often utilize the 3D region proposal network (RPN) for

© Springer Nature Switzerland AG 2021
I. Rekik et al. (Eds.): PRIME 2021, LNCS 12928, pp. 212–220, 2021.
https://doi.org/10.1007/978-3-030-87602-9_20

nodule screening, followed by a 3D classifier for false-positive reduction. Such solutions have achieved more than 95% in sensitivity with eight false-positives per scan [9].

However, there remain some limitations in the existing pulmonary nodule detection methods, which hamper their performance in the actual clinical applications. (1) They do not introduce domain-specific knowledge in medical images, such as vessels and ribs, which is the essential difference between natural images and medical images. (2) A proposal bounding box is appropriate for nodule location regression but is usually not enough to distinguish a true positive in hard cases, such as some normal tissues which have similar morphological appearances as nodules. (3) A single learning-based model is usually not robust enough in the face of input shift. It may produce various results on almost the same input, which lacks feasibility in clinical practice.

To address the aforementioned issues, here we attempt to integrate the human anatomical structures as prior information, and provide a novel method to ensure the robustness in nodule detection. We first extract the coarse segmentation of the anatomical structures from CT, and then sample image patches that may contain false-positive candidates based on anatomical structure complexity. Motivated by related works on false-positive reduction [4,7,11], we propose a parallel multi-task region-of-interest (RoI) head for anatomical structure-aware nodule classification. We define an auxiliary task, named as the context category classification, to classify whether the nodule is located near the anatomical structures based on context feature maps around the nodule. We also use three parallel sub-heads to imitate the diagnosis of a case by more than one doctor, and force the three sub-heads to learn from different nodules' features but generate the same results. The final model benefits from both the anatomical structures and parallel multi-task RoI head, and achieves 98.1% in sensitivity at one false-positive per scan and 99.4% in sensitivity at two false-positives per scan.

2   Method

Figure 1 shows the overall framework of the novel pulmonary nodule detection method, which is based on Faster-RCNN [10] with the feature pyramid network (FPN) [8] as the main architecture. The network can be separated into feature extractors, RPN head, and RoI head. We use a modified 3D ResNet18 to extract multi-scale feature maps (C1-C4) from the input CT volume patches. Then we use C2, C3, C4 as FPN inputs and generate P2, P3 from backbone features. Next we use a shared RPN head on multi-scale feature maps to propose candidate 3D bounding boxes across different scales or levels. We use non-maximum suppression (NMS) to integrate these proposals to get RoIs for further classification and bounding boxes regression. While in RoI head, we propose a parallel multi-task RoI head which has three sub-heads to classify nodule or non-nodule using both nodule features and context features.

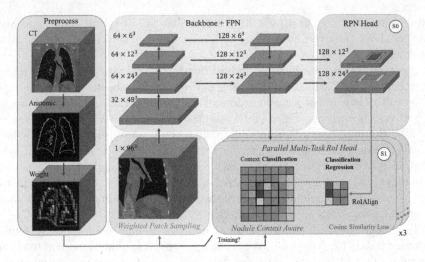

Fig. 1. The proposed framework. A whole CT volume is fed into the preprocess module to extract coarse anatomical structures and weights. Then the input image patch is sampled via the weights. The backbone and FPN extract multi-scale feature maps and fed into the RPN head. Next, the RPN head proposes RoIs to the parallel multi-task RoI head. Finally, the RoI head classify whether the proposal is a nodule based on the features of nodule and context.

2.1   Anatomical Structure-Awareness

We use thresholding to extract coarse anatomical structures. We define a voxel outside the official provided lung mask as rib when the HU value is higher than 400, and a voxel inside the lung mask as vessel when the value is higher than −400. Finally, we can extract coarse anatomical structures as Fig. 2 shows.

Fig. 2. A coarse anatomical structure extracted from CT. The above shows the 3D rendering of the masks of lungs, ribs and vessels.

Weighted Patch Sampling. In 3D detection tasks, we usually crop patches that contain one or more ground-truth bounding boxes, called positive patches, as training inputs. Sometimes, we also crop negative patches, which do not contain detection targets, as part of the training inputs. In this paper, we propose a weighted patch sampling method to sample negative patches. While we are calculating the distance between each voxel and the anatomical structure, we can easily get the sampling weight of each voxel by calculating the reciprocal distance. Therefore, some hard patches which may contain hard negative examples will have a higher weight to be sampled than easy patches.

Nodule Context-Awareness. According to [5], 10% of the 209 lung cancers are located in the contours of the lungs (pleural-attached nodules), 62.7% are located in the periphery of the lungs (peripheral nodules), and only 27.3 are located in the central or middle of the lungs. According to [2], the distribution of pulmonary nodules has a certain pattern. There are more lung nodules in the upper lobe than in the lower lobe. Larger pulmonary nodules are mainly distributed in the lower lobe.

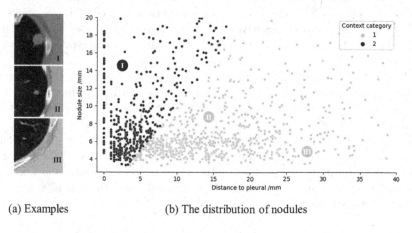

(a) Examples   (b) The distribution of nodules

Fig. 3. The distribution of nodules. Some nodules, such as a.I, are located in pleural, of which the distances are zero. Some nodules (a.II) are also located around and may grow to approach the pleural. Some small nodules (a.III) are located far away.

To clarify the distribution of nodules, we count the sizes of 1186 nodules and their distances from the periphery of the lung. The distribution is shown in Fig. 3b. We find that the sizes of lung nodules are mainly distributed in 3–10 mm, and the distances from the periphery of the lung are mainly distributed in 0–15 mm. We argue that, by keeping aware of the lung's periphery, the nodule detection can be improved in sensitivity. So, we define a **Context Category** for fine-grained classification to find hard nodules which locate near the anatomical structures. We define a nodule as a hard example (context category equals to 2), such as Juxta-pleural, when the minimum distance between the nodule and the nearest anatomical structure is less than the diameter of the nodule. Note that the remaining nodules are defined as an easy example (context category equals to 1).

2.2 3D Region Proposal Network.

The positive and negative anchors for training RPN are defined as follow: If an anchor overlaps a ground truth bounding box with the intersection over union (IoU) higher than 0.7^3, we consider it as a positive anchor (nodule category equals to 1). On the other hand, if an anchor has IoU with all ground-truth boxes less than 0.2^3, we consider it as a negative anchor (nodule category equals to 0). To avoid an extreme large number of negative anchors, the maximum number of negative anchors is 800. The left anchors are considered as ignored.

Due to the extreme imbalance between positive anchors and negative anchors, a sampler is needed to balance the numbers of anchors from both positive and negative. Here we use all positive anchors and only sample as many negative anchors as positive anchors from a total of 800 negative anchors. Finally, the classification loss of RPN head can be defined as the mean of BCE loss of positive anchors and BCE loss of negative anchors. The regression loss can be defined as L1 loss between ground-truth bounding boxes and corresponding positive anchors.

To generate nodule proposals for the following RoI head, two $3 \times 3 \times 3$ convolutional layers are applied to the feature map, followed by two parallel $1 \times 1 \times 1$ convolutional layers to generate the classification probability. The same network architecture is used to generate six regression terms: central z-, y-, x- coordinates, depth, height, and width, associated with each anchor at each voxel on the feature map. We chose a cube of size 4, 5, 10, 20 and 40 as the 5 anchors in this work. To limit the cost caused by too many proposals, we do non-maximum suppression (NMS) on the top 2000 proposals and select the top 1000 proposals as the following inputs.

2.3 3D Parallel Multi-task RoI Head

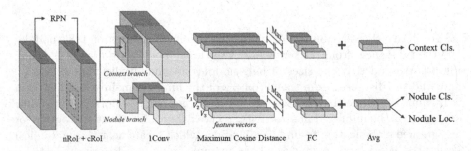

Fig. 4. Parallel Multi-Task RoI Head. For each proposed nodule RoI (nRoI), we use two times scaled nRoI to generate context RoI (cRoI). The context branch only focus on the context category classification. For each branch, we use three parallel sub-heads to do the same task. In order to avoid the high similarity between each sub-heads, we maximum cosine distance to constrain the similarity.

We use a parallel multi-task RoI head to generate nodule RoI (nRoI) and context RoI (cRoI) feature maps, which is the main novelty. For each nRoI (colored green in Fig. 4), We use an RoIAlign layer to crop and resize the feature maps from multi-scale feature maps generated from FPN. We use two times scaled proposal bounding boxes as cRoI (colored orange in Fig. 4) to do the same operation to generate context feature maps.

To generate nodule proposals' classification targets and regression targets, we define as follows. If a proposal overlaps a ground-truth bounding box with IoU higher than 0.7^3, we consider it as a positive proposal. Then the left is considered as negative proposals. To avoid too many negative proposals from dominating the loss, we sample three times as many negative proposals as positive proposals. In context task, we use pre-defined context category as classification target and the regression task is not turned on.

To imitate the diagnosis of a case by more than one doctor, we use three parallel sub-heads (generating three feature vectors in Fig. 4) to classify the same proposals. To make the three sub-heads as different as possible, we use the cosine distance to constrain the similarity between each head output. The similarity loss function is defined as Eq. 1. So that, different heads can learn from different features but produce the same results, which is called different routes to the same goal.

$$L_{sim} = 1 - Dist_{cos} = \frac{1}{3}(Sim(V_1, V_2) + Sim(V_1, V_3) + Sim(V_2, V_3)) \quad (1)$$

Here, Sim represents cosine similarity function and V_i represents the feature vector of the ith sub-head.

2.4 Implementation and the Training Strategy

In the training, for each model, we use 150 epochs in total with SGD optimization and the momentum as 0.9. The batch size parameter is limited by GPU memory and we set batch size as 8 (float16) on our Nvidia GTX 1080Ti GPU. We use weight decay as 0.0001. The initial learning rate is 0.01, 0.001 after half the total number of epochs, and 0.0001 after epoch 120. We also use a linear warming-up learning rate schedule in the first epoch. We use a hard-negative example mining (HNEM) sampler in positive anchors and negative anchors sampling for classification balance. For all training steps, we use a random shifted ground-truth bounding box as additional proposals.

3 Experiments

3.1 Datasets and Experiment Setting

The performance of the proposed framework is evaluated on the most popular LUNA16 challenge dataset which consists of 1186 nodules in the size between 3–30 mm from 888 CT scans and agreed by at least 3 out of 4 radiologists.

It is divided into 10 subsets. To conduct a fair comparison with other methods, we follow the same process to conduct cross-validations by using 9 subsets for training and the remaining 1 subset for testing, then obtain the final results by averaging the 10 experiments.

While processing the official segmentation results of the lung provided by the LUNA16 challenge, we have found serious errors in more than ten cases. So, we manually process these cases and segment the lungs. We resample the CT volume to (1, 1, 1) mm and crop the image inside the lung mask after a dilation, and fill the value outside. Data augmentation is applied by flipping and resizing the CT scans. Except for that, we also use elastic transformation, random rotation, random shift and random noise.

Same as other methods, the Free-Response Receiver Operating Characteristic (FROC) analysis and Competition Performance Metric (CPM) of detection sensitivity and the corresponding false-positives at 1/8,1/4,1/2,1,2,4,8 per scan are employed to measure the performance. The CPM score is calculated by the average sensitivity for all the levels of false-positives per scan.

3.2   Experimental Results

Table 1 shows the FROC evaluation results with 1/8,1/4,1/2,1,2,4 and 8 false-positive levels of our proposed method compared with state-of-the-art methods. The blob numbers in the table indicate the best performance within each column. All the methods are tested on LUNA16 dataset followed by the same FROC evaluation. As shown in the table, our framework obtains over 98% sensitivity at 1 FPs/Scan and almost 100% sensitivity at 2 FPs/Scan. We believe that the parallel multi-task RoI head forces different sub-heads to detect nodules from different features. Therefore, our method tends to detect all nodules and almost never miss them.

Table 1. FROC Performance comparison with the state-of-the-arts: sensitivity (recall) and the corresponding false-positives at 1/8,1/4,1/2,1,2,4,8 per scan.

Method	1/8	1/4	1/2	1	2	4	8	CPM
Zhu et al. [16]	0.692	0.769	0.824	0.865	0.893	0.917	0.933	0.842
Wang et al. [15]	0.676	0.776	0.879	0.949	0.958	0.958	0.958	0.878
Song et al. [12]	0.723	0.838	0.887	0.911	0.928	0.934	0.948	0.881
Ding et al. [3]	0.748	0.853	0.887	0.922	0.938	0.944	0.946	0.891
Khosravan et al. [6]	0.709	0.836	0.921	0.953	0.953	0.953	0.953	0.897
Cao et al. [1]	**0.868**	**0.900**	0.913	0.915	0.916	0.931	0.932	0.911
Liu et al. [9]	0.848	0.876	0.905	0.933	0.943	0.957	0.970	0.919
Ours	0.710	0.840	**0.942**	**0.981**	**0.994**	**0.995**	**0.995**	**0.922**

4   Conclusion

In this paper, we have proposed an effective Parallel Multi-task RoI Head by integrating the human anatomical structures as auxiliary tasks and combining several weak heads to produce a powerful ensemble. This framework provides a more similar method to clinical diagnosis and is a novel attempt to integrate domain expert knowledge in medical images into deep learning. The proposed framework also outperforms the state-of-the-art methods and has achieved high sensitivity which has a great potential in routine clinical practice.

Acknowledgements. This work was supported by the National Key Research and Development Program of China (2018YFC0116400).

References

1. Cao, H., et al.: A two-stage convolutional neural networks for lung nodule detection. IEEE J. Biomed. Health Inform. **24**(7), 2006–2015 (2020)
2. Chen, L., et al.: An artificial-intelligence lung imaging analysis system (alias) for population-based nodule computing in ct scans. Comput. Med. Imaging Gr. **89**, 101899 (2021)
3. Ding, J., Li, A., Hu, Z., Wang, L.: Accurate pulmonary nodule detection in computed tomography images using deep convolutional neural networks. In: Descoteaux, M., Maier-Hein, L., Franz, A., Jannin, P., Collins, D.L., Duchesne, S. (eds.) MICCAI 2017. LNCS, vol. 10435, pp. 559–567. Springer, Cham (2017). https://doi.org/10.1007/978-3-319-66179-7_64
4. Dou, Q., Chen, H., Yu, L., Qin, J., Heng, P.A.: Multilevel contextual 3-d CNNs for false positive reduction in pulmonary nodule detection. IEEE Trans. Biomed. Eng. **64**(7), 1558–1567 (2016)
5. Horeweg, N., et al.: Characteristics of lung cancers detected by computer tomography screening in the randomized nelson trial. Am. J. Respir. Crit. Care Med. **187**(8), 848–854 (2013)
6. Khosravan, N., Bagci, U.: S4ND: single-shot single-scale lung nodule detection. In: Frangi, A.F., Schnabel, J.A., Davatzikos, C., Alberola-López, C., Fichtinger, G. (eds.) MICCAI 2018. LNCS, vol. 11071, pp. 794–802. Springer, Cham (2018). https://doi.org/10.1007/978-3-030-00934-2_88
7. Kim, B.C., Yoon, J.S., Choi, J.S., Suk, H.I.: Multi-scale gradual integration CNN for false positive reduction in pulmonary nodule detection. Neural Netw. **115**, 1–10 (2019)
8. Lin, T.Y., Dollár, P., Girshick, R., He, K., Hariharan, B., Belongie, S.: Feature pyramid networks for object detection. In: Proceedings of the IEEE Conference on Computer Vision and Pattern Recognition, pp. 2117–2125 (2017)
9. Liu, J., Cao, L., Akin, O., Tian, Y.: 3DFPN-HS^2: 3D feature pyramid network based high sensitivity and specificity pulmonary nodule detection. In: Shen, D., et al. (eds.) MICCAI 2019. LNCS, vol. 11769, pp. 513–521. Springer, Cham (2019). https://doi.org/10.1007/978-3-030-32226-7_57
10. Ren, S., He, K., Girshick, R., Sun, J.: Faster r-cnn: Towards real-time object detection with region proposal networks. arXiv preprint arXiv:1506.01497 (2015)

11. Setio, A.A.A., et al.: Pulmonary nodule detection in CT images: false positive reduction using multi-view convolutional networks. IEEE Trans. Med. Imaging **35**(5), 1160–1169 (2016)
12. Song, T., et al.: CPM-Net: a 3d center-points matching network for pulmonary nodule detection in CT scans. In: Martel, A.L., et al. (eds.) MICCAI 2020. LNCS, vol. 12266, pp. 550–559. Springer, Cham (2020). https://doi.org/10.1007/978-3-030-59725-2_53
13. Sung, H., et al.: Global cancer statistics 2020: globocan estimates of incidence and mortality worldwide for 36 cancers in 185 countries. CA Cancer J. Clin. **71**(3), 209–249 (2021)
14. Team, N.L.S.T.R.: Reduced lung-cancer mortality with low-dose computed tomographic screening. N. Engl. J. Med. **365**(5), 395–409 (2011)
15. Wang, B., Qi, G., Tang, S., Zhang, L., Deng, L., Zhang, Y.: Automated pulmonary nodule detection: high sensitivity with few candidates. In: Frangi, A.F., Schnabel, J.A., Davatzikos, C., Alberola-López, C., Fichtinger, G. (eds.) MICCAI 2018. LNCS, vol. 11071, pp. 759–767. Springer, Cham (2018). https://doi.org/10.1007/978-3-030-00934-2_84
16. Zhu, W., Liu, C., Fan, W., Xie, X.: Deeplung: deep 3d dual path nets for automated pulmonary nodule detection and classification. In: 2018 IEEE Winter Conference on Applications of Computer Vision (WACV), pp. 673–681. IEEE (2018)

Towards Cancer Patients Classification Using Liquid Biopsy

Sebastian Cygert[1]([✉])(ID), Franciszek Górski[1], Piotr Juszczyk[1],
Sebastian Lewalski[1], Krzysztof Pastuszak[1,2](ID), Andrzej Czyżewski[1](ID),
and Anna Supernat[1,2](ID)

[1] Faculty of Electronics, Telecommunications and Informatics,
Gdańsk University of Technology, Gdańsk, Poland
`sebcyg@multimed.org`
[2] Intercollegiate Faculty of Biotechnology, University of Gdańsk and Medical
University of Gdańsk, Gdańsk, Poland

Abstract. Liquid biopsy is a useful, minimally invasive diagnostic and monitoring tool for cancer disease. Yet, developing accurate methods, given the potentially large number of input features, and usually small datasets size remains very challenging.

Recently, a novel feature parameterization based on the RNA-sequenced platelet data which uses the biological knowledge from the Kyoto Encyclopedia of Genes and Genomes, combined with a classifier based on the Convolutional Neural Network (CNN), allowed significantly improving the classification accuracy. In this work, we take a closer look at this approach and find that similar results can be obtained using significantly smaller models. Additionally, competitive results were achieved using gradient boosting. Since it has another advantage of adding interpretability to the model, we further analyze it in this work.

Keywords: Image-based classification · Tumor educated platelets · RNA sequencing · Liquid biopsy

1 Introduction

Liquid biopsies offer a minimally invasive sample collection instead of tissue biopsies of solid tumors, traditionally used in cancer evaluation. The most common material for this type of analysis is blood: the source of circulating tumor DNA,

This work has been partially supported by Statutory Funds of Electronics, Telecommunications and Informatics Faculty, Gdansk University of Technology. This work was supported in part through the European Regional Development Fund as part of the Project entitled: Academy of Innovative Applications of Digital Technologies under Grant The Operational Programme "Digital Poland" 2014-2020 number POPC.03.02.00-00-0001/20-00. This research was supported by the SONATA grant of the National Science Centre (2018/31/D/NZ5/01263) and Medical University of Gdańsk statutory work (ST-23, 02-0023/07).

© Springer Nature Switzerland AG 2021
I. Rekik et al. (Eds.): PRIME 2021, LNCS 12928, pp. 221–230, 2021.
https://doi.org/10.1007/978-3-030-87602-9_21

circulating tumor cells, miRNAs, exosomes and, lately, tumor-educated platelets (TEPs). The introduction of high-throughput sequencing techniques allowed for the unprecedented resolution of the analysis. However, generated data complexity and a multitude of features created the need for more advanced approaches than assuming a simple cut-off for final result interpretation. The utility of Support Vector Machine (SVM) and Particle Swarm Optimization-enhanced SVM, known as throbmoSeq classifier, applied to sequenced RNA of tumor educated platelets has already been demonstrated for cancer detection (e.g., non-small cell lung cancer, breast cancer, sarcoma) [3].

Recent work [20] further improved the classification accuracy by implementing biological knowledge on the sequenced RNA molecules from the Kyoto Encyclopedia of Genes and Genomes [10]. Features obtained from an RNA-sequenced platelet, were converted into images and classified by custom-built CNN architecture, resulting in a significant improvement. Their approach introduced two main novelties:

1. using novel feature extraction step.
2. using the CNN-based model with a custom architecture for classification.

However, from the paper, it is unknown whether the improvements come from the new feature extraction step, using a CNN-based model, or using custom architecture. Therefore, we take a step-by-step approach in this work, starting with the standard CNN architectures and detailed data analysis. To improve CNN classification accuracy, standard techniques such as ImageNet pretraining, Dropout [27] and mixup data augmentation [28] are applied.

Finally, other machine learning approaches such as k-nearest neighbors (kNN) and gradient boosting [11] were applied, to compare their accuracies with a CNN-based approach. To sum up, the contributions of this work are as follows:

- We performed an ablation study on the CNN-based classifier using different architectures and regularization strategies to improve model accuracy.
- It was shown that the CNN models are not crucial to the final model performance and similar accuracy can be obtained by using gradient tree boosting. It has another advantage of adding interpretability to the model, which is briefly analyzed in our work.

2   Method

2.1   Parameterization

Briefly, raw RNA-sequencing data encoded in FASTQfiles were subjected to a standardized RNA-sequencing alignment pipeline, as described in Thromboseq protocol [2]. The expression data for each sample were then normalized using DESeq2 package [17] with Variance Stabilizing Transformation [14]. Gencode v19 GRCh37 annotation [10] was used for annotation. Transcripts that could not be mapped to a transcript with Gencode status "known" were excluded.

Filtered expression profiles were then used to build images. Each row corresponds to a signaling pathway from the KEGG database [10]. Each pixel in a row corresponds to the expression level of a single transcript from the pathway. Pathways from the KEGG database corresponding to three aspects: cancer, metabolism and signaling processes, were selected. R package gage was used to gather KEGG pathway data [18]. As a result, a feature vector is a two-dimensional array with 345 rows (number of signaling pathways) and 243 columns (length of the longest pathway).

In [20] such parameterized data is then directly fed into the CNN. However, because each row contains a different number of values, more than half of the values in the array are empty (filled with 0s). As such, we experiment with another input variant where all values are put into a square of minimal size (by simply removing empty values in the original arrays). Only the last few values in the last row are empty. However, now each row may contain data from different signaling pathways. It allowed us to reduce the size of the array to 143 rows by 143 columns and reduce input dimensionality from 83835 to 20449 values, which could make the task easier for the classifier (especially given the limited data). In the experiment section, this variant is named *reduced*.

2.2   Methods

In this section our methods for liquid biopsy data classification are described.

While CNNs were originally developed for computer vision, they were used in a much larger set of applications, including analysis of EEG signals [5], calling genetic variants [23], or text classification [7]. It is because CNNs are parameter-efficient algorithms exploiting local feature patterns. However, for many applications labelled data are scarce, which makes the training very challenging. In such a scenario it was shown that usually smaller architectures are already very efficient [24]. In this work, standard CNN architecture, namely ResNet [12] is used for liquid biopsy classification, and due to the small data regime, smaller variants of the ResNet architecture are used. Additionally, it was tested whether ImageNet pretraining helps in our scenario.

A standard approach to prevent model overfitting is to use some form of data augmentation. While many forms of data augmentation exist for images (e.g., rotations, translations, changes in color), none of the standard forms of data augmentation can be applied to liquid biopsy data. As such, in this work a data-agnostic *mixup* [28] augmentation routine is applied to the data. It creates new data samples by means of linear interpolation between existing data:

$$\tilde{x} = \lambda x_i + (1 - \lambda)x_j$$
$$\tilde{y} = \lambda y_i + (1 - \lambda)y_j$$

where (x_i, y_i) and (x_j, y_j) are randomly selected training pairs of input vectors and the corresponding label, and $\lambda \in [0, 1]$ is the interpolating factor. In the original paper λ is drawn from a symmetric Beta distribution and its α value is a hyperparameter. Despite its simplicity, *mixup* is a powerful technique that works as a strong regularizer on the model. Apart from *mixup*, other experiments were conducted using another regularization technique, namely Dropout [27], which works by randomly zeroing output from some of the neurons during training.

Another popular machine learning technique is a gradient tree boosting [11], which often performs very well on diverse applications such as ranking problems [4], recent COVID-19 patient deterioration prediction [25] and many others [19]. A great advantage of gradient boosting algorithms is their interpretability, especially important for medical applications. Additionally, a number of well-maintained open-source libraries are available (e.g., XGBoost[6], LightGBM [16]. A popular XGBoost library is used for data processing in this work, and standard hyperparameter search is applied over selected parameters.

Finally, a simple k-nearest neighbors classifier was applied to the problem. While we do not expect it to perform better than previous methods, it is an important baseline for comparing results to. It could be possible that the classification improvements in [20] were possible mainly due to the novel parameterization, and in such a case even a simple k-nearest neighbors algorithm would perform well.

3   Experiments

3.1   Evaluation

Table 1. Datasets used for experimentation.

Name	Train samples	Test samples	Imbalance ratio	Details
OC [20]	158	104	8.36	ovarian cancer
NSCLC [3]	157	447	1.96	non-small cell lung cancer
Sarcoma [13]	118	56	1.8	sarcoma

Three publicly available datasets were used to test the classifier (Table 1): 401 non-small cell lung cancer patients (NSCLC) and 203 healthy controls [3], 62 sarcoma and 37 former sarcoma patients who recovered at least 5 years ealier, now treated as healthy) and 75 healthy controls [13] and the original imPlatelet dataset consisting of 204 healthy controls and patients with ovarian cancer (28) or benign gynaecological conditions (30) [20].

For evaluation, the same setting as in [20] was followed, i.e., the model is evaluated on exactly the same held-out test set. Then, the rest of the data is split into train and validation parts and a 5-fold stratified validation is run.

Balanced accuracy is used to measure performance, which is defined as:

$$\text{Balanced accuracy} = \frac{\text{Sensitivity} + \text{Specificity}}{2}$$

where sensitivity is the true positive rate and the specificity is the true negative rate.

3.2 CNN Model

For the CNN approach, standard ResNet backbones are used for classification. Since the amount of data is limited, we focused on smaller variants of the ResNet, namely ResNet-18 and ResNet-34 from the PyTorch library [21] were tested. First, a detailed study of the influence of different factors is performed on the ovarian cancer dataset. To find the best model a grid search is executed, where the learning rate is sampled from the range [0.01, 0.1], Dropout rate $\in \{0.2, 0.5\}$ and L1-weights regularization $\in \{0.001, 0.0001\}$. During training standard cross-entropy classification loss it optimized. The model is being trained for 100 epochs and the model for testing is chosen based on the balanced accuracy. As there are still minor differences between the same runs, each experiment is repeated 3 times, and mean accuracy is reported.

In the first experiment, two parameterization approaches for the CNN model, as described in Sect. 2.1, were compared (Table 2). In general all models perform very well on the validation set (balanced accuracy ranging from 88% to 92%), and as expected, have a few percent accuracy drops on the test set. In general, there is no clear winner for one specific architecture or data parameterization. Note, that the model with the highest validation accuracy obtain the lowest score on the test set. For the next round of experiments, it was decided to utilize the *reduced* parameterization, as it performs similarly to the original parameterization, while using a much smaller input size.

Table 2. Accuracy of different backbones and parameterization on ovarian cancer classification. Balanced accuracy reported.

Backbone	Validation acc.	Test acc.
Standard parameterization [20]		
ResNet-18	0.9080	**0.8958**
ResNet-34	0.8793	0.8317
Reduced parameterization		
ResNet-18	0.8938	0.8563
ResNet-34	**0.9218**	0.8255

Further, it was tested whether using ImageNet pretraining can improve the final accuracy (Table 3). As it can be noticed, using ImageNet pretraining and mixup improved the validation and testing accuracy. Also the variance in classification accuracy was reduced, which shows that using the above methods helped to stabilize trained models. Balanced accuracy reported.

Finally, evaluation was performed on NSCLC and Sarcoma datasets. For the NSCLC dataset the best model obtained a balanced accuracy of 86,52% on the test set. The Sarcoma dataset turned out to be challenging, which might be because of the limited dataset size. When doing standard 5-fold validation, it turned out that the accuracy on the balanced dataset is poorly correlated with

Table 3. Effects of ImageNet pretraining and *mixup* data augmentation on the ovarian cancer classification.

Model	Validation acc.	Test acc.	Test std.
ResNet-18	0.8938	0.8563	0.0614
ResNet-34	0.9218	0.8255	0.0738
ImageNet pretraining			
ResNet-18	0.9236	0.8952	**0.0056**
ResNet-34	0.9236	0.8652	0.0229
mixup			
ResNet-18	**0.9379**	0.8798	0.0242
ResNet-34	0.9042	0.8221	0.0477
mixup + ImageNet pretraining			
ResNet-18	0.9343	**0.9043**	0.0328
ResNet-34	0.9343	0.8782	0.0185

the accuracy on the test set. As such, the models were trained for 100 epochs and simply the model from the last epoch was used for testing. It was possible since no signs of overfitting were noticed, which might be due to the used regularization techniques, i.e., dropout. In such a setting, the Sarcoma balanced accuracy was 94,09%.

3.3   Other Algorithms

In this section experiments with tree gradient boosting and kNN model are presented. For the XGBoost model the following parameters were used in the hyperparameter search: maximal depth of a tree $\in \{1, 2, 3, 4\}$, number of boosting stages $\in \{50, 100, 300, 500\}$ and learning rate $\in \{0.1, 0.01\}$, following insights from [19].

In the case of kNN algorithm, given the large dimensionality of the input space, first, a Principal Component Analysis (PCA) is performed using *scikit* library [22]. Then the grid search for the kNN algorithm is applied, searching for the number of neighbours ($n \in \{1, 2, 3\}$) and number of principal components (explained variance in $\{0.6, 0.7, 0.8, 0.9, 0.95, 0.99, 0.999\}$. For evaluation, the same procedure is applied as in the CNN model. Stratified 5-fold cross-validation is used for model selection (PCA is separately computed for each fold), and models with the best validation accuracy are used for testing (Table 4).

As expected the kNN algorithm is the worst performing algorithm. However, on the OC dataset it reached 69.06% of balanced accuracy which is a fair result. When comparing gradient boosting and CNN classifier, CNN scores similar on the OC dataset and better on remaining datasets, and CNN accuracy is the most stable across datasets. However, note that the CNN model is the only one that used data augmentation, so it is very likely that gradient boosting would benefit

Table 4. Accuracy comparison of different classification methods on all datasets. Balanced accuracy reported.

Aggregation method	Validation acc.	Test acc.
OC		
CNN	0.9343	**0.9043**
Boosting	1.0	0.8991
kNN	0.7556	0.6906
NSCLC		
CNN	0.9129	**0.8652**
Boosting	0.76	0.7343
kNN	0.61	0.5299
Sarcoma		
CNN	1.00	**0.9409**
Boosting	0.9818	0.6316
kNN	0.4522	0.3592

from it, especially on the Sarcoma dataset, on which the model is overfitting (large difference between validation and test accurracy).

3.4 Discussion

In this work various machine learning approaches were evaluated on the task of cancer patient classification using liquid biopsy. It was found that both standard CNN-based models and gradient boosting algorithms perform very well. However, all of the models are sensitive to the hyperparameters. It is because of the limited size of datasets and a high number of input features. At the same time, it is expected that ensembling results from different models will further increase and stabilize the performance.

Compared to the recent work [20] it was found that standard CNN backbones (i.e. ResNet architecture) can perform very well on the task (as opposed to the custom architecture). Further, it was shown that other models (i.e., gradient boosting) could also perform very well on the task, given the novel parameterization proposed in [20]. Our models performed better on the Sarcoma dataset and worse on the OC dataset and NSCLC datasets. At the same time, the CNN model used in our work is significantly smaller in terms of a number of parameters.

Various regularization techniques were tested for the CNN model (Dropout, *mixup* data augmentation, ImageNet pretraining, L1-weight regularization). At the same time, there is no combination of methods and hyperparameters that work the best on all datasets; in general, those methods allowed to improve and stabilize the performance.

The importance of features was calculated using XGboost built-in functionality and depended on the number of splits a particular feature was involved in. In the NSCLC dataset, RPL7A showed the highest importance. This gene has been studied only in relation to osteosarcoma [29]. Four genes demonstrated consistently high importance in the detection of ovarian cancer - SH3GL2 involved in breast [15] and lung cancer [8], PRPF6, which is involved in tumor growth in colon cancer [1], HLA-DRA, which expression levels are known to affect the prognosis of a number of malignancies [9] and UGT2B7, which mutations are known to increase the risk of breast and colorectal cancer [26]. UGT2B7 has not been studied in relation to ovarian cancer and the research on PRFR6 and SH3GL2 has been minimal. Since the performance of gradient boosting on sarcoma dataset was poor, no analysis of feature importance was performed. The analysis of feature importance may provide additional targets for further research on the biological background of studied cancers.

4   Conclusions

In this work an analysis of different machine learning approaches to patients classification using liquid biopsy data, was presented. It was found out that given the novel parameterization presented in [20], standard CNN-based models and gradient boosting methods are very effective. However, because of the limited datasets size, and significant size of the input space, different regularization techniques (such as dropout, mixup data augmentation) are crucial to the final performance of the model and its stability.

Gradient boosting allowed us to add interpretability to the model. Using data augmentation for the gradient boosting model to improve and stabilize its performance is essential for future work. It will also allow us to perform a more detailed analysis of the importance of the features returned by the model.

Acknowledgments. Authors would like to thank Tomasz Bączek, Myron Best, Jacek Bigda, Peter Grešner, Jacek Jassem, Tomasz Stokowy, Thomas Würdinger, Anna Żaczek for constant support.

References

1. Adler, A.S., et al.: An integrative analysis of colon cancer identifies an essential function for PRPF6 in tumor growth. Genes Dev. **28**(10), 1068–1084 (2014)
2. Best, M.G., In 't Veld, S.G.J.G., Sol, N., Wurdinger, T.: RNA sequencing and swarm intelligence-enhanced classification algorithm development for blood-based disease diagnostics using spliced blood platelet RNA. Nat. Protoc. **14**(4), 1206–1234 (2019)
3. Best, M.G., et al.: Swarm intelligence-enhanced detection of non-small-cell lung cancer using tumor-educated platelets. Cancer Cell **32**(2), 238–252 (2017)
4. Burges, C.J.: From ranknet to lambdarank to lambdamart: An overview. Technical report 23–581 (2010)

5. Cecotti, H., Graser, A.: Convolutional neural networks for p300 detection with application to brain-computer interfaces. IEEE Trans. Pattern Anal. Mach. Intell. **33**(3), 433–445 (2010)

6. Chen, T., Guestrin, C.: XGBoost: a scalable tree boosting system. In: Proceedings of the 22nd ACM SIGKDD International Conference on Knowledge Discovery and Data Mining, pp. 785–794. ACM (2016)

7. Conneau, A., Schwenk, H., Barrault, L., LeCun, Y.: Very deep convolutional networks for text classification. In: Proceedings of the 15th Conference of the European Chapter of the Association for Computational Linguistics, EACL (2017)

8. Dasgupta, S., et al.: SH3GL2 is frequently deleted in non-small cell lung cancer and downregulates tumor growth by modulating EGFR signaling. J. Mol. Med. **91**(3), 381–393 (2013)

9. Dunne, M.R., et al.: HLA-DR expression in tumor epithelium is an independent prognostic indicator in esophageal adenocarcinoma patients. Cancer Immunol. Immunother. **66**(7), 841–850 (2017). https://doi.org/10.1007/s00262-017-1983-1

10. Frankish, A., et al.: GENCODE reference annotation for the human and mouse genomes. Nucleic Acids Res. **47**(D1), D766–D773 (2019)

11. Friedman, J.H.: Greedy function approximation: a gradient boosting machine. Ann. Stat. 1189–1232 (2001)

12. He, K., Zhang, X., Ren, S., Sun, J.: Deep residual learning for image recognition. In: Proceedings of the IEEE Conference on Computer Vision and Pattern Recognition (2016)

13. Heinhuis, K.M., et al.: Rna-sequencing of tumor-educated platelets, a novel biomarker for blood-based sarcoma diagnostics. Cancers **12**(6) (2020)

14. Huber, W., von Heydebreck, A., Sültmann, H., Poustka, A., Vingron, M.: Variance stabilization applied to microarray data calibration and to the quantification of differential expression. Bioinformatics **18**(suppl_1), S96–S104 (2002)

15. Kannan, A., et al.: Mitochondrial reprogramming regulates breast cancer progression. Clin. Cancer Res. **22**(13), 3348–3360 (2016)

16. Ke, G., et al.: Lightgbm: a highly efficient gradient boosting decision tree. Adv. Neural. Inf. Process. Syst. **30**, 3146–3154 (2017)

17. Love, M.I., Huber, W., Anders, S.: Moderated estimation of fold change and dispersion for rna-seq data with deseq2. Genome Biol. **15**(12), 1–21 (2014)

18. Luo, W., Friedman, M.S., Shedden, K., Hankenson, K.D., Woolf, P.J.: Gage: generally applicable gene set enrichment for pathway analysis. BMC Bioinform. **10**(1), 1–17 (2009)

19. Olson, R.S., Cava, W.G.L., Mustahsan, Z., Varik, A., Moore, J.H.: Data-driven advice for applying machine learning to bioinformatics problems. In: Biocomputing 2018: Proceedings of the Pacific Symposium, pp. 192–203 (2018)

20. Pastuszak, K., et al.: Implatelet classifier: image-converted rna biomarker profiles enable blood-based cancer diagnostics. Molecular Oncology (2021)

21. Paszke, A., et al.: Pytorch: an imperative style, high-performance deep learning library. Adv. Neural Inf. Process. Syst. **32** (2019)

22. Pedregosa, F., et al.: Scikit-learn: machine learning in python. J. Mach. Learn. Res. **12**, 2825–2830 (2011)

23. Poplin, R., et al.: A universal SNP and small-indel variant caller using deep neural networks. Nat. Biotechnol. **36**(10), 983–987 (2018)

24. Raghu, M., Zhang, C., Kleinberg, J.M., Bengio, S.: Transfusion: understanding transfer learning for medical imaging. In: Annual Conference on Neural Information Processing Systems, NeurIPS (2019)

25. Shamout, F.E., et al.: An artificial intelligence system for predicting the deterioration of COVID-19 patients in the emergency department. arXiv preprint (2020). https://arxiv.org/abs/2008.01774
26. Shen, M.L., et al.: Associations between UGT2B7 polymorphisms and cancer susceptibility: a meta-analysis. Gene **706**, 115–123 (2019)
27. Srivastava, N., Hinton, G., Krizhevsky, A., Sutskever, I., Salakhutdinov, R.: Dropout: a simple way to prevent neural networks from overfitting. J. Mach. Learn. Res. **15**(1), 1929–1958 (2014)
28. Zhang, H., Cissé, M., Dauphin, Y.N., Lopez-Paz, D.: Mixup: beyond empirical risk minimization. In: 6th International Conference on Learning Representations, ICLR (2018)
29. Zheng, S.E., et al.: Down-regulation of ribosomal protein L7A in human osteosarcoma. J. Cancer Res. Clin. Oncol. **135**(8), 1025–1031 (2009)

Foreseeing Survival Through 'Fuzzy Intelligence': A Cognitively-Inspired Incremental Learning Based *de novo* Model for Breast Cancer Prognosis by Multi-Omics Data Fusion

Aviral Chharia[1,2(✉)] and Neeraj Kumar[2(✉)] [iD]

[1] Mechanical Engineering Department, Thapar Institute of Engineering
and Technology, Patiala, PB 147004, India
achharia_be18@thapar.edu
[2] Computer Science and Engineering Department, Thapar Institute of Engineering
and Technology, Patiala, PB 147004, India
neeraj.kumar@thapar.edu

Abstract. High-precision breast cancer prognosis is crucial for early disease identification, avoiding hazardous side-effects of unnecessary therapies, and decreasing mortality rates through personalized and tailored treatment regimens. However, designing a prognosis model continues to be challenging, given the intricate relationship between distinct genetic attributes, varied clinical results of drug therapies, the noisy nature of gene expressions, and the high-class imbalance seen in multimodal cancer data. Furthermore, because labeled omics data collection is costly and requires highly-trained experts, the data available is very limited. This makes the design of the conventional machine and deep learning models incredibly challenging as they require large quantities of data for learning the underlying intricate patterns and would otherwise overfit, decreasing model precision. Moreover, all present models suffer from a 'closed world assumption.' These models, once trained, cannot be updated in real-time (when more omics data is available in the future) without a complete re-training. The present study is the first to introduce the 'Fuzzy' way towards Breast cancer prognosis, framing the task as an incremental learning problem. The proposed approach allows the model to continually update its learned feature space on a non-stationary multimodal data stream emulating the human brain's remarkable quality to learn over time. We demonstrate the model's ability to learn complex relationships between different multimodal attributes, training on severely imbalanced and limited data by mapping it to a high-dimensional 'fused' feature space. The proposed model surpasses state-of-the-art machine learning (ML) models significantly. These results suggest that prediction through 'fuzzy intelligence' is a promising approach towards breast cancer prognosis.

Keywords: Breast cancer prognosis · Survival prediction ·
Multimodal feature fusion · Data-efficient learning · Machine learning

© Springer Nature Switzerland AG 2021
I. Rekik et al. (Eds.): PRIME 2021, LNCS 12928, pp. 231–242, 2021.
https://doi.org/10.1007/978-3-030-87602-9_22

1   Introduction

According to the American Society of Oncology, Breast Cancer or Breast Carcinoma is the most frequently occurring cancer amongst women. In 2020, breast cancer affected 276, 480 women in the United States alone. The development of a Breast cancer prognosis model can help oncologists offer personalized treatment plans, especially in the case of the more aggressive Invasive breast cancer, which spreads within the body, reducing 5-year survival rates to as low as 27%. Moreover, an effective prognosis can help to increase life expectancy, especially if the patient is a short-term survivor. Breast cancer prognosis is challenging due to its underlying heterogeneous nature and high complexity. Patients in the same stage and similar clinical characteristics often undergo different therapies, with disparate responses in most cases affecting overall survival. This makes the design of a co-relative prognosis model challenging and, therefore, of practical interest to Oncologists. Recent epidemiological and linkage research has established that mutations in genes and lack of 'alleles' in the BRCA1 locus increase susceptibility to breast cancer. Moreover, gene expression and DNA copy number alteration (CNA) data are quite noisy and large (order of 2×10^4 features). However, the number of training samples is very less, making patients' characterization as long-term (>5-year) and short-term survivors (<5-year) a challenging task.

Related Works. Several studies have focused on predicting survival rates in patients diagnosed with breast cancer but have certain limitations. [1] was the first to propose a prognosis model but used only gene expression profiles. Recent development in high throughput microarrays and gene expression technologies has shown that gene signatures are not the only contributing factor in breast cancer. Assuming different genes of a particular patient may have significant relations amongst themselves, [2] used support vector machine (SVM), while [3] used Random Forest coupled with efficient feature selection for high accuracy prognosis. The deep belief network proposed by [4] combined two independent microarray data, i.e., gene expression and clinical while using principal component analysis (PCA) for dimensionality reduction. The major limitation of these models was that they assumed that different modalities have the same feature representations. Recently, deep learning-based supervised feature extraction has gained immense attention [5]. Recent work by [6] used a score-level fusion of coefficients for integrating multi-modal data. However, the major disadvantage of the model was that these coefficients need to be manually determined, which is an iterative and challenging task in itself. Work by [7] combined CNN-based stacked feature extraction with various ML models. The major limitation of this work was that the proposed method used models which are quite data-intensive and cannot work on severely imbalanced datasets. Also, the study lacks cross-validation on larger datasets required for deep learning models. Furthermore, since these models are trained on limited data, they may be prone to overfitting.

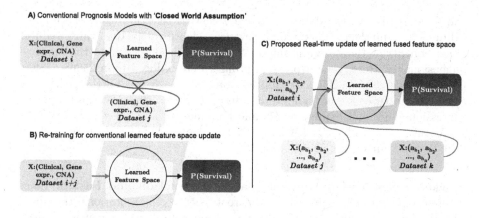

Fig. 1. Illustration of the **A.** Conventional prognosis models with 'closed world assumption' **B.** The required re-training for conventional leaned feature space update **C.** Proposed real-time update $\mathbf{X} : (a_{h_1}, a_{h_2}, ..., a_{h_n}) \rightarrow \mathbf{i}, \mathbf{j}, \mathbf{k}$ of learned fused feature space

Contribution. The significant contributions and novel aspects of this work are-

1. The present work is the *first* to formulate Cancer Prognosis as an incremental learning problem combining multimodal omics data and proposes a *de novo* cognitively-inspired 'fuzzy' network for Breast Cancer Prognosis. The proposed model outperforms other state-of-the-art ML models by a significant margin. Furthermore, as compared to prior techniques, real-time architecture update eliminates the need for model re-training if new labeled data is available in the future.
2. The model's other novelty lies in its capacity to attain high classification accuracy despite being trained on limited data samples, as proven experimentally, in contrast to previously proposed models that need large amounts of multi-omics data difficult and costly to obtain.
3. Another unique feature of the model lies in its robustness to high-class imbalance, commonly seen in real-world multi-omics datasets, as demonstrated in the experiments performed. On the contrary, as the class imbalance in the dataset increases, most ML and deep learning models exhibit a significant reduction in classification accuracy.

2 Proposed Methodology

Figure 2 illustrates the architecture of the proposed Breast cancer prognosis model. This section discusses in detail the proposed methodology.

Weighted K-NN and mRMR Feature Selection. *Firstly*, the Weighted nearest neighbor algorithm [8] is used for estimating the missing gene expression and CNA profiles in the dataset. *Secondly*, feature selection using the mRMR (Minimum redundancy maximum relevance) algorithm [9,10], which reduces dimensionality without significant data loss was used to escape the curse of dimensionality. Here, gene expression profiles were reduced from $24368 \rightarrow 400$, CNA from $26298 \rightarrow 200$ and clinical from $27 \rightarrow 25$ [6]. Further, the gene expression features are normalized and discretized into three categories: under-expression (-1), over-expression (1) and baseline (0), i.e., $\in \{-1, 0, 1\}$ [9]. The CNA features are discretized $\in \{-2, -1, 0, 1, 2\}$ and clinical data is normalized $\in [0, 1]$ using min-max normalization.

Stacked CNN for Multimodal Feature Extraction and Fusion. Since different data modalities may have different feature representations, the direct feature fusion of multi-sourced data to a deep neural net may not be ideal. Therefore separate CNN models [7] are used for each: clinical, gene expression and CNA. Each of the CNNs is trained on a single METABRIC modality with **AUC** value as the evaluation metrics. Binary-cross entropy is used as the loss function $\mathcal{L}$ with **L2** regularization and learning rate 10^{-3} for 8 mini-batches and 20 training epochs as,

$$\mathcal{L}(y_t, \hat{y}_t) = -\frac{1}{\mathcal{N}} \sum_{i=0}^{\mathcal{N}} \left[y_t(i) \log \hat{y}_t(i) - (1 - y_t(i)) \log(1 - \hat{y}_t(i)) \right] + \frac{1}{2} \lambda \sum_{k=1}^{\mathbf{K}} \sum_{j=1}^{\mathbf{n_k}} \sum_{i=1}^{\mathbf{m_k}} w_{ij}^{k^2}$$

where $\mathcal{N}$ is the batch size, $\mathbf{K}$ is the number of weight matrices in the CNN, $\mathbf{W}^k = (w_{ij}^k)_{(\mathbf{m_k} \times \mathbf{n_k})}$ is the k_{th} weight matrix, y_t and $\hat{y}_t$ are the actual and predicted labels. A feature map i.e., element-wise $\odot$ followed by addition between the filter matrix and corresponding values of input matrix is produced. Glorot normal initializer [12] is used for filter matrix initialization. It selects random numbers with mean $= 0$ and standard deviation in, $\left[-\sqrt{\frac{2}{n_i + n_o}}, \sqrt{\frac{2}{n_i + n_o}} \right]$, where n_i and n_o represents number of input and output units for selected layer, respectively. For the Convolution layer, 4 filters were used with size 15, and the stride size being 2. The introduced hidden layer had 150 hidden units. The obtained hidden feature vectors from each trained CNNs are fused to get stacked features.

Mapping n-dimensional 'Fused' Feature Space. The obtained stacked feature vector $\{a_h, C_i\}$ for each patient is passed to the input nodes $\{a_1, ..., a_h\}$ of the fuzzy classifier [13] after normalization $\in [0.01, 0.99]$. The fuzzy classifier works by creating hyperboxes $\mathcal{H}$ [14] which is a geometrical shape defined in the n-dimensional feature space. Parameters $V_j = (v_{j1}, v_{j2}, ..., v_{jn})$ and $W_j = (w_{j1}, w_{j2}, ..., w_{jn})$ are used to define the min and max-coordinates of $\mathcal{H}$, while 'θ' i.e., the hyperbox expansion coefficient $\in (0, 1)$ represents its size and 'γ' represents the fuzziness control parameter.

Incremental Learning Based Model Training. Fig. 1 illustrates the incremental learning approach, on which the proposed model is based. For each input feature vector, Classifying Neurons (CLN) performs the classification of learned

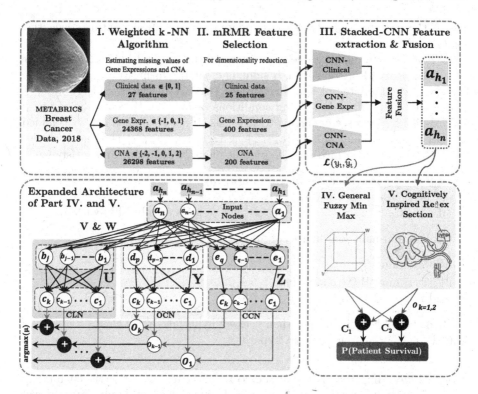

Fig. 2. Architecture of the proposed Breast cancer prognosis model based on 'Fuzzy' Incremental Learning for classifying patients as long and short-term survivors on multimodal omics (i.e., gene expressions, copy number alteration and clinical) data

data using min-max hyperboxes. In CLNs, neuron b_j represents hyperbox fuzzy set $B_j = A_h, V_j, W_j, f(A_h, V_j, W_j) \forall (A_h \in I^n)$. To compute class memberships, activation function by [15] is used to assign membership value = 1 when the test sample falls within $\mathcal{H}$. In other cases, when test sample lies outside $\mathcal{H}$, the membership value is calculated on the basis of its distance from extreme co-ordinates of $\mathcal{H}$. The classifying neuron activation function i.e., b_j is defined as,

$$b_j(a_h, V_j, W_j) = \min_{i=1..n} \left(\min \left[\left(1 - f(a_{hi} - W_{ji}, \gamma)\right), \left(1 - f(V_{ji} - a_{hi}, \gamma)\right) \right] \right)$$

where, $f(x, \gamma) = 0$ if $x\gamma < 0$; 1 if $x\gamma > 1$ and equal to $x\gamma$ if it lies $\in [0, 1]$. In the middle layer, the input nodes and the hyperbox ($\mathcal{H}$) nodes are connected together. These connections represents $\mathbf{V}$ and $\mathbf{W}$ of the n-dimensional hyperbox fuzzy set [16]. During training, the neurons in the middle layer are created dynamically and connection between the $\mathcal{H}$ node b_j to a class node C_j, is represented by matrix $\mathbf{U}$, where, $u_{ij} = 1$ if $b_j \in C_j$ else $u_{ij} = 0$. This real-time architecture update allows for the midway introduction of new labeled data. More labeled multimodal omics data, *if available*, in future may be directly passed through the trained model to update its learned fused feature space, thus enabling Incremental learning.

Whenever a training sample is encountered that doesn't belong to a class the model has learned previously, a $\mathcal{H}$ node is created. During training, the model tries to accommodate subsequent samples $\{a_h, C_i\}$ in the previously made $\mathcal{H}$ belonging to the same class using conditions below [13]. If expansion of any of the existing hyperboxes belonging to that class is not feasible, a new $\mathcal{H}$ is added; i.e., for a new training sample $\{a_h, C_i\}$, a hyperbox $\{b_j, C_j\}$ is found such that $C_j = C_i$ or $C_j = C_0$ which has the highest membership value and satisfying-

1. $\theta_{max} \geq \frac{1}{n} \sum_{i=1}^{n}(\max(w_{ji}, a_{hi}) - \min(v_{ji}, a_{hi}))$
2. b_j is not associated with any OCN/CCN
3. if $C_i = C_0$ or $C_j = C_0$ then $\mu_j > 0$, where μ_j is membership with b_j,

Adjust min-max coordinates of b_j, as,

$V_{ji}^{new} = \min(V_{ji}^{old}, a_{hi})$,
$W_{ji}^{new} = \max(W_{ji}^{old}, a_{hi})$, where $i = 1, 2, ..., n$
and if $C_j = C_0$ and $C_i \neq C_0$ then $C_j = C_i$

If no suitable b_j is present then a novel hyperbox $\mathcal{H}$ for class C_i is created with $V_j = W_j = a_h$; i.e., a point hyperbox.

Cognitively-Inspired Reflex Section. The Reflex section is cognitively-inspired [16] from the human brain and handles cases of $\mathcal{H}$ overlap and containment which may arise due to visual feature overlap in the high dimensional feature space. The Overlap Compensation Neuron (**OCN**) becomes active if the test data lies within the overlap space and generates two compensation outputs, one each for the two overlapping classes with activation function-

$$d_{j_p} = T(b_j(a_h, V_j, W_j) - 1) \times \left(-1 + \frac{1}{n} \sum_{i=1}^{n} \max\left(\frac{a_{hi}}{w_{p_{ji}}}, \frac{v_{p_{ji}}}{a_{hi}} \right) \right)$$

where, $T(x) = 0$ if $x < 0$ and 1 if $x \geq 0$. The Containment Compensation Neuron (**CCN**) overcomes $\mathcal{H}$ containment cases and has activation function-

$$e_j = -1 \times T(b_j(a_h, V, W) - 1)$$

The final class for each sample is computed as the **argmax** of the sum of the membership and compensation values.

3 Experiments and Results

3.1 Dataset and Evaluation Metrics

The Molecular Taxonomy of Breast Cancer International Consortium [17] data contains the clinical data, gene expression profiles and CNA profiles of 1980 breast cancer patients in the METABRIC trial. The pre-processed METABRICS data at (https://github.com/USTC-HIlab/MDNNMD) is used in this study for validating the proposed model. In datasets which are skewed towards a particular class, accuracy alone may be misleading while accessing model performance. Therefore, on imbalanced datasets, precision along with accuracy is considered as a better evaluation metric [18].

3.2   Exp. 01. Evaluation and Comparison of Model Performance on Limited Data Subset Configuration

In the first experiment, the model is trained on limited data subset configurations, i.e., $n = 200, 300, 400, 500$ (with no class-imbalance), where n is the number of samples. The obtained results are compared with state-of-the-art ML models trained on the same number of data samples (see Table 1). Here, confusion-matrix based evaluation metrics was used for performance comparison. Further, it is to be noted that the same feature selection and pre-processing techniques (Weighted k-NN algorithm and mRMR Feature selection) are used while implementing the ML models for a fair comparison of results. The proposed model outperforms all other models with a significant margin in all data subset configurations, i.e., for $n = 200, 300, 400, 500$. On $n = 300$ samples data, the proposed model surpasses the best-performed ML model by 16.81% on the accuracy, 16.31% on precision, and 15.93% on recall. This demonstrates the strong ability of the model to achieve high performance on limited datasets. Furthermore, it is seen that as the number of training samples varies, the second-best performing model loses consistency, i.e., for $n = 200$, the Naive Bayes is the second best-performed model, its performance decreases for $n = 300, 400, 500$, where Ridge classifier and Random forest are seen performing better. This illustrates that no single ML model exhibits robust and consistent performance while training on limited data. On the contrary, the proposed model, which outperforms other models, demonstrates consistency on all subset configurations of limited data.

3.3   Exp. 02. Evaluation and Comparison of Model Performance on Severely Imbalanced Data Configuration

Most multi-omics datasets available suffer from high-class imbalance (example, 75% − 25% class distribution in METABRICS trail) since positive samples are difficult to obtain compared to the negative ones. Most deep learning and ML models exhibit a reduction in performance on high class-imbalance. To demonstrate the model's robustness to class imbalance, we compare the model performance on two dataset subset configurations: *first* with no class imbalance (50% − 50% class distribution) and *second* with severely skewed class distribution of 75% − 25%. The obtained results (see Table 3) are compared with ML models trained on the same subset configuration. Here, precision is considered as better evaluation metric than accuracy alone [18]. It is seen that the proposed

Table 1. Comparative Results with state-of-the-art ML models on limited data subset configurations for 5–fold cross validation. In each column, the best results are typeset in **boldface** and the second best results are underlined. Here **QDA:** Quadratic Discriminant Analysis, **Ada:** Ada Boost Classifier, **K-NN:** K-Neighbours Classifiers, **LightBGM:** Light Gradient Boosting Machine, **LR:** Logistic Regression, **DT:** Decision Tree, **GBC:** Gradient Boosting Classifier, **LDA:** Linear Discriminant Analysis, **RF:** Random Forest, **SVM:** Support Vector Machine (Linear Kernel), **ET:** Extra-Trees Classifier, **NB:** Naive Bayes

Model	n = 200				n = 300				n = 400				n = 500			
	Acc	Prec	Rec	F1	Acc	Prec	Rec	F1	Acc	Prec	Rec	F1	Acc	Prec	Rec	F1
QDA	51.06	52.78	54.19	53.36	57.93	58.79	64.76	61.29	49.47	49.58	55.71	51.59	56.46	59.52	57.54	58.42
Ada	53.33	53.77	58.48	55.79	66.02	66.52	69.39	67.86	68.81	69.15	67.33	68.07	77.38	78.22	80.16	79.04
K-NN	54.71	61.73	35.14	43.23	61.23	70.45	44.46	53.63	60.95	69.33	40.66	50.72	59.03	71.97	41.37	51.81
LightGBM	61.88	63.60	62.38	62.39	64.13	65.98	64.72	65.04	71.69	70.59	74.68	72.27	77.08	79.70	77.44	78.31
LR	61.96	63.61	64.10	63.45	69.81	71.75	69.44	70.49	75.26	77.30	70.95	73.85	76.24	77.71	79.59	78.34
DT	61.98	65.97	63.90	64.15	63.61	64.47	65.63	64.49	70.60	71.88	68.10	69.57	72.19	74.09	74.74	74.23
Ridge	62.01	65.17	61.33	62.64	67.48	68.93	67.62	68.18	76.70	81.15	70.32	74.86	73.93	76.43	74.21	75.20
GBC	64.87	68.09	68.00	66.53	70.33	73.22	68.53	70.51	74.55	76.75	69.66	72.90	78.82	81.11	78.52	79.76
LDA	65.48	66.63	67.90	66.98	62.23	66.01	59.39	62.12	62.73	63.98	55.74	59.34	63.03	66.99	59.69	63.03
RF	66.24	72.59	62.57	65.75	67.44	69.60	66.62	67.57	70.95	71.79	68.10	69.72	78.82	80.47	80.11	80.20
SVM	66.27	70.69	66.57	67.09	66.49	66.83	71.30	68.76	73.10	71.00	78.23	73.97	74.24	76.87	76.34	76.23
ET	66.35	72.51	64.10	67.08	69.85	70.23	72.25	71.11	74.87	73.30	76.72	74.81	77.67	80.39	77.97	78.97
NB	73.41	73.68	77.90	75.01	69.86	69.40	74.98	71.70	68.81	68.15	70.26	68.65	74.23	74.04	79.54	76.53
Proposed	**77.50**	**85.0**	73.91	**79.07**	**86.67**	**85.71**	**90.91**	**88.24**	**83.75**	**83.33**	**81.08**	**82.19**	**83.00**	79.55	**81.40**	**80.46**

Table 2. Qualitative and Quantitative comparison of results on METABRIC trail data with proposed models in literature.

Model	Acc. ↑	Prec. ↑	Rec. ↑	F1 ↑	Dataset↓	Model Training ability data		Imbalance
						Limited	Incremental	
Deep Learning models								
Khademi et al., 2013 [19]	80.00	–	–	–	1980	Low	–	Low
Khademi et al., 2015 [4]	82.00	–	–	–	1980	Low	–	Low
Sun et al, 2019 [6]	82.60	74.90	45.00	–	1980	Low	–	Low
Arya et al., 2020 [7]	90.20	84.10	74.70	–	1980	–	–	Low
Machine Learning models								
Quadratic Disc. Analysis	53.22	53.36	53.88	53.46	1000	Low	–	Low
K-Neighbor Classifier	61.95	73.97	36.68	48.79	1000	Low	–	Low
Linear Discriminant Anal.	69.52	72.25	63.02	67.26	1000	Low	–	Low
Decision Tree	73.11	73.49	72.78	73.04	1000	Low	–	Low
Naive Bayes	74.10	83.48	60.16	69.88	1000	Low	–	Low
Random Forest	75.96	74.87	78.51	76.50	1000	Low	–	Low
Ridge Classifier	76.82	80.33	71.34	75.47	1000	Low	–	Low
Extra Trees Classifier	77.39	77.08	77.93	77.48	1000	Low	–	Low
Ada Boost Classifier	79.39	80.42	77.93	78.94	1000	Low	–	Low
Support Vector Machine	79.54	80.61	78.22	79.16	1000	Low	–	Low
Light Gradient Boosting	79.83	81.22	77.65	79.30	1000	Low	–	Low
Logistic Regression	80.83	82.43	78.51	80.31	1000	Low	–	Low
Gradient Boosting	81.26	81.29	81.37	81.30	1000	Low	–	Low
Proposed	87.00	93.01	89.26	91.09	1000	High	√	High

Table 3. Comparative Results with state-of-the-art ML models for 5−fold cross validation on imbalanced data subset configurations ($n = 1000$). In each column, the best results are typeset in **boldface** and the second best results are <u>underlined</u>

Model	50%–50%				25%–75%				Δ Prec.
	Acc	Prec	Rec	F1	Acc	Prec	Rec	F1	
QDA	53.22	53.36	53.88	53.46	53.36	26.50	47.63	33.94	−26.86
K-NN	61.95	73.97	36.68	48.79	72.96	39.55	09.68	14.56	−34.42
LDA	69.52	72.25	63.02	67.26	65.81	36.49	48.86	41.69	−35.76
DT	73.11	73.49	72.78	73.04	71.40	42.00	35.32	38.02	−31.49
NB	74.10	<u>83.48</u>	60.16	69.88	53.36	30.71	<u>68.21</u>	42.28	−52.77
RF	75.96	74.87	78.51	76.50	79.26	<u>71.98</u>	30.71	42.13	<u>−2.89</u>
Ridge	76.82	80.33	71.34	75.47	73.53	47.91	47.17	47.37	−32.42
ET	77.39	77.08	77.93	77.48	78.55	65.69	32.37	42.80	−11.39
Ada Boost	79.39	80.42	77.93	78.94	78.12	59.49	48.30	52.63	−20.93
SVM	79.54	80.61	78.22	79.16	77.83	59.58	50.62	53.58	−21.03
LightGBM	79.83	81.22	77.65	79.30	77.26	57.25	40.38	47.18	−23.97
LR	80.83	82.43	78.51	80.31	<u>80.11</u>	64.03	48.86	<u>55.23</u>	−18.40
GBC	<u>81.26</u>	81.29	<u>81.37</u>	<u>81.30</u>	78.98	63.24	38.63	47.78	−18.05
Proposed	**87.00**	**93.01**	**89.26**	**91.09**	**86.00**	**91.15**	**89.93**	**90.54**	**−1.86**

model has consistent performance (Δ Precision $= -1.86$) on both the class distributions, whereas all other models show a significant decrease in performance (large variation in Δ Precision observed).

3.4 Exp. 03. Quantitative and Qualitative Comparison of Overall Model Performance

For an in-depth quantitative and qualitative comparison, $n = 1000$ was taken as the subset configuration. The proposed model was trained for short and long-term patient survival prediction following the conventional $80 - 20\%$ train-test split, with hyperparameters $\theta = 0.7263$ and $\gamma = 2$, found experimentally during model hyperparameters tuning and parametric study (see Exp. 04). The number of $\mathcal{H}$ formed during training was found to be 12. The proposed model is compared with state-of-the-art ML and deep learning models on the METABRICS trail data, as shown in Table 2. The obtained results show that the proposed model outperforms all other ML models by a significant margin ($\approx 5.74\%$ improvement on accuracy, $\approx 9.53\%$ on precision). Moreover, the modal performance is at par with other deep learning approaches despite training on a limited dataset. It can be inferred that the proposed model can predict patient survival with comparable accuracy and higher precision while requiring a relatively small dataset to train. The primary reason for this is the data-intensive nature of majority of proposed models, which fail when applied to imbalanced datasets. Moreover, the proposed model preserves both the contrasting and similar characteristics of each class sample. This differs from deep learning models that learn mostly the contrasting features while minimizing a loss function $\mathcal{L}$. The results confirm that 'fuzzy' way is more suited towards imbalanced and limited data.

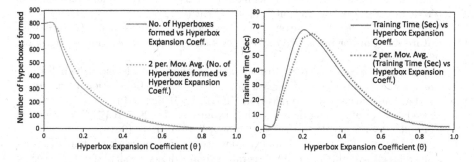

Fig. 3. Effect of variation in hyperbox ($\mathcal{H}$) expansion coefficient on **A.** Number of hyperbox ($\mathcal{H}$) formed during model training **B.** Model training time (*sec*)

3.5   Exp. 04. Parametric Study and Time Complexity Analysis

A parametric study was performed to analyze the effect of the variations of model hyperparameters i.e., θ and γ. From Fig. 3, we conclude that as θ increases, the number of hyperbox ($\mathcal{H}$) created during training shows an 'exponential' increase rather than a 'linear' one. In contrast, the training time first shows a sharp rise until $\theta = 0.2$, after which it decreases exponentially. The study quantifies that the proposed model has an significantly less total training time (≈ 5 s) compared to other models. However, the test time is comparatively higher (≈ 90 s/sample).

4   Discussion and Conclusions

The present study proposes a *de novo* approach towards the development of prognosis models framing the task as an 'incremental learning' problem. The proposed model addresses the problem of limited availability of high-throughput multi-omics datasets and the high-class imbalance seen in them. The obtained results establish the model's effectiveness and quantify that fuzzy classifier-based models are more suited towards problems where the dataset is highly imbalanced or limited, such as developing prognosis models combining multi-omics data. The proposed model surpasses state-of-the-art ML models significantly. These results suggest that prediction through 'fuzzy intelligence' is a promising approach towards breast cancer prognosis. In future work, we aim to expand the model for ovarian cancer, cervical cancer, fallopian tube cancers, etc., among others caused by BRCA1 and BRCA2 mutations. Future research may also integrate a fourth multi-modal data such as gene methylation, miRNA expression or pathology image dataset to improve classification performance further.

References

1. van de Vijver, M.J., et al.: A gene-expression signature as a predictor of survival in breast cancer. N. Engl. J. Med. **347**, 1999–2009 (2002)
2. Xu, X., Zhang, Y., Zou, L., Wang, M., Li, A.: A gene signature for breast cancer prognosis using support vector machine. In: 2012 5th International Conference on BioMedical Engineering and Informatics. IEEE (2012)
3. Nguyen, C., Wang, Y., Nguyen, H.N.: Random forest classifier combined with feature selection for breast cancer diagnosis and prognostic. J. Biomed. Sci. Eng. **06**, 551–560 (2013)
4. Khademi, M., Nedialkov, N.S.: Probabilistic graphical models and deep belief networks for prognosis of breast cancer. In: 2015 IEEE 14th International Conference on Machine Learning and Applications (ICMLA). IEEE (2015)
5. Szegedy, C., et al.: Going deeper with convolutions. In: 2015 IEEE Conference on Computer Vision and Pattern Recognition (CVPR). IEEE (2015)
6. Sun, D., Wang, M., Li, A.: A multimodal deep neural network for human breast cancer prognosis prediction by integrating multi-dimensional data. IEEE/ACM Trans. Comput. Biol. Bioinform. **16**, 841–850 (2019)
7. Arya, N., Saha, S.: Multi-modal classification for human breast cancer prognosis prediction: proposal of deep-learning based stacked ensemble model. IEEE/ACM Trans. Comput. Biol. Bioinform. 1 (2020)
8. Troyanskaya, O., et al.: Missing value estimation methods for DNA microarrays. Bioinformatics **17**, 520–525 (2001)
9. Gevaert, O., De Smet, F., Timmerman, D., Moreau, Y., De Moor, B.: Predicting the prognosis of breast cancer by integrating clinical and microarray data with Bayesian networks. Bioinformatics **22**, e184-90 (2006)
10. Ding, C., Peng, H.: Minimum redundancy feature selection from microarray gene expression data. J. Bioinform. Comput. Biol. **03**, 185–205 (2005)
11. Peng, H., Long, F., Ding, C.: Feature selection based on mutual information criteria of max-dependency, max-relevance, and min-redundancy. IEEE Trans. Pattern Anal. Mach. Intell. **27**, 1226–1238 (2005)
12. Glorot, X., Bengio, Y.: Understanding the difficulty of training deep feedforward neural networks. In: Teh, Y.W., Titterington, M. (eds.) Proceedings of the Thirteenth International Conference on Artificial Intelligence and Statistics, pp. 249–256. PMLR, Chia Laguna Resort, Sardinia, Italy (2010)
13. Simpson, P.K.: Fuzzy min-max neural networks. I. Classification. IEEE Trans. Neural Netw. **3**, 776–786 (1992)
14. Alpern, B., Carter, L.: The hyperbox. In: Proceeding Visualization 1991. IEEE Computer Society Press (2002)
15. Gabrys, B., Bargiela, A.: General fuzzy min-max neural network for clustering and classification. IEEE Trans. Neural Netw. **11**, 769–783 (2000)
16. Nandedkar, A.V., Biswas, P.K.: A general reflex fuzzy min-max neural network. Eng. Lett. **14**, 195–205 (2007)
17. Curtis, C., et al.: The genomic and transcriptomic architecture of 2,000 breast tumours reveals novel subgroups. Nature **486**, 346–352 (2012)
18. Saito, T., Rehmsmeier, M.: The precision-recall plot is more informative than the ROC plot when evaluating binary classifiers on imbalanced datasets. PLoS One **10**, e0118432 (2015)
19. Khademi, M.: Probabilistic graphical models for prognosis and diagnosis of breast cancer (2014)

Improving Across Dataset Brain Age Predictions Using Transfer Learning

Lara Dular[(✉)], Žiga Špiclin,
and The Alzheimer's Disease Neuroimaging Initiative

Faculty of Electrical Engineering, Laboratory of Imaging Technologies,
University of Ljubljana, Tržaška 25, 1000 Ljubljana, Slovenia
lara.dular@fe.uni-lj.si
http://lit.fe.uni-lj.si/

Abstract. Brain age has shown it's potential as a biomarker of healthy or accelerated neurological ageing. Utilizing deep learning methods, brain age predictions for healthy individuals have become increasingly more accurate, but seem adversely affected when applied to new scanner data or differently preprocessed scans. We thus focused on transfer learning methods for convolutional neural network based brain age prediction models. The four models were trained and evaluated on a large multi-site dataset ($N = 2543$) and new site longitudinal dataset ($N = 5632$). Next, we assessed the ability of three transfer learning approaches, namely bias correction (BC), domain adaptation (DA) and full transfer (FT), to generalize the brain age prediction performance across the datasets. Our results indicate that models using transfer learning outperform models trained from scratch in similar studies. We further show that simpler and less expensive transfer learning approaches, such as BC or DA, perform better than FT and generalize well across datasets and preprocessing procedures, with mean absolute error and mean absolute (longitudinal) difference error as low as 3.3 and 1.1 years respectively, increasing their potential to practically deliver brain age biomarker to aid in diagnosis of neurodegenerative diseases and/or monitoring of their progression.

Keywords: Brain age · Deep learning · Transfer learning · Domain adaptation · Longitudinal predictions

1 Introduction

Brain age involves the prediction of chronological age from structural brain MR scans, such as T1-weighted (T1w) images, using machine learning algorithms. The difference between the predicted and chronological age shows great potential as a biomarker of healthy brain ageing. Numerous studies reported increased

Electronic supplementary material The online version of this chapter (https://doi.org/10.1007/978-3-030-87602-9_23) contains supplementary material, which is available to authorized users.

© Springer Nature Switzerland AG 2021
I. Rekik et al. (Eds.): PRIME 2021, LNCS 12928, pp. 243–254, 2021.
https://doi.org/10.1007/978-3-030-87602-9_23

brain age in patient cohorts with neurodegenerative diseases, such as Alzheimer's dementia [6]. Namely, as a result of neurodegeneration, the predicted age is generally much higher than the chronological age indicating accelerated ageing.

In order to apply brain age analysis to detect accelerated ageing and thus diagnose neurodegenerative diseases and/or monitor their progression, it is first necessary to train a model for brain age prediction on a dataset of healthy subjects. With the use of deep learning models, specifically convolutional neural networks (CNN), recent studies reported increasingly accurate brain age predictions [4,9,13]. However, the performance of models is generally adversely affected when applied to new site dataset, unseen in model training, or differently preprocessed scans, thus limiting the practical value of brain age as a biomarker.

A recent brain age study by Jonsson et al. [9] shows the use of transfer learning approach for adapting a trained CNN for an accurate brain age prediction on a new site dataset, previously not used in model training. They achieve this by retraining a CNN model by first freezing the model weights of the convolutional layers, so that only the weights of fully connected layers are trainable. Second, full model was fine-tuned, by retraining all layers of CNN on the new site dataset.

Besides the aforementioned full transfer (FT), computationally less demanding approaches of transfer learning are found in the literature. For instance, Karani et al. [10] proposed training deep learning models on multi-site neuroimaging datasets by using a separate batch normalization for each site/dataset. They hypothesized that the trained weights of convolutional filters are domain-agnostic, and only the batch normalization layers should be adapted on each new dataset. Hence, for transfer learning on a new site data, only weights of batch normalization layers were retrained, keeping other layers fixed. However, this so-called domain adaptation (DA) has not yet been evaluated for brain age regression task. Further, for traditional machine learning models, Franke and Gaser [6] notice that predictions on new site dataset exhibit a systematic offset. They correct this bias by a simple linear regression fit.

Our aim was to investigate the prediction accuracy of four state-of-the-art CNN-based brain age prediction models using three transfer learning approaches of different complexities and thus show that simple and computationally less expensive approaches preform better (or at least comparably) well to full transfer, currently established in brain age literature [9]. In addition, we assessed the effectiveness of transfer learning for brain age on new site dataset, which was preprocessed by a pipeline different from the one applied to the training data. Such a scenario, while common in practice, was not yet assessed in similar studies. Furthermore, the four brain age prediction models were applied on new (unseen) site MRI dataset that included longitudinal scans (acquired at different timepoints) and their ability to estimate the brain age (longitudinal) difference, with and without transfer learning, was evaluated.

2   Methods

2.1   Data

For the purpose of this study, we aggregated T1w MR images from seven publicly available datasets to insure a large enough dataset for CNN model training. Upon visual quality control discarding scans with motion artifacts, the training datasets consisted of 2543 T1w images of healthy adults aged between 18 and 96 years. The dataset was split into train ($N = 2040$), validation ($N = 253$) and test dataset ($N = 250$). Information about the datasets is given in Table 1.

Table 1. Dataset information with dataset name [reference], number of subjects, number of T1w scans and age span, according to the study aim.

Aim: Train, Validation, Test (Multi-site scanner data)			
Dataset	$N_{subjects}$	$N_{samples}$	Age span [years]
IXI [8]	472	472	20.0–86.3
CamCAN [14,17][a]	628	628	18.0–88.0
CC-359 [16]	349	349	29.0–80.0
FCON 1000 [1]	607	607	18.0–85.0
OASIS: Longitudinal [11][b]	78	78	60.0–96.0
ADNI [2][c]	248	248	55.0–92.0
ABIDE I [3]	161	161	18.0–56.2
Total	2543	2543	18.0–96.0
Aim: Test (Unseen scanner data)			
Dataset	$N_{subjects}$	$N_{samples}$	Age span [years]
UK Biobank [12]	2816	5632	47.2–82.7

[a]Data collection and sharing for this project was partially provided by the Cambridge Centre for Ageing and Neuroscience (CamCAN). CamCAN funding was provided by the UK Biotechnology and Biological Sciences Research Council (grant number BB/H008217/1), together with support from the UK Medical Research Council and University of Cambridge, UK.

[b]Data were provided in part by OASIS Longitudinal: Principal Investigators: D. Marcus, R, Buckner, J. Csernansky, J. Morris; P50 AG05681, P01 AG03991, P01 AG026276, R01 AG021910, P20 MH071616, U24 RR021382

[c]Data used in preparation of this article were in part obtained from the Alzheimer's Disease Neuroimaging Initiative (ADNI) database (adni.loni.usc.edu). As such, the investigators within the ADNI contributed to the design and implementation of ADNI and/or provided data but did not participate in analysis or writing of this report. A complete listing of ADNI investigators can be found at http://adni.loni.usc.edu/wp-content/uploads/how_to_apply/ADNI_Acknowledgement_List.pdf

Transfer learning approaches were evaluated on new site dataset, specifically a subset of UK Biobank (UKB) dataset, which was not included in model training and validation. The UKB subset included only the subjects with two T1w scans; we thus identified 2816 subjects with 5632 T1w scans. The average time between scans was 2.25 ± 0.12 years. Approximately 10% of the dataset (264 subjects and 528 T1w scans) were assigned at random into the training dataset for transfer learning.

2.2 Preprocessing

Multi-site scanner data from the seven publicly available datasets (Table 1, top) was uniformly preprocessed for model training and validation as follows. First T1w scan was converted to Nifti format and denoised using Adaptive non-local means. Next, spatial intensity bias was corrected on the denoised image using N4 bias correction algorithm without mask, implemented in the ANTs package[1]. Denoised and bias corrected scans were co-registered to MNI152 atlas (ICBM 2009c Nonlinear Symmetric) using affine transformation. The denoised T1w scan was sinc resampled to size $193 \times 292 \times 193$ and spacing $1\,\text{mm}^3$. For image registration and resampling we used NiftyReg software[2]. The obtained resampled T1w scan was grayscale windowed, by saturating intensity outliers below 5-th and above 99-th intensity distribution percentile. Finally, the N4 bias correction was applied, using MNI152 atlas mask dilated by 3 voxels.

The UKB dataset preprocessing did not follow the steps described above. We instead used the materials already provided in the dataset, specifically the raw defaced T1w scans in patient space and the linear transformation matrices for registration to MNI nonlinear 6th generation atlas space ($T_{sub \to 6gMNI}$). To perform affine registration of each T1w scan to MNI152 7th generation atlas space, we precomputed the affine co-registration between the 6th and 7th generation MNI152 atlases (i.e. $T_{6gMNI \to 7gMNI}$) using NiftyReg and composed it with the scan-specific $T_{sub \to 6gMNI}$ matrix. Final test scans were obtained by cubic resampling. This dataset served to assess the capability of transfer learning on across scanner data and on differently preprocessed scans. The obtained preprocessed images were cropped to $170 \times 189 \times 157$ voxels around the original image's center.

2.3 Model Architectures

For comparison of transfer learning methods, we implemented four CNN brain age estimation models based on recent published works. Each architecture was implemented following the associated research paper, however, hyperparameters were fine-tuned to assure best model predictions.

Main differences between the four models were in the size and representation of the input images. Both Model 1, proposed by Cole et al. [5] and Model 4,

---

[1] N4 bias field correction: https://manpages.debian.org/testing/ants/N4BiasFieldCorrection.1.en.html.

[2] NiftyReg Software: http://cmictig.cs.ucl.ac.uk/wiki/index.php/NiftyReg.

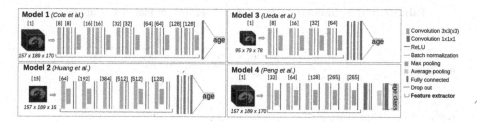

Fig. 1. Architectures of four implemented models for brain age prediction.

proposed by Peng et al. [13], were trained on full resolution 3D T1w MR images. Model 2, proposed by Huang et al. [7], was trained on 2D images, where 15 axial image slices were used as channel inputs of 2D CNN. For Model 3, proposed by Ueda et al. [18], Gaussian smoothing and downsampling to size $95 \times 79 \times 78$ were applied to the preprocessed T1w scans, using SimpleITK library.

All four models were composed of a feature extraction part, further composed of convolutional blocks consisting of one or more convolutional layers, batch normalization, max pooling layer and a ReLU activation function, appearing in different orders and with different hyperparameters. Feature extraction was followed by one or more fully connected layers for Models 1, 2 and 3, and by a global pooling layer for Model 4. The detailed model architectures are shown in Fig. 1.

Brain age estimation is typically formulated as a regression task, where a model outputs a single continuous value, i.e., the predicted age. Alternatively we may split the real line into disjoint age subintervals and reformulate age regression as a classification problem. For a given subject we thus output a vector, in which each value represents the probability that the subject's age is in a particular subinterval. The final age is computed as the weighted sum of the mean subinterval age and the associated probability, $y' = \sum_j p_j age_j$, where p_j denotes the probability of class j and age_j the center of the class interval. In our case, Models 1, 2, and 3 formulate age prediction as a regression task and Model 4 as a classification task.

2.4   Model Training

All four models were trained using SGD optimizer with momentum 0.9, whereas the L1 loss was used for Models 1, 2 and 3 (formulated as regression models), and Kullback-Leibler divergence (KLD) loss between discretized true age and prediction vector for Model 4 (classification model).

The learning rate decay schedule for each model was implemented as proposed in the respective original studies. Namely, for Model 1, the decay of the initial learning rate was set to 3% after each epoch, for Model 2 and 3 the learning rate lr_i on i-th epoch was computed as $lr_i = \frac{lr_0}{1+(i\lambda)}$, where lr_0 denotes the starting learning rate and λ the learning rate decay. Finally, for Model 4 the learning rate was multiplied by 0.3 on every 30 epochs.

Table 2. Hyperparameter values for each of four implemented models.

	Model 1	Model 2	Model 3	Model 4
Input size	$157 \times 189 \times 170$	$157 \times 189 \times 15$	$95 \times 79 \times 78$	$157 \times 189 \times 170$
Batch size	16	32	8	9
Learning rate	1×10^{-4}	1×10^{-3}	5×10^{-5}	1×10^{-2}
Lr decay	3%	1×10^{-4}	1×10^{-4}	$\times 0.3$ every 30 ep.
Weight decay	5×10^{-5}	1×10^{-3}	5×10^{-4}	1×10^{-3}
Momentum	0.9	0.9	0.9	0.9
Parameters	$\approx$900 000	$\approx$6.6 mio	$\approx$11 mio	$\approx$3 mio
MAE_{10} $med[min, max]$	3.57 [3.52, 3.61]	4.23 [4.14, 4.67]	3.57 [3.52, 4.26]	3.35 [3.29, 3.42]

Hyperparameter values were chosen based on grid search in parameter space, set around the values proposed in the respective original papers. For instance, batch size was set to 4, 8, 16, 32 and 64 for Models 2 and 3. Due to GPU constraints we trained Model 1 with batch size 4, 8, 16 and 24 (maximal sample size) and Model 4 with batch size 4 and 9 (maximal sample size). Learning rate values were set to $\{10^{-3}, 10^{-4}, 5 \cdot 10^{-5}, 10^{-5}, 10^{-6}\}$ for models with the L1 loss and $\{10^{-2}, 10^{-3}, 10^{-4}\}$ for models with the KLD loss function.

To ensure training convergence, Models 1 and 4 were trained for 110 and Models 2 and 3 for 400 epochs. To assure the optimal choice of hyperparameters with respect to both mean absolute values (MAE), as defined in Sect. 3.2, and convergence, we computed the median MAE across last 10 training epochs and chose the hyperparameter setting that achieved the minimal value. The selected hyperparameter values are reported in Table 2.

The benefit of augmentation is well argued in brain age literature [13,18]. All models were trained using the same data augmentation procedure: (i) random shifting along all major axes with probability of 0.3 for an integer sampled from $[-s, s]$, where s was 3 for Model 3, and s was 5 for Models 1, 2, and 4; (ii) random padding with probability of 0.3 for an integer from $[0, p]$, where p was 2 for Model 3, and p was 5 for Models 1, 2, and 4; (iii) flipping over mid sagittal plane with probability of 0.5. Postprocessing involved bias correction on the obtained predictions, as proposed for Model 4 [13] and first described by Smith et al. [15], and aimed to correct for age overestimation in younger individuals and underestimation in older individuals, which is a result of regression dilution. Linear correction was applied by fitting a regression line $\hat{y} = \beta_1 y + \beta_0$ on the validation set, where y denotes true and $\hat{y}$ predicted value. The estimated coefficients β_0 and β_1 were then used for correcting the predicted brain age on the test set as $y' = (\hat{y} - \beta_0)/\beta_1$.

Experiments were run using Intel Core i7-8700K CPU and three NVIDIA GeForce RTX 2080 Ti GPUs. Models were implemented in PyTorch 1.4.0 for Python 3.6.8. Each model was trained five times, using different random weight initialization. Model achieving minimal MAE values on validation dataset was chosen as the final model. All four models achieved MAE values on the test dataset comparable to results of studies using similar number of sites, dataset size and age span of training data (cf. Table 3).

Table 3. Mean error (ME) and mean absolute error (MAE) computed for four final models on the test dataset (multi-site scanner data).

	Model 1		Model 2		Model 3		Model 4	
	ME	MAE	ME	MAE	ME	MAE	ME	MAE
Test dataset	−0.39	3.30	0.21	3.70	−0.12	3.41	0.09	3.23

3 Experiments

3.1 Transfer Learning

We tested and compared three different transfer learning approaches for tuning the pretrained model performance on new (unseen) site dataset. For the new site dataset, a subset of UKB dataset was used such that each subject had at least two T1w scans. Our aim was to assess the brain age models' performances and the ability of each transfer learning approach for prediction of longitudinal age difference. To apply transfer learning, a subset of 528 scans of 264 subjects was used as training dataset.

Baseline: Predictions using transfer learning approaches were compared to model predictions without any additional training. For these age predictions, bias was linearly corrected using parameters β_0 (offset) and β_1 (slope) computed on the validation dataset.

Bias Correction: The simplest approach was implemented as a postprocessing procedure. Each of the four models was used for age prediction on UKB dataset. Then, bias correction was applied using linear regression, similarly as described in Sect. 2.4. Here, 528 images of training dataset were used for fitting the regression line. The estimated coefficients β_0 and β_1 were then used for correcting the predicted brain age on the remaining subject images of the UKB dataset.

Domain Adaptation: Transfer learning using domain adaptation [10] was performed by retraining batch normalization layers, while keeping weights of other layers fixed. In contrast to Karani et al. [10], initial models were trained with shared batch normalization weights for all seven datasets.

Full Transfer: In accordance with recent brain age literature, the models were fine-tuned using a two step procedure [9]: (i) keeping the weights of feature extractor fixed and retraining only the layers following feature extractor, (ii) retraining all model weights.

3.2 Metrics

For comparison we report metrics typically reported in brain age literature, namely mean absolute error (MAE)

$$MAE = \frac{1}{N} \sum_{i=1}^{N} |y_i' - y_i|,$$

and mean error (ME)

$$ME = \frac{1}{N} \sum_{i=1}^{N} (y_i' - y_i),$$

where y_i denotes true age and y_i' predicted age of i-th subject.

To assess whether the models correctly predict the difference between the baseline and follow-up scan in the UKB longitudinal data, we compute the age difference between scans of i-th subject, $d_i = y_i^F - y_i^B$, and its estimation, $d_i' = y_i^{F'} - y_i^{B'}$, where $y_i^{F'}$ and $y_i^{B'}$ denote the respective follow-up and baseline age prediction.

Similarly as above, we compute mean difference error (MdE) and mean absolute difference error (MAdE) as

$$MdE = \frac{1}{N} \sum_{i=1}^{N} (d_i - d_i'),$$

$$MAdE = \frac{1}{N} \sum_{i=1}^{N} |d_i - d_i'|.$$

4 Results

We compare the results of three transfer learning methods, i.e. bias correction (BC), domain adaptation (DA), full transfer (FT) with the baseline prediction without transfer learning, for four brain age prediction models on new site longitudinal dataset. The predictions for 80 random subjects are pictured in Fig. 2.

Table 4. ME and MAE baseline results and the results of three transfer learning approaches: bias correction (BC), domain adaptation (DA), and full transfer (FT).

	Model 1		Model 2		Model 3		Model 4	
	ME	MAE	ME	MAE	ME	MAE	ME	MAE
Baseline	−4.334	5.494	−10.751	10.981	−7.770	8.103	0.796	3.841
BC	−0.503	3.491	−0.342	4.178	−0.579	3.650	−0.334	3.578
DA	−0.054	3.720	−0.300	4.134	−0.295	3.699	−0.087	3.281
FT	−1.408	3.802	0.290	4.146	−0.754	3.953	−0.944	3.839

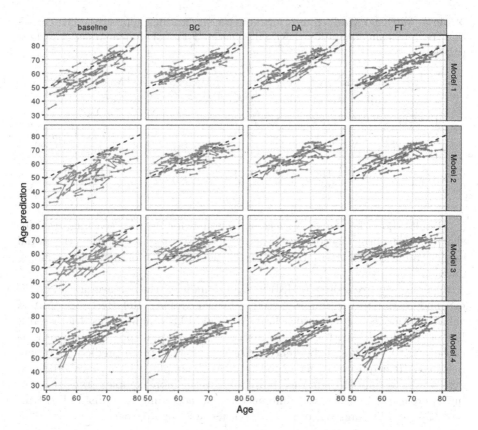

Fig. 2. Longitudinal predictions of 80 UKB subjects chosen at random. Each row corresponds to one of four models, each column corresponds to baseline and three transfer learning approaches: bias correction (BC), domain adaptation (DA), and full transfer (FT).

Baseline predictions exhibit a large prediction error, resulting in MAE of almost 11 years for Model 2 (Table 4). Large MAE and negative ME values for Models 2 and 3 indicate a systematic age underestimation across the whole age span. An exception in this regard is Model 4 with ME values within the interval $[-1, 1]$ for baseline and transfer methods. For comparison, in a study by Peng et al. [13], Model 4 trained from scratch on approximately 500 images of UKB achieves MAE of approximately 3.5 years, which is higher than 3.28 achieved with DA transfer learning.

When comparing transfer learning methods we observed that DA approach resulted in lower MAE than FT for all four models. Moreover, the simplest BC method performed best for Model 1 and comparable to DA for Models 2 and 3, but resulted in ME further from zero for all models. In Fig. 2 we observe that the DA approach significantly improved the prediction of Model 4 for those subjects, for which other methods overestimated the time between MRI acquisitions.

Table 5. MdE and MAdE results for three transfer learning approaches: bias correction (BC), domain adaptation (DA), full transfer (FT) and without transfer learning (baseline).

	Model 1		Model 2		Model 3		Model 4	
	MdE	MAdE	MdE	MAdE	MdE	MAdE	MdE	MAdE
Baseline	−0.644	1.526	−0.876	2.040	−0.129	1.654	−0.253	1.341
BC	−1.121	1.384	−1.387	1.696	−0.774	1.317	−0.693	1.226
DA	−0.849	1.516	−1.141	1.603	−0.678	1.496	−0.436	1.139
FT	−0.794	1.418	−1.206	1.647	−1.125	1.237	−0.195	1.451

Results show that DA is a successful method for transfer learning, even for shared batch normalization layers across multi-site datasets during the initial model training. We attribute this to the heterogeneity of multi-site training dataset.

Overall negative scores of MdE in Table 5 indicate that on average all methods underestimate the time between scan acquisitions. Relatively low MAdE values for Models 2 and 3 despite large ME and MAE values suggest that regardless of the global bias, the models are consistent in their predictions and thus capture the longitudinal difference quite well. For accurate predictions of gap between MR acquisitions, measured by MAdE, simple BC approach produced best results for Model 1 and was comparable to the more complex transfer learning approaches. Longitudinal differences were, according to MdE and MAdE, best captured by Model 4, using DA transfer learning approach.

5   Conclusion

The aim of this study was to compare three transfer learning approaches and evaluate their ability to predict brain age on new site dataset. For this purpose we trained and validated four CNN model architectures on multi-site dataset, trained on 2D, downsampled 3D and full resolution 3D T1w MR images. Our results on new (unseen) site dataset, with different T1w preprocessing, showed that simpler transfer learning approaches, such as bias correction and domain adaptation perform at least comparable, if not better to full transfer learning on all four tested model architectures. Furthermore, transfer learning with domain adaptation resulted in lower MAE values for Model 4 as compared to training from scratch on a dataset with the same preprocessing in a related study [13].

When assessing age difference prediction, the MAdE error, estimating the absolute predicted error between two consecutive scans of 2816 subjects, ranged from approximately 1.1 (best, Model 4 with DA transfer learning) and up to 2 years (worst, baseline Model 2). Relative to the average time of 2.25 years between the consecutive scan acquisitions the observed errors were rather high. Nevertheless, the MAdE was generally reduced using transfer learning, most consistently using simple bias correction approach.

Our results indicate that even with simple transfer learning approaches, such as bias correction or domain adaptation, the brain age analysis may be generalized across datasets and on datasets with new preprocessing, thereby increasing its potential to become a common biomarker to diagnose neurodegenerative diseases and/or monitor their progression.

Acknowledgements. This study was supported by the Slovenian Research Agency (Core Research Grant No. P2-0232 and Research Grants Nos. J2-8173 and J2-2500).

References

1. Functional Connectomes (FCON 1000). http://fcon_1000.projects.nitrc.org/indi/enhanced/neurodata.html. Accessed 27 July 2021
2. Alzheimer's Disease Neuroimaging Initiative (ADNI). http://adni.loni.usc.edu/. Accessed 27 July 2021
3. Autism Brain Imaging Data Exchange I (ABIDE I). http://fcon_1000.projects.nitrc.org/indi/abide/abide_I.html. Accessed 27 July 2021
4. Cole, J.H., et al.: Brain-predicted age in Down syndrome is associated with beta amyloid deposition and cognitive decline. Neurobiol. Aging **56**, 41–49 (2017). https://doi.org/10.1016/j.neurobiolaging.2017.04.006
5. Cole, J.H., et al.: Predicting brain age with deep learning from raw imaging data results in a reliable and heritable biomarker. Neuroimage **163**, 115–124 (2017). https://doi.org/10.1016/j.neuroimage.2017.07.059
6. Franke, K., Gaser, C.: Longitudinal changes in individual BrainAGE in healthy aging, mild cognitive impairment, and Alzheimer's disease. GeroPsych **25**(4), 235–245 (2012). https://doi.org/10.1024/1662-9647/a000074
7. Huang, T., et al.: Age estimation from brain MRI images using deep learning. In: 2017 IEEE 14th International Symposium on Biomedical Imaging (ISBI 2017), pp. 849–852, April 2017. https://doi.org/10.1109/ISBI.2017.7950650
8. IXI Dataset. https://brain-development.org/ixi-dataset. Accessed 27 July 2021
9. Jonsson, B.A., et al.: Brain age prediction using deep learning uncovers associated sequence variants. Nat. Commun. **10**(1), 5409 (2019). https://doi.org/10.1038/s41467-019-13163-9
10. Karani, N., Chaitanya, K., Baumgartner, C., Konukoglu, E.: A lifelong learning approach to brain MR segmentation across scanners and protocols. arXiv:1805.10170 [cs, stat], May 2018
11. Marcus, D.S., Fotenos, A.F., Csernansky, J.G., Morris, J.C., Buckner, R.L.: Open access series of imaging studies: longitudinal MRI data in nondemented and demented older adults. J. Cogn. Neurosci. **22**(12), 2677–2684 (2010). https://doi.org/10.1162/jocn.2009.21407
12. Miller, K.L., et al.: Multimodal population brain imaging in the UK Biobank prospective epidemiological study. Nat. Neurosci. **19**(11), 1523–1536 (2016). https://doi.org/10.1038/nn.4393
13. Peng, H., Gong, W., Beckmann, C.F., Vedaldi, A., Smith, S.M.: Accurate brain age prediction with lightweight deep neural networks. Med. Image Anal. **68**, 101871 (2021). https://doi.org/10.1016/j.media.2020.101871

14. Shafto, M.A., et al.: The Cambridge Centre for ageing and neuroscience (Cam-CAN) study protocol: a cross-sectional, lifespan, multidisciplinary examination of healthy cognitive ageing. BMC Neurol. **14** (2014). https://doi.org/10.1186/s12883-014-0204-1

15. Smith, S.M., Vidaurre, D., Alfaro-Almagro, F., Nichols, T.E., Miller, K.L.: Estimation of brain age delta from brain imaging. Neuroimage **200**, 528–539 (2019). https://doi.org/10.1016/j.neuroimage.2019.06.017

16. Souza, R., et al.: An open, multi-vendor, multi-field-strength brain MR dataset and analysis of publicly available skull stripping methods agreement. Neuroimage **170**, 482–494 (2018). https://doi.org/10.1016/j.neuroimage.2017.08.021

17. Taylor, J.R., et al.: The Cambridge Centre for ageing and neuroscience (Cam-CAN) data repository: Structural and functional MRI, MEG, and cognitive data from a cross-sectional adult lifespan sample. Neuroimage **144**(Pt B), 262–269 (2017). https://doi.org/10.1016/j.neuroimage.2015.09.018

18. Ueda, M., et al.: An age estimation method using 3D-CNN from brain MRI images. In: 2019 IEEE 16th International Symposium on Biomedical Imaging (ISBI 2019), pp. 380–383, April 2019. https://doi.org/10.1109/ISBI.2019.8759392

Uncertainty-Based Dynamic Graph Neighborhoods for Medical Segmentation

Ufuk Demir[1](✉)(ID), Atahan Ozer[1](ID), Yusuf H. Sahin[1](ID), and Gozde Unal[2](ID)

[1] Faculty of Computer and Informatics Engineering, Computer Engineering,
Istanbul Technical University, 34469 Istanbul, Turkey
{demiruf17,ozera17,sahinyu}@itu.edu.tr
[2] Faculty of Computer and Informatics Engineering, AI and Data Engineering,
Istanbul Technical University, 34469 Istanbul, Turkey
gozde.unal@itu.edu.tr
https://vision.itu.edu.tr

Abstract. In recent years, deep learning based methods have shown success in essential medical image analysis tasks such as segmentation. Post-processing and refining the results of segmentation is a common practice to decrease the misclassifications originating from the segmentation network. In addition to widely used methods like Conditional Random Fields (CRFs) which focus on the structure of the segmented volume/area, a graph-based recent approach makes use of certain and uncertain points in a graph and refines the segmentation according to a small graph convolutional network (GCN). However, there are two drawbacks of the approach: most of the edges in the graph are assigned randomly and the GCN is trained independently from the segmentation network. To address these issues, we define a new neighbor-selection mechanism according to feature distances and combine the two networks in the training procedure. According to the experimental results on pancreas segmentation from Computed Tomography (CT) images, we demonstrate improvement in the quantitative measures. Also, examining the dynamic neighbors created by our method, edges between semantically similar image parts are observed. The proposed method also shows qualitative enhancements in the segmentation maps, as demonstrated in the visual results.

Keywords: Segmentation · Graph neural networks · Refinement

1 Introduction

Deep convolutional neural networks (CNN) have proven to be powerful for computer vision tasks including classification, segmentation, and retrieval [1]. Considering the time spent on the manual segmentation by the medical experts for making quantitative measurements, medical image segmentation by CNNs

---

U. Demir and A. Ozer—The authors have equal contribution.

© Springer Nature Switzerland AG 2021
I. Rekik et al. (Eds.): PRIME 2021, LNCS 12928, pp. 255–265, 2021.
https://doi.org/10.1007/978-3-030-87602-9_24

attained wide usage especially in recent years [9,15,16,18]. However, despite the fact that many of these models create a rough and useful segmentation, especially for segmented organ borders or organs with similar tissue (e.g. pancreas parenchyma has a similar contrast with the bowel in CT imaging [2]), they can produce unreliable results. Those regions can be presumed by their pixel-wise uncertainty values at the test time. Considering the Monte Carlo Drop Out [10] (MCDO) uncertainty of these areas in the segmentation network output, Ding et al. developed an uncertainty-aware training procedure that focuses on segmenting the relatively certain parts correctly and reserving the remaining uncertain parts for expert decision [5]. Although their method improved the segmentation for certain areas, it is still challenging to manually segment the uncertain regions.

To improve the CNN's results in medical image segmentation, post-processing methods like conditional random fields (CRF) [13] could be directly applied to improve the network predictions [6,9]. However, the CRF depends strongly on the shape priors and is an independent process from the network features. In [4], a region growing algorithm to refine the network output focusing on the uncertain pixels is used and better results than a CRF process are obtained. Similarly, to benefit from the network uncertainties in the post-processing step, Soberanis-Mukul et al. developed a graph convolutional network (GCN) Refinement technique [20]. The main contributions of this work focus on two advancements over the GCN Refinement procedure as follows:

- A new dynamic neighbor selection mechanism is defined.
- The dynamic neighbor selection mechanism is applied in two setups: inter-graph and intra-graph.

Then, these methods on the neighbor selection are applied for an uncertainty-aware training strategy in which the GCN and the segmentation network are trained end-to-end. Our method increases the performance of the segmentation in a selected application, which is the pancreas segmentation from CT images.

2   Related Work

Uncertainty Estimation: Uncertainty estimation is a critical task in automated medical imaging. In general, uncertainty can be modeled in two ways, aleatoric and epistemic uncertainty [22]. Aleatoric uncertainty is related to noisy observations of the distribution. On the other hand, epistemic uncertainty is related to deficient observations of the distribution [10]. The second one can be reduced given enough data however, the first one can not be reduced without removing the source of the noise. Kendal and Gal [10] proposed it is possible to estimate epistemic uncertainty by using drop-out layers of the model and named this process as Monte Carlo Dropout (MCDO). This metric can provide uncertainty estimation per pixel during segmentation and hence it is possible to find possible erroneous predictions using pixels with high uncertainty.

Graph Neural Networks: The prominence of Graph Neural Networks (GNNs) is increasing due to the latest advancements in the area [3,21]. One of the works that accelerated the research is Graph Convolutional Neural Networks (GCN) by Kipf and Weil [12]. They presented a convolution-based layer propagation network that can directly work in a graph structure. The capabilities of various GNN [8,14] models in terms of aggregation schemes is investigated by Xu et al. [24] in a mathematical frame to characterize the expressive power of GNNs. It is shown that single aggregators may fail to distinguish representations for node classification. The recent work of Corso et al. [3] shows multiple aggregation functions in GNNs are required to maximize information extraction from the network and presented Principle Neighborhood Aggregation (PNA) blocks.

Dynamic Graph Neural Networks: For point cloud segmentation and classification tasks with graph neural networks, nearest spatial neighbors according to point coordinates could be selected to create the edges between the points [25]. However, in Dynamic Graph CNNs (DG-CNN) [23], after every network layer, different edges are created according to feature distances between each point. Thus, at the beginning of the network, edges indicate the spatially close points, whereas at the end, they indicate semantically close points.

Our work utilizes uncertainty-aware CNN training by using graphs that are constructed by the MCDO process during segmentation network training. Uncertainty graphs are refined by employing multiple aggregator functions and dynamic edge calculations.

3   Method

3.1   GCN Refinement

We take the GCN Refinement work of Soberanis-Mukul et al. [20] as our baseline. In GCN Refinement, a graph is created by selecting the uncertain voxels and some certain voxels next to the uncertain ones as graph nodes. Then, for every node, 6 edges are created according to a 6-neighborhood, and 16 edges are created randomly. Using the prediction outputs of certain nodes with low uncertainty, a GCN model is trained in a semi-supervised manner.

In this paper, we focus on two main issues of the GCN Refinement procedure. First, the random neighbor selection process is problematic since using random neighbors, edges could be created between unrelative nodes, and the reproducibility of the technique could be decreased. Second, the network's contribution to the GCN is limited since the training process is not executed in an end-to-end manner. To address those limitations, we devise a dynamic neighbor selection, which is investigated through both an intra-graph and inter-graph edge selection procedure.

Some related definitions that are used throughout the paper are presented next. The uncertainty value $\mathbb{U}(x)$ for the voxel at coordinate x of a segmentation output is calculated using the entropy,

$$\mathbb{U}(x) = -\sum_{c=1}^{M} P(x)^c \log P(x)^c, \tag{1}$$

where $P(x)^c$ is the probability of the voxel belonging to the class c. This probability is estimated by using the expectation of the MCDO process with T passes as

$$\mathbb{E}(x) = \frac{1}{T} \sum_{t=1}^{T} g(V(x), \theta_t), \tag{2}$$

where g represents the segmentation network, $V(x)$ represents the voxel intensity and θ_t represents the network parameters. For each node of the graph, expectation $\mathbb{E}(x)$, entropy $\mathbb{U}(x)$, and voxel value $V(x)$ are used as node features. Weights for the edges are calculated as weighted summation of three different metrics such as expectation diversity $div(x_i, x_j)$ [26] (Eq. 3), relative intensity $int(x_i, x_j)$ (Eq. 4) and relative 3-D position $pos(x_i, x_j)$ (Eq. 4), as follows:

$$div(x_i, x_j) = \sum_{c=1}^{M} (P^c(x_i) - P^c(x_j)) \log \frac{P^c(x_i)}{P^c(x_j)}, \tag{3}$$

$$int(x_i, x_j) = \exp\left(-\frac{\|V(x) - V(x_j)\|^2}{2\sigma_1}\right), \tag{4}$$

$$pos(x_i, x_j) = \exp\left(-\frac{\|x_i - x_j\|^2}{2\sigma_2}\right). \tag{5}$$

In our work, we keep uncertainty calculation and graph features the same as the baseline work. Our novelty lies in our connectivity structure and uncertainty-aware CNN training as described in the next section.

3.2 Dynamic Edge Selection and Uncertainty Aware Training

The input CT image slices are fed to a 2D segmentation network, which is a standard U-Net [18] model. The network is trained until it converges, then it continues training with the graph-based method. At this stage, for each iteration, the uncertainty analysis is performed on the input volume. Then, a graph model for the volume, its edges, and graph features are created to use in the GCN. Figure 1 illustrates the proposed method.

In [20], two types of edge creation mechanisms are used. First, for every node, connections are created to the 6-neighbors. These local connections bring regional information about each node, however, as the neighbor nodes have nearly the same features with a selected node, the global graph topology is not discovered until the last layers of the network. Second, in order to tolerate the aforementioned problem, a set of randomly selected 16 neighbors are added to

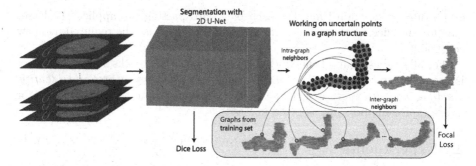

Fig. 1. In our uncertainty-aware training procedure, the GCN is trained combined with the segmentation network. In the GCN, in addition to the edges created using 6-neighborhood, intra-graph neighbors or inter-graph neighbors are also selected. Neighbors for only one node are shown in the figure. (Color figure online)

each node, and the corresponding edges are created. The random selection process helps to improve the quantitative results, however, it lacks the reproducibility of the results. Our hypothesis is that instead of choosing random neighbors, neighbors that are chosen considering feature distances could improve the quantitative results and their interpretability. Thus, inspired by the neighbor-creation mechanism in DG-CNN, we define two different types of neighbors in the feature space: intra-graph and inter-graph neighbors.

Intra-graph Neighbors: As represented by the red lines in Fig. 1, after the patient graph is constructed, the k-nearest neighbors algorithm is applied to select the nearest 5 new nodes from the same patient graph for each node.

Inter-graph Neighbors: For each training sample used to train the segmentation network, graphs are created and graph features are calculated individually. Then, for each node of the test graph, a total of 5 new neighbors are selected as illustrated by the purple lines in Fig. 1, according to dynamic feature distances obtained from graphs of training samples. To decrease the memory need, farthest point sampling [17] with ratio $\frac{1}{40}$ is applied on the train graphs.

Both inter-graph and intra-graph neighbor selection mechanisms can be applied to two different procedures: refinement and uncertainty-aware training (UAT)[1].

Refinement: The refinement procedure is similar to the one presented in GCN Refinement [20]. All voxels under a segmentation mask of the test object are refined in a semi-supervised manner. The segmentation network is not affected by this procedure.

Uncertainty-Aware Training (UAT): We combined the graph network with the segmentation network and trained using the training set in a supervised manner. Thus wrongly labeled voxels have the chance to be corrected by the

---

[1] Our usage of the term Uncertainty-Aware Training is different from Ding et al. [5].

graph during the train time. For each backward passing, we applied the losses only for one slice for the segmentation network to decrease the required memory constraint.

For all studies, a simple network containing two PNA blocks and a GCN layer is used. For the PNA blocks, *mean, min, max, std* aggregators; and *identity, amplification, attenuation* scalers are used. The graph model and its usage for inter-graph neighbors are as given in Fig. 2.

4   Experiments

For a fair evaluation, the same pancreas CT dataset from NIH [19] and the U-Net [18] segmentation model officially shared for the GCN Refinement are used. Considering the hyper-parameters and training procedure from the baseline, the GCN part is trained using an Adam [11] optimizer with a learning rate $1e^{-2}$. To ensure the balance between the GCN part and the U-Net, a learning rate of the U-Net is selected as $1e^{-5}$. The number of nodes and edges are selected heuristically. The other hyper-parameters are selected as the same as the baseline. The U-Net model is trained alone for more than 50 epochs until its performance converged. For the U-Net model and GCN, dice loss and focal loss are used respectively. For all experiments, we used the PyTorch 1.7.1 framework and PyTorch Geometric [7] library. We trained the models on a device having Titan RTX.

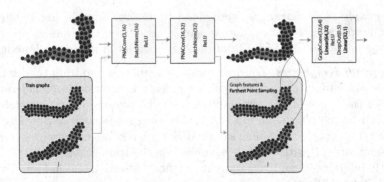

Fig. 2. Graph network model used for the "Inter-graph Neighbors" setup. Graph features are calculated in the PNA blocks for both the graph to refine and the train graphs. Then, new edges are created using graph features.

To evaluate our method, we designed the experiments given in Table 1. We execute 4 modes of our method: our own base methods "Intra-graph" and "Inter-graph"; as well as the "Intra-graph UAT" and "Inter-graph UAT" with the uncertainty-aware components.[2]

---

[2] https://github.com/ituvisionlab/Uncertainty-Based-Dynamic-Graph-Neighborh oods.

Table 1. The list of experiments to evaluate our method on pancreas segmentation on CT images.

Experiment	6-neighbors	Intra-graph neighbors	Inter-graph neighbors	Uncertainty-aware training
6-Connectivity	✓			
Intra-graph	✓	✓		
Inter-graph	✓		✓	
Intra-graph UAT	✓	✓		✓
Inter-graph UAT	✓		✓	✓

4.1   Quantitative Results

For both refinement and UAT setups, the Dice scores for the test set are given in Table 2 and 3 respectively. In both setups, our method overperforms the baseline GCN refinement. Detailed explanations for each experiment's results are as below.

6-Connectivity: Since connectivity is dramatically reduced, the GCN model suffered from deficient information. In fact, it decreases the segmentation network's performance. We can interpret that even random neighbors are useful to keep the contextual information as in the original GCN Refinement.

Inter-graph: Since connectivity is increased quantitatively and semantically meaningful edges are created, our method overperformed the baseline.

Intra-graph: Using the connections inside the same patient's graph, better results than Inter-graph refinement are obtained. The best refinement performance in terms of the Dice score is achieved with this setup.

Inter-graph UAT: Improved results are obtained compared to the Inter and Intra-graph models in terms of Dice scores and their standard deviations. These improvements show that our UAT procedure is quantitatively better than refinement.

Intra-graph UAT: This experiment yielded the best scores among all setups. Our deductions for the *Intra-graph* experiment is also valid for this experiment.

Table 2. Dice score results for Uncertainty Aware Training results.

	Dice score
GCN refinement (Baseline)	77.81 ± 6.3
6-Connectivity	76.11 ± 7.81
Inter-graph	78.32 ± 6.41
Intra-graph	$\mathbf{78.87 \pm 6.24}$

Table 3. Dice score results for refinement setups.

	Dice score
GCN refinement (Baseline)	76.9 ± 6.6
Inter-graph UAT	78.84 ± 5.84
Intra-graph UAT	$\mathbf{79.26 \pm 5.78}$

In Fig. 3, visual results of our refinement method are compared with baseline results. To demonstrate the correctly refined parts in uncertain regions, uncertainty maps are also included. As stated in the previous works [5, 20], uncertain regions frequently occur in the border areas. As it can be seen from the visualizations, the refinement method improved the results of uncertain regions both for inter-graph and intra-graph neighborhoods compared to baseline work. The same results are also obtained for the UAT method in Fig. 4. In UAT visualizations, We observed the best visual results coherently with Dice scores given in Table 2 and 3.

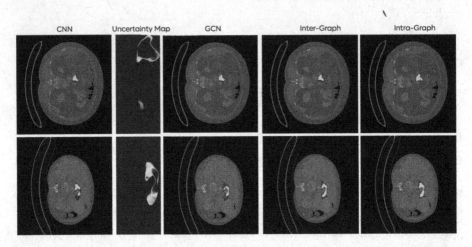

Fig. 3. Comparison of the GCN refinement. Red represents false positive, green represents true positive, and blue represent false negative regions. Each row corresponds to another 2D slice. The columns correspond to the CNN model results, the uncertainty map, the GCN Refinement results, the Inter-Graph and the Intra-Graph results respectively. (Color figure online)

4.2   Neighboring Results

The adequateness of our neighbor selection mechanism can be investigated by checking the selected neighbors. In Fig. 5, some voxels from test slices and label maps of their selected neighbors are shown. According to the results, we can conclude that the selected neighbors demonstrated a semantical similarity as argued by Wang et al. for the DG-CNN [23]. Also for the roughly certain voxels like the ones at the center of the image, the found neighbors are at the more certain positions of the pancreas.

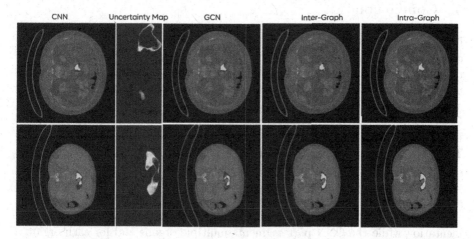

Fig. 4. Visualization of the UAT method. Same coloring and row order with Fig. 3 is applied. The columns correspond to the CNN model results, the uncertainty map, the GCN Refinement results, the Inter-Graph with UAT and the Intra-Graph with UAT results respectively.

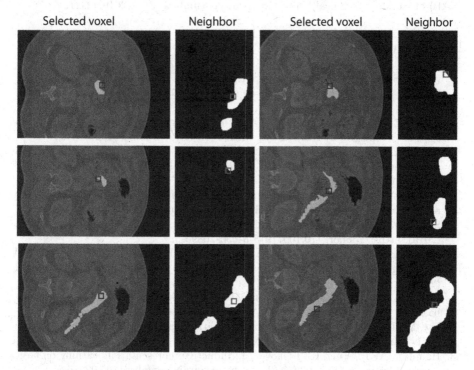

Fig. 5. First and third columns show some voxels from pancreas slices, second and fourth columns show pancreas label maps of one of their neighbors.

5   Conclusion

In this study, we introduced an uncertainty-aware CNN model training procedure and a dynamic edge selection method. Although our graph generation relies on that of the baseline GCN Refinement method, unlike the baseline, we utilize the graph network also for CNN training, allowing the model to learn about uncertain regions in the segmentation. Additionally, inspired by the DG-CNN, we implemented two different neighbor selection methods: Intra-graph and Inter-graph. In our best setup using Intra-graph neighbors for Uncertainty-Aware Training, we obtained an increase of ∼1.45% over GCN Refinement. Investigating the quantitative results for both refinement and UAT, we can infer that using these neighbor selection mechanisms and hence providing more contextual information about the pancreas caused a general improvement over the results.

As future work, the proposed method could be extended to multi-organ segmentation while the CNN part segments multiple organs and for each organ, a different GCN is trained. Also, the method could be applied to larger datasets to investigate whether the quality of the inter-graph neighborhood is dependent on the variety of the dataset.

Acknowledgement. This work is supported by the Scientific Research Project Unit (BAP) of Istanbul Technical University, Project Number: MOA-2019-42321.

References

1. Anwar, S.M., Majid, M., Qayyum, A., Awais, M., Alnowami, M., Khan, M.K.: Medical image analysis using convolutional neural networks: a review. J. Med. Syst. **42**(11), 1–13 (2018)
2. Boers, T., et al.: Interactive 3D U-Net for the segmentation of the pancreas in computed tomography scans. Phys. Med. Biol. **65**(6), 065002 (2020)
3. Corso, G., Cavalleri, L., Beaini, D., Liò, P., Velickovic, P.: Principal neighbourhood aggregation for graph nets. CoRR abs/2004.05718 (2020). https://arxiv.org/abs/2004.05718
4. Dias, P.A., Medeiros, H.: Semantic segmentation refinement by Monte Carlo region growing of high confidence detections. In: Jawahar, C.V., Li, H., Mori, G., Schindler, K. (eds.) ACCV 2018. LNCS, vol. 11362, pp. 131–146. Springer, Cham (2019). https://doi.org/10.1007/978-3-030-20890-5_9
5. Ding, Y., et al.: Uncertainty-aware training of neural networks for selective medical image segmentation. In: Medical Imaging with Deep Learning, pp. 156–173. PMLR (2020)
6. Feng, N., Geng, X., Qin, L.: Study on MRI medical image segmentation technology based on CNN-CRF model. IEEE Access **8**, 60505–60514 (2020)
7. Fey, M., Lenssen, J.E.: Fast graph representation learning with pytorch geometric. arXiv preprint arXiv:1903.02428 (2019)
8. Hamilton, W.L., Ying, R., Leskovec, J.: Inductive representation learning on large graphs. CoRR abs/1706.02216 (2017). http://arxiv.org/abs/1706.02216
9. Kamnitsas, K., et al.: Efficient multi-scale 3D CNN with fully connected CRF for accurate brain lesion segmentation. Med. Image Anal. **36**, 61–78 (2017)

10. Kendall, A., Gal, Y.: What uncertainties do we need in Bayesian deep learning for computer vision? CoRR abs/1703.04977 (2017). http://arxiv.org/abs/1703.04977
11. Kingma, D.P., Ba, J.: Adam: a method for stochastic optimization. arXiv preprint arXiv:1412.6980 (2014)
12. Kipf, T.N., Welling, M.: Semi-supervised classification with graph convolutional networks. CoRR abs/1609.02907 (2016). http://arxiv.org/abs/1609.02907
13. Krähenbühl, P., Koltun, V.: Efficient inference in fully connected CRFs with Gaussian edge potentials. In: Advances in Neural Information Processing Systems, vol. 24, pp. 109–117 (2011)
14. Meng, Y., et al.: Regression of instance boundary by aggregated CNN and GCN. In: Vedaldi, A., Bischof, H., Brox, T., Frahm, J.-M. (eds.) ECCV 2020. LNCS, vol. 12353, pp. 190–207. Springer, Cham (2020). https://doi.org/10.1007/978-3-030-58598-3_12
15. Milletari, F., et al.: Hough-CNN: deep learning for segmentation of deep brain regions in MRI and ultrasound. Comput. Vis. Image Underst. **164**, 92–102 (2017)
16. Milletari, F., Navab, N., Ahmadi, S.A.: V-Net: fully convolutional neural networks for volumetric medical image segmentation. In: 2016 Fourth International Conference on 3D Vision (3DV), pp. 565–571. IEEE (2016)
17. Qi, C.R., Yi, L., Su, H., Guibas, L.J.: PointNet++: deep hierarchical feature learning on point sets in a metric space. arXiv preprint arXiv:1706.02413 (2017)
18. Ronneberger, O., Fischer, P., Brox, T.: U-Net: convolutional networks for biomedical image segmentation. In: Navab, N., Hornegger, J., Wells, W.M., Frangi, A.F. (eds.) MICCAI 2015. LNCS, vol. 9351, pp. 234–241. Springer, Cham (2015). https://doi.org/10.1007/978-3-319-24574-4_28
19. Roth, H., Farag, A., Turkbey, E.B., Lu, L., Liu, J., Summers, R.M.: Data from pancreas-CT (2016)
20. Soberanis-Mukul, R.D., Navab, N., Albarqouni, S.: An uncertainty-driven GCN refinement strategy for organ segmentation. In: Machine Learning for Biomedical Imaging (MELBA) (2020). mIDL 2020 Special Issue
21. Veličković, P., Cucurull, G., Casanova, A., Romero, A., Liò, P., Bengio, Y.: Graph attention networks (2018)
22. Wang, G., Li, W., Aertsen, M., Deprest, J., Ourselin, S., Vercauteren, T.: Aleatoric uncertainty estimation with test-time augmentation for medical image segmentation with convolutional neural networks. CoRR abs/1807.07356 (2018). http://arxiv.org/abs/1807.07356
23. Wang, Y., Sun, Y., Liu, Z., Sarma, S.E., Bronstein, M.M., Solomon, J.M.: Dynamic graph CNN for learning on point clouds. ACM Trans. Graph. (ToG) **38**(5), 1–12 (2019)
24. Xu, K., Hu, W., Leskovec, J., Jegelka, S.: How powerful are graph neural networks? CoRR abs/1810.00826 (2018). http://arxiv.org/abs/1810.00826
25. Zhang, Y., Rabbat, M.: A graph-CNN for 3D point cloud classification. In: 2018 IEEE International Conference on Acoustics, Speech and Signal Processing (ICASSP), pp. 6279–6283. IEEE (2018)
26. Zhou, Z., Shin, J., Zhang, L., Gurudu, S., Gotway, M., Liang, J.: Fine-tuning convolutional neural networks for biomedical image analysis: actively and incrementally. In: Proceedings of the IEEE Conference on Computer Vision and Pattern Recognition, pp. 7340–7351 (2017)

FLAT-Net: Longitudinal Brain Graph Evolution Prediction from a Few Training Representative Templates

Guris Özen[1], Ahmed Nebli[1,2], and Islem Rekik[1]($\boxtimes$)

[1] BASIRA Lab, Faculty of Computer and Informatics, Istanbul Technical University, Istanbul, Turkey
irekik@itu.edu.tr
[2] National School for Computer Science (ENSI), Mannouba, Tunisia
http://basira-lab.com

Abstract. Diagnosing brain dysconnectivity disorders at an early stage amounts to understanding the evolution of such abnormal connectivities over time. Ideally, without resorting to collecting more connectomic data over time, one would predict the disease evolution with high accuracy. At this point, generative learning models from limited data can come into play to predict brain connectomic evolution over time from a single acquisition timepoint. Here, we aim to bridge the gap between data scarcity and brain connectomic evolution prediction by proposing our novel Few-shot LeArning Training Network (FLAT-Net), the *first* framework leveraging the few-shot learning paradigm for brain connectivity evolution prediction from baseline timepoint. To do so, we introduce the concept of learning from *representative connectional brain templates (CBTs)*, which encode the most centered and representative features (i.e., connectivities) for a given population of brain networks. Such CBTs capture well the data heterogeneity and diversity, hence they can train our predictive model in a frugal but generalizable manner. More specifically, our FLAT-Net starts by clustering the data into k clusters using the renowned K-means method. Then, for each cluster of *homogenous* brain networks, we create a CBT, which we call cluster specific-CBT (cs-CBT). We solely use *each cs-CBT* to train a distinct geometric generative adversarial network (gGAN) (i.e., for k clusters, we extract k cs-CBTs, and we train k gGANs (sub-model) each for a distinct cs-CBT) to learn the cs-CBT evolution over time. At the testing stage, we compute the Euclidean distance between the testing subject and each cs-CBT, and we select the gGAN model trained on the closest cs-CBT to the testing subject for prediction. A series of benchmarks against variants and excised interpretations of our framework showed that the proposed FLAT-Net, training strategy, and sub-model selection are promising strategies for predicting longitudinal brain alterations *from only a few representative templates*. Our FLAT-Net code is available at https://github.com/basiralab/FLAT-Net.

Keywords: Brain graph evolution prediction · Connectional brain template · Few-shot learning · Adversarial graph neural networks

© Springer Nature Switzerland AG 2021
I. Rekik et al. (Eds.): PRIME 2021, LNCS 12928, pp. 266–278, 2021.
https://doi.org/10.1007/978-3-030-87602-9_25

1 Introduction

Personalized treatments have become the default curative technique for many brain disorders [1]. As reported in [2], there is a correlational relationship between the quality of brain disease evolution prediction over time and the chances of patient recovery, disease reversal (e.g., mild cognitive impairment (MCI)), and healing costs reduction. As a result, a few recent studies relied on the nascent breakthroughs in machine learning techniques to predict brain connectivity evolution over time from a single timepoint [3,4].

For instance, [3] proposed Learning-Guided Infinite Network Atlas (LINAs) framework, the first graph-based machine learning method to predict the longitudinal brain network evolution from a base timepoint. Specifically, LINAs models the brain as a graph holding pairwise connectivities in-between regions of interest (ROIs). However, LINAs learning paradigm is severely dichotomized, resulting in defaulted feedback loops within its building learning blocks. To overcome this limitation, [4] proposed EvoGraphNet as a framework that leverages a graph-based Generative Adversarial Network (gGAN) [5] composed of many sequentially-trained gGANs to foresee brain graph evolution over time in an end-to-end fashion. Specifically, each gGAN aims to learn a transformation function linking a prior brain graph distribution at a timepoint t_i to its subsequent distribution at timepoint t_{i+1}. Notably, GANs are a pair of neural networks composed of a generator and a discriminator originally proposed by [6]. While the generator's purpose is to learn how to produce imitations of the ground truth data, the discriminator's purpose is to learn how to discriminate between the ground-truth data and the fake brain graph data produced by the generator.

With such good predictions delivered by [3] and [4], such frameworks still hold numerous limitations. For instance, these studies have only focused on predicting the longitudinal brain alterations overlooking the challenges coming with the scarcity of longitudinal neuroimaging datasets. Therefore, one can see that LINAs [3] and EvoGraphNet [4] might not deliver decent results, especially in scenarios where large datasets are not available due to the high medical image acquisition cost. As a result, these limitations raise the following question: *given a small set of training examples, how can we accurately and efficiently predict brain graph evolution over time?*

To address this question, we propose our Few-shot LeArning Training Network (FLAT-Net), a novel training and testing scheme for brain graph prediction using few representative examples. Now, it is easy to see how few-shot learning schemes are heavily prone to data heterogeneity and overfitting. Here we introduce the concept of learning from *representative connectional brain templates (CBTs)*, which encode the most centered and representative features (i.e., connectivities) for a given population of brain networks [5,7]. Such CBTs capture well the data heterogeneity and diversity, hence they can train our predictive model in a frugal but generalizable manner. To handle connectomic data heterogeneity, we first start by grouping together homogenous brain networks at baseline timepoint by clustering our dataset into k homogenous set of clusters by using the renowned K-means [8] method. Next, to reduce the effect of overfitting, we learn a cluster-specific CBT, which holds the most centered set of pair-wise connectivities across the cluster population at baseline timepoint. To this aim, we further extend the work of [7] to propose a Single Deep Graph Normalizer (sDGN) to map the single-view brain networks (SVBNs) of the subject population into a normalized population-representative connectional brain template (CBT). We remind the reader that a CBT [5, 7] is a brain connectivity graph that selectively captures the most common pair-wise features across a population of input brain graphs (i.e., for each pair-wise connectivity, CBT captures the most common connectivities representing a population of brain graphs). We also note that, for each cluster, we extract a distinct CBT that we call cluster-specific CBT (cs-CBT), each presenting one representative template (shot) to train our model on.

Given k cs-CBTs corresponding to k clusters, we propose to train k distinct geometric deep learning frameworks each on a single cs-CBT in an end-to-end fashion. Specifically, we draw inspiration from the landmark EvoGraphNet [4] to predict the brain graph evolution over time given the longitudinal cs-CBTs as training samples. We also refer to each of the k EvoGraphNets as *sub-model*. Finally, we propose to personalize our testing phase by enhancing its prediction scheme. To do so, we propose to compute the Euclidean distance between each cs-CBT and a given baseline test connectome. Next, our FLAT-Net selects the sub-model trained on the cs-CBT that scored the lowest Euclidean distance following the assumption that the lowest Euclidean distance indicates the most suitable sub-model.

To summarize our framework, there are three main building blocks in our FLAT-Net: (i) preprocessing (i.e., clustering and cs-CBT extraction), (ii) training k geometric deep learning models, and (iii) sub-model selection. Below, we list the primary contributions of our work:

1. *On a conceptual level.* FLAT-Net is the first geometric deep learning approach that predicts brain graph evolution over time **using only a few representative training templates.**
2. *On a methodological level.* FLAT-Net is a prediction method that handles data heterogeneity as well as reduces overfitting by training a set of geometric deep learning sub-models (i.e., EvoGraphNet) using population-representative brain networks (i.e., cs-CBTs).
3. *On clinical level.* FLAT-Net can be used for clinical applications independently from the availability of a high amount of data to formalize adequate personalized treatments.

2   Proposed Method

In this section, we explain the key building steps of our proposed FLAT-Net to predict brain graph evolution over time using representative shots. Table 1 displays the key mathematical notations used throughout this paper.

FLAT-Net Overview. Here, we aim to predict brain graph evolution using a few training connectional templates. However, according to [9], it is easy to

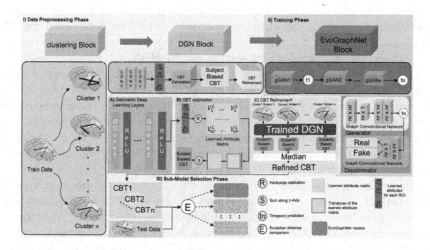

Fig. 1. *Proposed three-step brain network evolution prediction framework using a few representative templates.* **(I)** *Data preprocessing phase.* We cluster the training data to obtain k clusters using the K-Means clustering algorithm. Next, we propose sDGN to acquire normalized cluster-specific CBTs (cs-CBT). sDGN is a three-step framework: **(A)** *Geometric deep learning layers.* We propose a three-layer geometric deep learning network. These layers aim to learn a feature matrix V^L which consists of embeddings for each ROI. **(B)** *CBT acquisition.* We propose the CBT acquisition phase as a set of serial operations on the learned V^L to obtain a subject-biased CBT. For $\mathbf{V}^L = \{V_1^L, V_2^L, ..., V_{n_r}^L\}$ first, we replicate the matrix horizontally n_r times to obtain $R \in \mathbb{R}^{n_r \times n_r \times d_l}$. Next, we compute the absolute difference of R, and its transpose R^T. Finally, we sum the result along the z-axis to estimate the final subject-biased CBT. **(C)** *CBT refinement.* The CBT refinement phase is proposed to eliminate the mentioned bias in cs-CBTs and accomplish a more representative cs-CBT for each cluster population. **(II)** *Training phase.* We propose the training scheme to generate a prediction model from our previously learned k cs-CBTs. We train k EvoGraphNet sub-models with *each cs-CBT* as a baseline timepoint. For each EvoGraphNet, an initial cs-CBT goes through a chain of n gGANs depending on the desired timepoint t_n to produce a graph distribution prediction at timepoint t_n. **(III)** *Sub-model selection phase.* We introduce the sub-model selection phase where we use Euclidean distance measurements between each cs-CBT and each testing baseline connectome to select the most appropriate sub-model for prediction.

see how outliers can dramatically alternate the prediction course especially given that brain connectomes (i.e., graphs) are often seen as rather *unique fingerprints* for each individual. Hence, we aim to address this problem via our proposed FLAT-Net framework. To do so, we design FLAT-Net with the mindset of training with a few population-driven brain connectivity templates. We propose combining three machine-learning fundamental concepts: data preprocessing, model training, and sub-model selection, respectively. While we detail each building block separately in the following sub-sections, here, we present an overview of the need and the conduct for each step in our FLAT-Net.

The preprocessing step consists in clustering our input brain graphs into k clusters in order to segregate subject groups with similar traits. This is especially valid for few-shot learning schemes as it handles data heterogeneity [10]. First, we choose to leverage the renowned K-means clustering algorithm [11] for its reliability and to facilitate our study reproducibility. Next, for each cluster, we learn a normalized population-representative CBT, which encodes the most common and centered connectivity scheme representing the *centroid* of the clustered training population. To obtain cs-CBTs, we draw inspiration from the DGN method [7] (i.e., the current state-of-the-art study in CBT genesis). We note that, here, we adapt the original implementation to our model. We term our new implementation sDGN. We remind the reader that DGN estimates CBTs for a given population of multi-view brain networks where each view characterize certain features of the brain structure, while sDGN uses single-view clustered brain networks for CBT generation.

Our model training consists of learning using a few representative brain templates where we train k EvoGraphNets, each on a distinct *time-dependent* cs-CBT at timepoints $\{t_i\}_{i=0}^m$. Specifically, each EvoGraphNet aims to learn how to map cs-CBT at a timepoint t_i to its following cs-CBT at timepoint t_{i+1} as shown in Fig. 1 in an end-to-end fashion. We note that since we have k EvoGraphNets, each one is optimized for one distinct cs-CBT. Lastly, we introduce the sub-model selection step. For each testing subject at baseline timepoint t_0, we aim to select the most suitable EvoGraphNet that guarantees the most *personalized* brain graph evolution prediction. To do so, we compute the Euclidean distance measurements between each cs-CBT and each validation data. Then, our FLAT-net decides to use the most appropriate sub-model (i.e., EvoGraphNet) based on the lowest Euclidean distance measurement. We provide in-depth details of each building block in the following sub-sections.

Data Preprocessing. Given a population of n subjects, each with m brain graphs observed at m timepoints, we propose to first disentangle the baseline connectomic data heterogeniety using K-means method [11]. According to [11], K-Means clustering is the most reproducible, computationally inexpensive and accessible clustering method. Next, we leverage our adapted sDGN algorithm to learn one CBT at baseline timepoint for each cluster. We justify the choice of using cs-CBTs to train our FLAT-Net by the reliable *cluster-representativeness* offered by these CBTs. We train sDGN using a training set of SVBNs $\mathbb{X_k} = \{\mathbf{X}_k^0, \mathbf{X}_k^1, ..., \mathbf{X}_k^i, ..., \mathbf{X}_k^m\}$ for each cluster k, where $\mathbf{X}_k^i$ denotes the single-view brain graph tensor stacking all connectivity matrices of the cluster k at a given timepoint t_i. Each single-view brain graph tensor $\mathbf{X}_k^i$, passes through three geometric deep learning layers to generate the cs-CBTs at each timepoint t_i. Specifically, each layer uses an edge-conditioned filter learner [12] and is separated by ReLU non-linearity. These layers aim to learn a feature matrix $\mathbf{V}_L \in \mathbb{R}^{n_r \times n_r \times d_l}$ which consists of embeddings for each ROI. Following the graph convolutional layers, we obtain a subject-biased CBT by following a set of serial operations on the learned $\mathbf{V}_L$. However, such CBT does not fully serve our aim. Therefore, we refine these learned CBTs to eliminate the referred bias and attain a more representative cs-CBT. To do so, we compute the element-wise median of subject-biased CBTs. We propose a subject-specific normalization loss (SNL) to evaluate the representativeness of the generated CBT. We use a random subset of the tensor $\mathbf{X}_k^i$. We define our SNL loss for training cluster subject tr as follows:

$$\mathcal{SNL}_{n_k^s} = \sum_{i=1}^{n_k^s} \|\mathbf{T}_k - \mathbf{X}_k^i\|_F \times \lambda_{C_k} \tag{1}$$

$$\min_{W_1,b_1,\dots W_l,b_l} \frac{1}{|\mathbf{X}_k^i|} \sum_{i=1}^{|\mathbf{T}_k|} \mathcal{SNL}_{n_k^s} \tag{2}$$

n_k^s denotes the training subject index in a cluster k where $n_k^s \in C_k$. Here C_k is the set of training subjects in cluster k. Namely, n_k denotes total number of training subjects and S is a random subset of cluster subjects. Note that we estimate the cluster-specific CBT at each timepoint t_i. The λ_{C_k} is a view specific normalization term that is defined as:

$$\lambda_{C_k} = \frac{\frac{1}{\mu_{C_k}}}{\max\left\{\frac{1}{\mu_j}\right\}_{j=1}^{n_k}}, \tag{3}$$

where μ_{C_k} is the mean of brain graph connectivity weights of cluster C_k and $\max\left\{\frac{1}{\mu_j}\right\}_{j=1}^{n_k}$ is denoted as the maximum of mean reciprocals $\frac{1}{\mu_1}$ and $\frac{1}{\mu_{n_k}}$.

Proposed Training Scheme. We train k EvoGraphNets [4] with the previously acquired k refined cs-CBTs to successfully predict brain graph evolution from a single observation in an end-to-end fashion. Specifically, EvoGraphNet is a set of sequentially trained gGANs that aim to predict a given data distribution evolution from an initial timepoint. Each gGAN consists of two neural networks: a generator G_i and a discriminator D_i. Specifically, the generator and the discriminator collectively learn how to generate an accurate prediction of the following timepoint t_{i+1} from the previous timepoint t_i prediction obtained by the preceding gGAN. Below is the adversarial loss function for one pair of

Table 1. Major mathematical notations

Mathematical notation	Definition
m	Total number of timepoints
n	Total number of subjects
n_r	Total number of regions of interest in brain
t_i	Index of timepoint
C_k	Set of samples in cluster k
n_k	Total number of training subjects for a given cluster
k	Total number of clusters
$\mathbb{X}_k$	Baseline timepoint tensor representation $\in \mathbb{R}^{n_k \times n_r \times n_r}$ of training subjects in cluster k
$\mathbf{X}_k^i$	The single view brain graphs of the training population at timepoint t_n
λ_{C_k}	Normalization term for SNL loss
μ_{C_k}	The mean of graph connectivity weights of cluster C_k
μ_k	Mean of the connected edge weights for node k in a ground-truth brain graph
σ_k	Standard deviation of the connected edge weights for node k in a ground-truth brain graph
$\hat{\mu}_k$	Mean of the connected edge weights for node k in a predicted brain graph
$\hat{\sigma}_k$	Standard deviation of the connected edge weights for node k in a predicted brain graph
p_k	Normal distribution of the connected edge weights for a node k in a ground-truth brain graph
p_k	Normal distribution of the connected edge weights for a node k in the ground truth brain graph
$\mathbf{T}_{t_i}^k$	Ground-truth training cs-CBT $\in \mathbb{R}^{n_r \times n_r}$ at t_i
$\hat{\mathbf{T}}_{t_i}^k$	Predicted cs-CBT $\in \mathbb{R}^{n_r \times n_r}$ at t_i
$\mathbf{X}_{t_i}^{tst}$	Brain graph connectivity matrices $\in \mathbb{R}^{n_r \times n_r}$ of test data at t_i
G_i	GAN generator at timepoint t_i
D_i	GAN discriminator at timepoint t_i
$\mathcal{L}_{full}$	Full loss function
$\mathcal{L}_{adv}$	Adversarial loss function
$\mathcal{L}_{L_1}$	l_1 loss function
$\mathcal{L}_{KL}$	KL divergence loss function
λ_1	Coefficient of adversarial loss
λ_2	Coefficient of l_1 loss
λ_3	Coefficient of KL divergence loss
$\mathbf{V}_L$	Learned attributes matrix $\in \mathbb{R}^{n_r \times n_r \times d_l}$
T_k	Set of cs-CBTs

generator G_i and discriminator D_i composing an elementary gGAN at a given timepoint t_i:

$$argmin_{G_i} max_{D_i} \mathcal{L}_{adv}(G_i, D_i) = \mathbb{E}_{\mathbf{T}^k}[log D_i(\mathbf{T}_{t_i}^k)] + \mathbb{E}_{\mathbf{T}^k}[log(1 - D_i(G_i(\mathbf{T}_{t_{i-1}}^k))] \quad (4)$$

$\hat{\mathbf{T}}_{t_{i-1}}^k$ denotes the predicted cs-CBT by a subsequent generator G_{i-1} at timepoint t_{i-1}. The generator G_i takes $\hat{\mathbf{T}}_{t_{i-1}}^k$ as an input and aims learn how to generate a prediction of $\mathbf{T}_{t_i}^k$, the ground-truth brain connectivity matrix for a timepoint t_i. The discriminator D_i takes the prediction $\hat{\mathbf{T}}_{t_i}^k$ produced by the generator G_i, and the ground-truth data $\mathbf{T}_{t_i}^k$ to learn how to better differentiate between ground-truth cs-CBT and the predicted cs-CBT.

Since we expect the brain connectivity alterations to be sparse over time, we hypothesize that the l_1 distance between successive timepoints t_{i-1} and t_i is considerably small. Therefore, in addition to the adversarial loss, we add l_1 loss to further minimize the distance between $\hat{\mathbf{T}}_{t_{i-1}}^{k'}$ and $\mathbf{T}_{t_i}^k$. Below is the l_1 loss function for each template (i.e., cs-CBT) $\mathbf{T}_k$ for an elementary timepoint t_i:

$$\mathcal{L}_{l1}(G_i, \mathbf{T}_k) = \|\hat{\mathbf{T}}_{t_{i-1}}^k - \mathbf{T}_{t_i}^k\|_1 \tag{5}$$

In addition to l_1 loss, we propose to use KL-divergence loss to eliminate the inconsistency between ground-truth $\mathbf{T}_{t_i}^k$ and the predicted $\hat{\mathbf{T}}_{t_i}^k$ connectivity weight distributions. To compute our KL loss, we first calculate the mean μ_k and the standard deviation $\sigma_{\mathbf{k}}$ of connected edge weights for each ROI in ground-truth cs-CBT $\mathbf{T}_{t_i}^k$. We define p_k as the normal distribution for the node k in the ground-truth cs-CBT $\mathbf{T}_{t_i}^k$. Then, we calculate the mean $\hat{\mu}_k$ and the standard deviation $\hat{\sigma}_{\mathbf{k}}$ of connected edge weights for each ROI in the predicted cs-CBT $\hat{\mathbf{T}}_{t_i}^k$. We define q_k as the normal distribution for the same node k in the predicted cs-CBT $\hat{\mathbf{T}}_{t_i}^k$. Below is the normal distributions expression:

$$p_k = N(\mu_k, \sigma_k), \tag{6}$$

$$q_k = N(\hat{\mu}_k, \hat{\sigma}_k) \tag{7}$$

The KL-divergence between the previously calculated normal distributions p_k and q_k for each cs-CBT $\mathbf{T}_{\mathbf{k}}$ is expressed as follows:

$$\mathcal{L}_{KL} = \sum_{k=1}^{n_r} KL(q_k\|p_k), \tag{8}$$

Namely, each node k's KL divergence is expressed as:

$$KL(q_k\|p_k) = \int_{-\infty}^{+\infty} q_k(x) log \frac{q_k(x)}{p_k(x)} dx \tag{9}$$

Eventually, we obtain the full loss function to optimize by adding all the above-mentioned loss functions as follows:

$$\mathcal{L}_{Full} = \sum_{i=1}^{m} \left(\lambda_1 \mathcal{L}_{adv}(G_i, D_i) + \lambda_2 \mathcal{L}_{l1}(G_i, \mathbf{T}_i^k) + \lambda_3 \mathcal{L}_{KL} \right) \tag{10}$$

λ_1, λ_2, and λ_3 are hyperparameters controlling the influence of each loss function in the overall purpose.

Sub-model Selection. Our testing strategy consists of selecting the most appropriate sub-model (i.e., trained EvoGraphNet) for a given testing subject. To do so, we calculate the Euclidean distances between each cs-CBT $\mathbf{T}_k$, and the ground-truth test data $\mathbf{X}_{t_i}^{tst}$ for each timepoint t_i. Then, we choose the sub-model trained for the cs-CBT $\mathbf{T}_k$ having the lowest distance to ground-truth test data $\mathbf{X}_{t_i}^{tst}$. Below we express, the absolute difference function to successfully choose the most compatible method:

$$\mathcal{D}(\mathbf{X}_{t_i}^{tst}, \mathbf{T}_k) = \|\mathbf{X}_{t_i}^{tst} - \mathbf{T}_k\|_1 \tag{11}$$

Next, we used the selected sub-model to predict the follow-up connectomes $\{t_i\}_{i=1}^m$ for the baseline testing connectome at t_0.

3   Results and Discussion

Evaluation Dataset. We used 58 subjects from the OASIS-21 longitudinal dataset [13]. This set consists of a longitudinal collection of 150 subjects aged 60 to 96. Each subject was scanned on two visits, separated by at least one year. We construct a cortical morphological network for each subject derived from cortical thickness measurement using structural T1-w MRI as proposed in [14,15]. The left cortical hemispheres are parcellated into 35 ROIs using Desikan-Killiany cortical atlas [16] for the given dataset. For our evaluation, we use the cortical morphological networks to predict brain network evolution for two consecutive timepoints t_1 and t_2 from a baseline observation t_0. We built our geometric deep learning layer architecture for sDGN, and gGAN architecture for EvoGraphNet with the PyTorch Geometric library.

Parameter Setting. In Table 2, we report the averaged prediction mean absolute error (MAE) and best MAE. In addition, in Table 3 we report the node strength [17] and eigenvector centrality [17]. For preprocessing, we set the number of clusters k to 3. For sDGN, we fixed the geometric deep learning layer output embeddings to 36, 24, and 5, consecutively. We note that each layer is followed by the ReLU activation function. All models are trained using Adam optimizer [18]. Specifically, we set our optimizer learning rate to 0.0005. The number of random samples in our SNL function is fixed to 10. The sDGN is trained for 300 epochs with 3-fold cross-validation. For training, we set each pair of gGAN hyperparameters as follows: $\lambda_1 = 2$, $\lambda_2 = 2$, and $\lambda_3 = 0.001$. Also, we chose AdamW [18] as our default optimizer. To do so, we set the exponential decay rate for the first moment estimates (i.e., beta 1) to 0.5, and the exponential decay rate for the second-moment estimates (i.e., beta 2) to 0.999 for the AdamW optimizer [18]. We set the learning rate of the optimizer as 0.01 for each generator and as 0.0002 for each discriminator. Finally, we trained our sub-models (i.e., EvoGraphNets) for 300 epochs using a single Tesla V100 GPU (NVIDIA GeForce GTX TITAN with 32 GB memory).

Comparison Methods and Evaluation. To evaluate the reliability of our FLAT-Net, we used a 3-fold cross-validation strategy for training and testing. Since there are no other existing works on few-shot learning brain graph evaluation prediction, we proposed to assess our method against three ablated approaches. For the first comparison method: FLAT-Net (w/o clustering), we simplified the preprocessing phase to observe the influence of our proposed clustering method on overall FLAT-Net performance. To do so, we eliminated the clustering operation and used solely our proposed sDGN to obtain one CBT of the training dataset at each timepoint independently. Specifically, the training is completed by using the longitudinal CBT to train one EvoGraphNet model. The second comparison method: FLAT-Net (w/o sub-model selection), we removed the sub-model selection phase and used one model for testing to evaluate the impact of our proposed sub-model selection process on overall FLAT-Net performance. Particularly, we create an average model by simply averaging the weights of sub-models learned by training of k cluster-specific EvoGraphNets at each follow-up timepoint independently. In the final comparison method FLAT-Net (w hierarchical clustering), we evaluated a different clustering approach. Basically, we used hierarchical clustering [19] in the preprocessing step prior to the model training.

Table 2. Prediction accuracy using mean absolute error (MAE) and best MAE of our proposed method and comparison methods at t_1 and t_2 timepoints.

Method	t_1		t_2	
	Mean MAE ± std	Best MAE	Mean MAE ± std	Best MAE
FLAT-Net (w/o clustering+w/o sub-model selection)	0.06559 ± 0.0005705	0.06621	0.06764 ± 0.0000381	0.06769
FLAT-Net (w/o sub-model selection)	0.06509 ± 0.0004667	0.06557	0.06555 ± 0.0002728	0.06593
FLAT-Net (w hierarchical clustering)	0.60295 ± 0.0003447	0.06076	0.0652 ± 0.00005244	0.065281
FLAT-Net	**0.05687 ± 0.0008025**	**0.057631**	**0.06021 ± 0.0002091**	**0.060499**

Table 3. Prediction accuracy using node centrality and eigenvector centrality mean absolute error (MAE) of our proposed method and comparison methods at t_1 and t_2 timepoints.

Method	t_1		t_2	
	Node Strength MAE	Eigenvector Centrality MAE	Node Strength MAE	Eigenvector Centrality MAE
FLAT-Net (w/o clustering+w/o sub-model selection)	1.01825	0.029177	1.146202	0.033098
FLAT-Net (w/o sub-model selection)	0.952191	**0.024133**	1.252288	**0.031682**
FLAT-Net (w hierarchical clustering)	1.056528	0.03066	1.195618	0.03361
FLAT-Net	**0.772809**	0.025313	**1.076591**	0.033308

Table 2 shows that our proposed framework outperformed each comparison method in terms of mean MAE (averaged across 3 folds) and in terms of the best MAE for prediction at $t1$ and t_2. In other words, using our few representative training templates, our FLAT-Net reduces the risk of model divergence caused by outliers. Our proposed framework achieved the minimum mean MAE in terms of node strength [17] (averaged across 3 folds) for both t_1 and t_2 in comparison with other methods. Particularly, node strength is a powerful indicator for analyzing the brain longitudinal evolution in both healthy and disordered brain networks. Alterations in node strength might be explained by the progression of undesired neurodegenerative diseases such as Alzheimer's disease [20]. However, our proposed framework could not display the best performance in terms of the mean eigenvector centrality (averaged across 3 folds). An increase in eigenvector centrality might be due to the fact that our proposed sDGN has a centralization effect making the framework less prone to recognizing localized differences between brain graph populations [21]. Overall, our proposed FLAT-Net achieved the best performance in terms of mean MAE, best MAE, and node strength mean MAE in comparison to benchmark methods and its variants (e.g., using hierarchical clustering). Thus, to some extent one can confirm that our method is reliable to simultaneously provide a precision diagnosis to patients regardless of the amount of the input data.

Limitations and Future Work. Despite the good performance displayed by our FLAT-Net, it has a few limitations. So far, our FLAT-Net only operates on single-edge connections between brain ROIs. Therefore, in future work, we aim to fully represent the complexity of the brain with distinct topological attributes. To do so, we further aim to generalize our proposed preprocessing, training, and sub-model selection phases to handle hypergraphs. Specifically, hypergraphs consist of hyperedges (i.e., high-order connections) between each ROI that captures the high-order relationships. Furthermore, we notice that the inclusion of the sub-model selection phase produced a considerable advancement in the longitudinal brain graph evolution prediction with few-shot learning. Consequently, in our future work, we aim to optimize the sub-model selection phase.

4   Conclusion

In this paper, we propose a novel brain graph evolution prediction framework using a few representative shots. Our FLAT-Net comprises three steps: preprocessing, training, and sub-model selection. We clustered the training data into a set of k clusters to overcome the prediction errors caused by outliers and to handle data heterogeneity. By learning a population-driven connectional template for each cluster and at each timepoint, we train an EvoGraphNet using each connectional template evolution trajectory. For trajectory inference, we select the EvoGraphNet trained on the cluster-specific CBT that is most similar to the given testing connectome at baseline timepoint t_0. Our results showed that FLAT-Net could remarkably improve the prediction accuracy compared to its excised variants. Our FLAT-Net can be used to predict both typical and disordered brain evolution trajectories using a few training brain samples. In our future research, we aim to investigate the prediction accuracy of our FLAT-Net on multimodal brain graphs where the relationship between two brain regions is capturde by multiple edge weights (e.g., neural correlation and morphological dissimilarity).

Acknowledgments. This work was funded by generous grants from the European H2020 Marie Sklodowska-Curie action (grant no. 101003403, http://basira-lab.com/normnets/) to I.R. and the Scientific and Technological Research Council of Turkey to I.R. under the TUBITAK 2232 Fellowship for Outstanding Researchers (no. 118C288, http://basira-lab.com/reprime/). However, all scientific contributions made in this project are owned and approved solely by the authors.

References

1. Price, R.B., Paul, B., Schneider, W., Siegle, G.J.: Neural correlates of three neurocognitive intervention strategies: a preliminary step towards personalized treatment for psychological disorders. Cogn. Ther. Res. **37**, 657–672 (2013)
2. Tan, L., Jiang, T., Tan, L., Yu, J.T.: Toward precision medicine in neurological diseases. Ann. Transl. Med. **4**, 104 (2016)
3. Ezzine, B.E., Rekik, I.: Learning-guided infinite network atlas selection for predicting longitudinal brain network evolution from a single observation. In: Shen, D., et al. (eds.) MICCAI 2019. LNCS, vol. 11765, pp. 796–805. Springer, Cham (2019). https://doi.org/10.1007/978-3-030-32245-8_88
4. Nebli, A., Kaplan, U.A., Rekik, I.: Deep EvoGraphNet architecture for time-dependent brain graph data synthesis from a single timepoint. In: Rekik, I., Adeli, E., Park, S.H., Valdés Hernández, M.C. (eds.) PRIME 2020. LNCS, vol. 12329, pp. 144–155. Springer, Cham (2020). https://doi.org/10.1007/978-3-030-59354-4_14
5. Bessadok, A., Mahjoub, M.A., Rekik, I.: Graph neural networks in network neuroscience (2021)
6. Goodfellow, I.J., et al.: Generative adversarial networks (2014)
7. Gurbuz, M.B., Rekik, I.: Deep graph normalizer: a geometric deep learning approach for estimating connectional brain templates. In: Martel, A.L., et al. (eds.) MICCAI 2020. LNCS, vol. 12267, pp. 155–165. Springer, Cham (2020). https://doi.org/10.1007/978-3-030-59728-3_16
8. Macqueen, J.: Some methods for classification and analysis of multivariate observations. In: 5th Berkeley Symposium on Mathematical Statistics and Probability, pp. 281–297 (1967)
9. Mejia, A.F., Nebel, M.B., Eloyan, A., Caffo, B., Lindquist, M.A.: PCA leverage: outlier detection for high-dimensional functional magnetic resonance imaging data. Biostatistics **18**, 521–536 (2017)
10. Snell, J., Swersky, K., Zemel, R.S.: Prototypical networks for few-shot learning. arXiv preprint arXiv:1703.05175 (2017)
11. Kolay, S., Ray, K.: K+ means: an enhancement over k-means clustering algorithm (2017)
12. Simonovsky, M., Komodakis, N.: Dynamic edge-conditioned filters in convolutional neural networks on graphs (2017)
13. Marcus, D., Fotenos, A., Csernansky, J., Morris, J., Buckner, R.: Open access series of imaging studies: longitudinal MRI data in nondemented and demented older adults. J. Cogn. Neurosci. **22**, 2677–2684 (2010)
14. Mahjoub, I., Mahjoub, M., Rekik, I.: Brain multiplexes reveal morphological connectional biomarkers fingerprinting late brain dementia states. Sci. Rep. **8** (2018)
15. Nebli, A., Rekik, I.: Gender differences in cortical morphological networks. Brain Imaging Behav. **14**(5), 1831–1839 (2019). https://doi.org/10.1007/s11682-019-00123-6

16. Fischl, B., et al.: Automatically parcellating the human cerebral cortex. Cereb. Cortex **14**, 11–22 (2004)
17. Oldham, S., Fulcher, B., Parkes, L., Arnatkeviciūtė, A., Suo, C., Fornito, A.: Consistency and differences between centrality measures across distinct classes of networks. PLOS ONE **14**, e0220061 (2019)
18. Loshchilov, I., Hutter, F.: Fixing weight decay regularization in adam (2017)
19. Murtagh, F., Contreras, P.: Methods of hierarchical clustering (2011)
20. Koelewijn, L., et al.: Alzheimer's disease disrupts alpha and beta-band resting-state oscillatory network connectivity. Clin. Neurophysiol. **128**, 2347–2357 (2017)
21. Binnewijzend, M., et al.: Brain network alterations in Alzheimer's disease measured by eigenvector centrality in fMRI are related to cognition and cerebrospinal fluid biomarkers. Alzheimer's Dementia **9**, P684 (2013)

Author Index

Ahn, June Hong   37
Amit, Mika   117
An, Sion   1
Andrearczyk, Vincent   147

Bai, Xiaoyu   47
Baltatzis, Vasileios   201
Bessadok, Alaa   11
Bintsi, Kyriaki-Margarita   201
Butskova, Anastasia   83

Cai, Jinzheng   47
Castelli, Joel   147
Chaudhary, Aashish   83
Chharia, Aviral   231
Chikontwe, Philip   1, 37
Cygert, Sebastian   221
Czyżewski, Andrzej   221

Demir, Ufuk   255
Depeursinge, Adrien   147
Desai, Sujal   201
Du, Xiaping   93
Dular, Lara   243

Ellis, Sam   201

Folgoc, Loïc Le   201
Fontaine, Pierre   147

Gharsallaoui, Mohammed Amine   25, 104
Glocker, Ben   201
Górski, Franciszek   221
Guvercin, Umut   25

Harrison, Adam   47
Hexter, Efrat   117
Hong, Kyung Soo   37
Hu, Heping   47
Huang, Lingyun   47
Huang, Tingting   128
Huo, Jiayu   93, 128
Huo, Yuankai   47

Jiang, Haoxiang   128
Jreige, Mario   147
Juhl, Rain   83
Jung, Euijin   1
Juszczyk, Piotr   221

Kang, Myeongkyun   37
Kim, Chanho   181, 192
Kim, Hye Jung   181, 192
Kim, Jaeil   181, 192
Kim, Won Hwa   181, 192
Kumar, Neeraj   231
Kuo, Chang-Fu   138

Lai, Bolin   47
Lewalski, Sebastian   221
Li, Gang   11
Lin, Weili   11
Liu, Heng   128
Liu, Mingxia   59, 157
Liu, Yun   59
Liu, Yunbi   59
Lu, Le   47, 138
Luna, Miguel   37

Mahjoub, Mohamed Ali   11
Makin, Stephen   168
Martinez Manzanera, Octavio E.   201
Maulana, Rizal   168
Mhiri, Islem   70
Miao, Shun   138

Nair, Arjun   201
Nebli, Ahmed   11, 266

Oreiller, Valentin   147
Özen, Guris   266
Ozer, Atahan   255

Pala, Furkan   70
Park, Sang Hyun   1, 37
Pastuszak, Krzysztof   221
Perek, Shaked   117

Pohl, Kilian M.   83
Prior, John O.   147

Qiao, Yuanfang   93, 212
Qin, Genggeng   59

Rachmadi, Muhammad Febrian   168
Rekik, Islem   11, 25, 70, 104, 266

Sahin, Yusuf H.   255
Schnabel, Julia A.   201
Shen, Dinggang   11
Shin, Ho Kyung   192
Skibbe, Henrik   168
Špiclin, Žiga   243
Sui, Jing   157
Sun, Li   157
Supernat, Anna   221

Tao, Haoyi   212
Tornaci, Furkan   104

Unal, Gozde   255

Valdés-Hernández, Maria del C.   168

Wang, Fakai   138
Wang, Peng   47

Wang, Qian   93, 128, 212
Wang, Sheng   128
Wang, Yirui   138
Wang, Zhuochen   128
Wardlaw, Joanna   168
Won, Dongkyu   1
Wu, Fan   128
Wu, Min   138
Wu, Yuhsuan   47

Xia, Yong   47
Xiao, Bin   128
Xiao, Jing   47, 138
Xue, Zhong   128, 212

Yang, Erkun   157
Yang, Jian   128
Yang, Wei   59
Yao, Dongren   157

Zhan, Yiqiang   212
Zhang, Lichi   93, 212
Zhao, Qingyu   83
Zheng, Kang   138
Zhou, Xiang   128
Zhou, Xiaoyun   138
Zhou, Xiao-Yun   47
Zukić, Dženan   83

Printed in the United States
by Baker & Taylor Publisher Services